Introduction to Fluid Dynamics in Physics and Astrophysics

This textbook provides an accessible and self-contained introduction to the physics behind fluid dynamics; exploring the laws of nature describing three out of four fundamental states of matter (liquids, gases, and plasmas). Based on years of teaching of fluid dynamics theory and computation at advanced undergraduate level, it provides readers with the tools to understand and model fluid dynamical systems across a wide range of applications, from dense liquids to dilute space plasmas. The book covers the principles of fluid dynamics for an audience without prior exposure to fluid dynamics specifically. Discussion of relevant vector algebra, thermodynamics and electromagnetism is included, to ensure that the book is accessible both to readers experienced in these topics and readers starting from a basic understanding.

Example applications are drawn both from astrophysics and physics, touching upon topical research such as relativistic blast waves from neutron star mergers and the implications of plasma nature of the atmosphere for present and future radio observatories. These are contrasted and complemented with examples from general physics (e.g. contrasting the incompressible nature of water with the dilute interstellar medium). It is an ideal textbook for advanced undergraduates studying the topic and will provide a solid foundation for further (postgraduate) studies into fluid dynamics in physics or astrophysics.

Key Features:
- Introduction to fluid dynamics pitched at an advanced undergraduate level, accessible to students who are still learning relevant mathematical techniques.
- Includes over 60 exercises and selected worked solutions, in addition to timely examples and easily accessible numerical demonstrations written in C and python for readers to experiment with (*https://github.com/hveerten/code_fluid_dynamics_book*).
- Up-to-date selection of topics including fluid dynamics in special relativity and computational fluid dynamics, written by an expert in the field. The book covers all that is needed to independently write a finite-volume solver for Euler's equations and/or reproduce the provided Python and C software.
- Covers a wide range of applications in astrophysics, including first-order Fermi acceleration in shocks, accretion discs, self-similarity in cosmic explosions and interstellar turbulence

Dr Hendrik J. van Eerten is a Reader in Theoretical and Computational Astrophysics at the University of Bath, United Kingdom. His area of research is transients in high-energy astrophysics, with a focus on theoretical and computational fluid dynamics of relativistic jets and cosmic explosions, in particular gamma-ray bursts.

Introduction to Fluid Dynamics in Physics and Astrophysics

Hendrik J. van Eerten

CRC Press
Taylor & Francis Group
Boca Raton London New York

CRC Press is an imprint of the
Taylor & Francis Group, an informa business

A CHAPMAN & HALL BOOK

Designed cover image: NASA/CXC/SAO/J.Hughes et al (X-rays) and NASA/ESA/Hubble Heritage Team (STScI/AURA) (optical).

First edition published 2024
by CRC Press
2385 NW Executive Center Drive, Suite 320, Boca Raton FL 33431

and by CRC Press
4 Park Square, Milton Park, Abingdon, Oxon, OX14 4RN

CRC Press is an imprint of Taylor & Francis Group, LLC

ISBN: 978-0-367-55777-5 (hbk)
ISBN: 978-0-367-55235-0 (pbk)
ISBN: 978-1-003-09508-8 (ebk)

DOI: 10.1201/9781003095088

Typeset in Nimbus Roman
by KnowledgeWorks Global Ltd.

Publisher's note: This book has been prepared from camera-ready copy provided by the authors.

Contents

Preface

Fluid dynamics is an integral part of physics and underpins a large fraction of astrophysical theory. While it tends to play a less central role in the undergraduate physics curriculum at universities than, say, classical mechanics, electrodynamics or thermodynamics, it is nevertheless one of those topics that introduces a key concept in physics students are not likely to have already seen elsewhere (in this case, continuum conservation laws and their implications). It is also where students are inevitably going to be tested on their true level of understanding of some of their core skills as physicists, in particular, their fluency in vector algebra and thermodynamics. In the context of astrophysical research and education, fluid dynamics plays a significant role in many fields, including stellar structure, galaxy dynamics, cosmology and high-energy astrophysics to name but a few. The list of astrophysical phenomena that require a measure of understanding of fluid dynamics is almost endless, including but not at all limited to stars, the interstellar medium, cosmic jets, atmospheres and the cosmic web.

Various books already exist covering facets of astrophysical fluid dynamics, some of these having achieved a well-deserved status as classic texts in the field. It is therefore imperative for any new arrival on this scene to make clear its target audience and its rationale for including the particular topics within fluid dynamics (a vast field in its own right) that it does. The current book is an outgrowth of two year-three undergraduate courses that I have taught over the past six years at the University of Bath in the UK, *Fluid Dynamics in Physics and Astrophysics* and *Computational Astrophysics*. The target audience is therefore the advanced undergraduate student who will bring into their study of the topic some experience with vector algebra, classical electrodynamics, classical mechanics, thermodynamics, statistical physics and special relativity but not necessarily a complete fluency or full mastery of these. Additionally, I have aimed to provide a book that can be used as a reference work or refresher text for more advanced students and researchers.

To make for a book that is both accessible to year-three students while still retaining its value when revisited or first encountered later on, I have aimed for the discussion to be self-contained where possible in a manner that is not too intrusive. A first chapter (re-)introduces the mathematical tools needed for the approach taken to the treatment of fluid dynamics in this book and can be skimmed by the more advanced reader. Similarly, the core concepts from thermodynamics that make an appearance are included but deferred to an appendix. The treatment does not shy away from some mathematical formalism in places but tries to foreground the physics as much as possible. Whether or not to make use of tensors and index notation was an easy question to answer, given the inclusion of special relativistic fluid dynamics, but I believe the other chapters also benefit from the choice to do so even if students have not yet encountered this previously when starting in this book. For example, viscosity also becomes easier to describe using index notation than in vector notation, in my view.

Most books on fluid dynamics will probably include core topics such as the conservation laws in Eulerian and Lagrangian form, some hydrostatics, fluid parcel kinematics, sound and shock waves, viscous flow, some on fluid instabilities and at least a passing mention of magnetohydrodynamics. Books with an astrophysical flavour will add to this inevitable topics such as accretion flow and some stellar structure theory. All are included here as well. Additionally, I have included fluid dynamics in special relativity (present, for example, in the classical text books *Fluid Mechanics* by Landau & Lifshitz and *Foundation of radiation hydrodynamics* by Mihalas & Mihalas but not in e.g. *Gas Dynamics* by Shu; see the bibliography for full references). With the advent of multi-messenger astrophysics, providing us with more channels to observe cosmic extreme plasma flows, this is once again a timely topic. Self-similar solutions and shell models both for relativistic and non-relativistic explosions are provided in some detail. This allows students to, for example, connect to the topic of gamma-ray bursts (my own area of research expertise) in a relatively easy manner.

Another substantial addition is an extended treatment of computational fluid dynamics using the finite volume method. There is not enough room in the book to include alternative approaches such as smoothed-particle hydrodynamics or finite element methods. A choice had to be made and the finite volume method serves well as an introduction to various well-known astrophysical hydrodynamics codes currently used in research. After reading this chapter, the student should have all the tools to write their own solver. An implementation in the C programming language of the methods discussed in this chapter is also available for download[1]. I believe it is empowering to students to be able to actually write such a code, rather than be presented with a generic overview for passive consumption, and that this helps students better understand key fluid behaviours such as shock waves, sound waves and instabilities. A code written on the basis of Chapter 13 might not on its own be competitive but could well serve as the basis for such a code.

The default unit system is *cgs* units, although references are made to other unit systems where useful. The motivation for this choice is that the actual units only really matter whenever specific examples of hydrodynamic systems are given. Since these are mostly drawn from astrophysics, where cgs units remain a standard, this will make for an easier connection to astrophysical literature. In various cases, ad-hoc units remain the most obvious choice and will be used instead (a solar mass, a parsec, a litre, and so forth).

When I teach astrophysical fluid dynamics to students in the first semester of their third year at two lecture hours per week, I do not cover the entire book. Instead, I roughly include the following sections: 1.1, 1.2, 1.4.1 (week 1); 2.1 (week 2); 2.2, 2.4 (week 3); 3.1, 3.2, 3.4, but not 3.4.1 (week 4); 4.1, 5.1, 5.2 (week 5); 6.1, 6.2, 6.3, 6.4, 6.5, 6.6, 6.7 (week 6); 7.1, 7.2, 7.3, 7.4 (week 7); 8.1, 8.2, 8.3, 8.4 (week 8); 9.1, 9.2, 9.3, 9.4, 9.5 (week 9); 2.4, 10.1, 11.1, 11.2 (week 10), leaving the final week to either

---

[1]See https://github.com/hveerten/code_fluid_dynamics_book for details.

consolidate earlier material or offer a survey of topics not already covered, such as plasma physics (Chapter 12). Other selections of material are certainly possible, and hopefully, the book is written in a manner that is sufficiently readable to be useful for self-study. In my own teaching, the material on computational fluid dynamics is all covered in a three-week block including practice sessions in PC labs.

I am indebted to many people with regard to this textbook, including all the Bath physics students who have helped to weed out the typos and mistakes in the lecture notes over the years. I gratefully acknowledge the many hours that my colleagues Geoffrey Ryan and David Tsang and my PhD student John Hope have invested in proofreading the text book and providing valuable feedback. The latter half of the discussion on accretion flow owes a lot to teaching notes prepared by Geoff that he has kindly allowed me to draw from directly.

Finally, I am grateful to my wife Lisette and children Merel, Esther and Erik for their helpful support and equally welcome distraction during the writing of this book.

1 Preliminaries

Fluid dynamics by its nature involves a substantial amount of vector algebra and differentiation. Hopefully, most of the techniques from these areas of maths that we will need for the purpose of this text book will have been encountered before by the reader. But to avoid getting lost in the details of vector notation and the subtler points about derivatives at a stage when we want to focus on other matters, we will start off with discussing the maths of fluid dynamics before introducing any of the physics. Even if most (if not all) of this chapter serves to refresh one's memory rather than introduce anything new, it is worthwhile to go through everything here at least once to ensure that the notation conventions to be followed in this book are clear.

1.1 VECTORS AND THEIR BASES

Vectors are objects that have a magnitude and a direction. They offer a convenient means of grouping quantities together that are sufficiently similar that one can conceive of a continuous rotation between them—the obvious example being directions in (three-dimensional) space. In a coordinate-independent manner, we can write

$$\mathbf{v} = v^x \hat{\mathbf{e}}_x + v^y \hat{\mathbf{e}}_y + v^z \hat{\mathbf{e}}_z. \tag{1.1}$$

Here v^x, v^y and v^z are the projected magnitudes of $\mathbf{v}$ along *basis* vectors $\hat{\mathbf{e}}_x$, $\hat{\mathbf{e}}_y$ and $\hat{\mathbf{e}}_z$. We start our discussion of vectors by considering Cartesian basis vectors, using a hat symbol to denote that they each have length unity. Cartesian basis vectors are also orthogonal. We define an *inner product* or *dot product*[1] that maps two vectors onto a scalar number. Denoting with a dot, this is defined to obey

$$\hat{\mathbf{e}}_x \cdot \hat{\mathbf{e}}_x = \hat{\mathbf{e}}_y \cdot \hat{\mathbf{e}}_y = \hat{\mathbf{e}}_z \cdot \hat{\mathbf{e}}_z = 1, \tag{1.2}$$

$$\hat{\mathbf{e}}_x \cdot \hat{\mathbf{e}}_y = \hat{\mathbf{e}}_y \cdot \hat{\mathbf{e}}_x = \hat{\mathbf{e}}_x \cdot \hat{\mathbf{e}}_z = \hat{\mathbf{e}}_z \cdot \hat{\mathbf{e}}_x = \hat{\mathbf{e}}_y \cdot \hat{\mathbf{e}}_z = \hat{\mathbf{e}}_z \cdot \hat{\mathbf{e}}_y = 0. \tag{1.3}$$

The dot product of two vectors is defined to be symmetric with respect to the interchange of the two vectors and to be linear under vector addition. The dot product of a vector with itself produces the square of the length of the vector, that is $\mathbf{v} \cdot \mathbf{v} = |\mathbf{v}|^2$ if the magnitude of a vector is indicated by surrounding it with two vertical bars. The *position vector* $\mathbf{r}$ is the vector that, at any given point in space, denotes the direction from the origin to that point and the length r of the distance between the origin and that point. In Cartesian coordinates we have

$$\mathbf{r} = x \hat{\mathbf{e}}_x + y \hat{\mathbf{e}}_y + z \hat{\mathbf{e}}_z, \tag{1.4}$$

and therefore

$$\mathbf{r} \cdot \mathbf{r} = x^2 + y^2 + z^2 = r^2, \tag{1.5}$$

---

[1] We will not need to distinguish between the inner product and dot product, even though the latter is formally a special case of the first.

DOI: 10.1201/9781003095088-1

providing an example demonstration of the connection between the inner product and the length of a vector.

Throughout this book we will work nearly exclusively in *Euclidean space*, or at least space that is not curved by gravity as allowed for by general relativity[2]. In this context, we do not need to worry about caveats to the statements above about position vectors and inner products. However, we remain free to define our basis vectors as we see fit. Basis vectors summarize the possible directions from each point in space, and while Cartesian basis vectors are extremely simple in that they are orthogonal, of unit length and position-independent, these are not strict requirements of basis vectors. For example, *spherical coordinates* are often useful in astrophysical applications of fluid dynamics. But in these coordinates, the property of position-independence no longer holds, as shown in Figure 1.1 where at each position the basis vectors point in different directions. But what, then, are basis vectors? How should they be interpreted and defined?

Let us define the *natural* basis vectors $\mathbf{e}_u$, $\mathbf{e}_v$, $\mathbf{e}_w$ for the coordinate basis of a given parametrization of coordinates u, v, w (note the absence of the hat symbol, the vectors do not need to have length unity). At each position (u, v, w) we can define a set of coordinate curves by varying one coordinate while leaving the others constant. The basis vectors are then defined to lie along these curves, so if we express our route along points on such a curve in terms of the position vector, we can take

$$\mathbf{e}_u = \frac{\partial \mathbf{r}}{\partial u}, \qquad \mathbf{e}_v = \frac{\partial \mathbf{r}}{\partial v}, \qquad \mathbf{e}_w = \frac{\partial \mathbf{r}}{\partial w}. \tag{1.6}$$

Partial derivatives are defined exactly to 'leave the others constant', so moving in the direction of a basis vector as defined above will indeed mean embarking on a trip along the appropriate coordinate curve. As a trivial example, consider the Cartesian basis vector $\mathbf{e}_x$:

$$\mathbf{e}_x = \frac{\partial \mathbf{r}}{\partial x} = \frac{\partial}{\partial x} \left(x\hat{\mathbf{e}}_x + y\hat{\mathbf{e}}_y + z\hat{\mathbf{e}}_z \right) = \hat{\mathbf{e}}_x, \tag{1.7}$$

with the definition, in this case, leading to a set of basis vectors of length unity.

As a less trivial example, consider spherical coordinates (r, θ, ϕ) with angles defined as shown in Figure 1.2, which obey

$$x = r\sin\theta\cos\phi, \qquad y = r\sin\theta\sin\phi, \qquad z = r\cos\theta. \tag{1.8}$$

Taking partial derivatives of the position vector $\mathbf{r} = x\hat{\mathbf{e}}_x + y\hat{\mathbf{e}}_y + z\hat{\mathbf{e}}_z$, we can obtain the basis vectors:

$$\mathbf{e}_r = \frac{\partial x}{\partial r}\hat{\mathbf{e}}_x + \frac{\partial y}{\partial r}\hat{\mathbf{e}}_y + \frac{\partial z}{\partial r}\hat{\mathbf{e}}_z = \sin\theta\cos\phi\hat{\mathbf{e}}_x + \sin\theta\sin\phi\hat{\mathbf{e}}_y + \cos\theta\hat{\mathbf{e}}_z, \tag{1.9}$$

$$\mathbf{e}_\theta = \frac{\partial x}{\partial \theta}\hat{\mathbf{e}}_x + \frac{\partial y}{\partial \theta}\hat{\mathbf{e}}_y + \frac{\partial z}{\partial \theta}\hat{\mathbf{e}}_z = r\cos\theta\cos\phi\hat{\mathbf{e}}_x + r\cos\theta\sin\phi\hat{\mathbf{e}}_y - r\sin\theta\hat{\mathbf{e}}_z, \tag{1.10}$$

$$\mathbf{e}_\phi = \frac{\partial x}{\partial \phi}\hat{\mathbf{e}}_x + \frac{\partial y}{\partial \phi}\hat{\mathbf{e}}_y + \frac{\partial z}{\partial \phi}\hat{\mathbf{e}}_z = -r\sin\theta\sin\phi\hat{\mathbf{e}}_x + r\sin\theta\cos\phi\hat{\mathbf{e}}_y. \tag{1.11}$$

---

[2]For our purpose, we can typically assume other curved spaces such as the surface of the Earth to be embedded in a higher-dimensional Euclidean space when it comes to defining a position vector that can be interpreted in terms of distance from a coordinate origin.

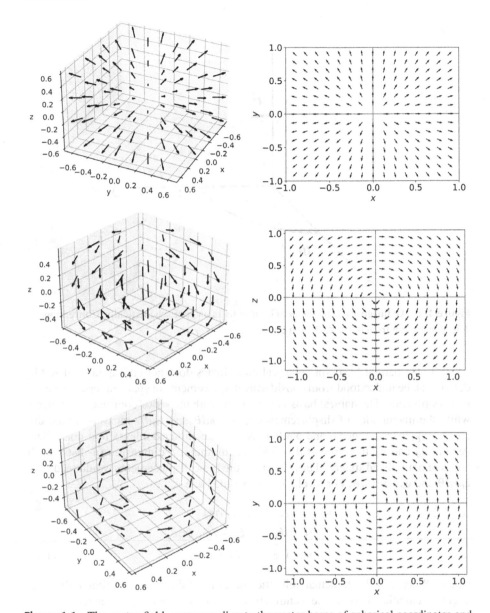

Figure 1.1 The vector fields corresponding to the vector bases of spherical coordinates and intersections in two dimensions. From top to bottom: $\hat{\mathbf{e}}_r$, $\hat{\mathbf{e}}_\theta$ and $\hat{\mathbf{e}}_\phi$. Top and bottom figure intersections show the xy-plane, and the middle shows the xz-plane.

In this case, the natural basis vectors are not yet normalized, except for $\mathbf{e}_r$, as shown by their inner products:

$$\mathbf{e}_r \cdot \mathbf{e}_r = 1, \qquad \mathbf{e}_\theta \cdot \mathbf{e}_\theta = r^2, \qquad \mathbf{e}_\phi \cdot \mathbf{e}_\phi = r^2 \sin^2 \theta. \qquad (1.12)$$

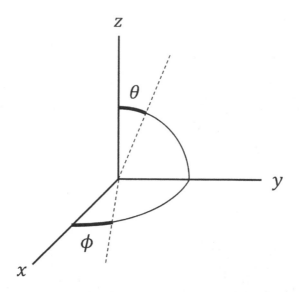

Figure 1.2 An image showing the definition of theta and phi schematically. Conventionally, ϕ is taken to run from 0 to 2π, θ from 0 to π.

Why the natural basis for spherical coordinates does not end up normalized by default can be understood from considering displacements in space in terms of spherical coordinates. The natural basis is associated with the set of coordinates $\{r, \theta, \phi\}$, while the magnitudes of displacements $\mathbf{e}_r dr$, $\mathbf{e}_\theta d\theta$ and $\mathbf{e}_\phi d\phi$ measure actual distances (e.g. in cm). Since θ and ϕ are in radians (i.e. dimensionless), non-normalized vectors $\mathbf{e}_\theta$ and $\mathbf{e}_\phi$ therefore still contain at minimum a dimension of distance. In terms of normalized spherical basis vectors, the same displacements would read $\hat{\mathbf{e}}_r dr$, $\hat{\mathbf{e}}_\theta r d\theta$ and $\hat{\mathbf{e}}_\phi r \sin\theta d\phi$. For a velocity, we would have

$$
\begin{aligned}
\mathbf{v} &= \dot{r}\mathbf{e}_r + \dot{\theta}\mathbf{e}_\theta + \dot{\phi}\mathbf{e}_\phi \\
&= \dot{r}\hat{\mathbf{e}}_r + r\dot{\theta}\hat{\mathbf{e}}_\theta + r\sin\theta\,\dot{\phi}\hat{\mathbf{e}}_\phi \\
&= v^r\hat{\mathbf{e}}_r + v^\theta\hat{\mathbf{e}}_\theta + v^\phi\hat{\mathbf{e}}_\phi,
\end{aligned}
\tag{1.13}
$$

where v^θ and v^ϕ refer to actual velocities in terms of distance over time rather than angular velocities $\dot{\theta}$ and $\dot{\phi}$ (i.e. centimetres per second, not radians per second). The overhead dots denote time derivatives, with $\dot{X} \equiv dX/dt$.

Even though we immediately expressed the position vector in terms of Cartesian basis vectors in equations 1.7 and 1.9–1.11, this is not compulsory and we could have postponed using a Cartesian basis or used a different basis altogether. Let us postpone Cartesian coordinates just by a little bit while re-establishing the spherical coordinate basis, in order to make this point explicit. Starting from the position vector in spherical coordinates $\mathbf{r} = r\hat{\mathbf{e}}_r$, we have

$$\mathbf{e}_r = \frac{\partial}{\partial r}(r\hat{\mathbf{e}}_r) = \hat{\mathbf{e}}_r, \tag{1.14}$$

consistent with Equation 1.9. Note that $\hat{\mathbf{e}}_r = \mathbf{e}_r$ is itself not a function of r and therefore not affected by the partial derivative. In the case of θ, ϕ, however, the situation is different, and we have

$$\mathbf{e}_\theta = \frac{\partial}{\partial \theta}(r\hat{\mathbf{e}}_r) = r\frac{\partial \hat{\mathbf{e}}_r}{\partial \theta} = r\hat{\mathbf{e}}_\theta. \tag{1.15}$$

(temporarily retreating into Cartesian coordinates and acting with the partial derivative on the sine and cosine terms of Equation 1.9 offers a conceptually straightforward way to confirm the last step above). Likewise, for ϕ we have

$$\mathbf{e}_\phi = \frac{\partial}{\partial \phi}(r\hat{\mathbf{e}}_r) = r\frac{\partial \hat{\mathbf{e}}_r}{\partial \phi} = r\sin\theta\hat{\mathbf{e}}_\phi. \tag{1.16}$$

This example therefore also serves to emphasize the position-dependent nature of the spherical coordinates basis. Rather than a single immutable set of directions, we have established a continuum of basis orientations throughout space. While spherical coordinates and curvilinear coordinates, in general, can be by far the most convenient coordinate system to use in certain cases, this position-dependence is not without its pitfalls, and we will return to this below.

Actually, what we have done so far is to define the local natural basis within a Euclidean *tangent space*. Formally, the tangent space intersects the space that we wish to map only at the point of interest and displacements within the tangent space away from this point are not necessarily physically meaningful. However, if the space we wish to map is itself a Euclidean space of equal dimension, then for all practical purposes the tangent space can be conflated with the space that we are mapping. But even in ordinary fluid mechanics outside of relativity theory, it is easy to conceive of situations where the two spaces are not equal. Consider, for example, a river running through a landscape of varying height. We only need two directions to map our position on the river, but the river coordinates map onto a threedimensional vector when the height of the surface is taken into account (see Problem 1.2). Another example is the curved surface of a sphere, e.g. the lower earth atmosphere. Again, we only need two coordinates to determine our position on this surface. In this case, the tangent surface is a plane (e.g. spanned by the basis vectors $\mathbf{e}_\theta$ and $\mathbf{e}_\phi$ from spherical coordinates). But as soon as we move within this plane we will leave the surface of the sphere, so therefore the curved surface cannot be conflated with the tangent plane. Both the river and the spherical surface are examples of curved spaces embedded in a space of higher dimension, unlike the intrinsically curved spaces that occur in the presence of strong gravity[3]. Further details can be curved coordinate systems found in Appendix C.

---

[3]For a less concise treatment of tangent spaces, see e.g. [13]. The term *natural basis* is drawn from this textbook, but other terminology, such as *tangent basis*, can also be found in the literature [17].

PROBLEM 1.1
The natural basis of cylindrical coordinates

Cylindrical coordinates ς, ϕ, z obey

$$x = \varsigma \cos \phi, \qquad y = \varsigma \sin \phi, \qquad z = z.$$

i) Compute the natural basis vectors $\mathbf{e}_\varsigma$, $\mathbf{e}_\phi$, $\mathbf{e}_z$ starting from the position vector in Cartesian coordinates x, y, z.

ii) Express the position vector in cylindrical coordinates.

iii) Sketch the three vector fields produced by these basis vectors, as in the spherical example Figure 1.1.

PROBLEM 1.2
Tangent planes

Consider a road running through a hilly landscape. The surface elevation level obeys $z = 1 + \sin(x)$. The road runs along the x-direction, covering a width from $y = -1$ to $y = 1$. Seen on a map, the road shows as a flat strip, and we define coordinates on this two-dimensional surface obeying a straightforward $\bar{x} = x$ and $\bar{y} = y$.

i) Give an expression for the normalized natural basis vectors $\hat{\mathbf{e}}_{\bar{x}}$ and $\hat{\mathbf{e}}_{\bar{y}}$ spanning the tangent plane, in terms of standard Cartesian unit vectors $\hat{\mathbf{e}}_x$, $\hat{\mathbf{e}}_y$ and $\hat{\mathbf{e}}_z$.

ii) To demonstrate that the natural basis defines a tangent plane that only instantaneously coincides with the surface it is tangent to, move a distance $\pi/6$ along the (un-normalized) vector $\mathbf{e}_{\bar{x}}$ in the tangent plane starting at the origin $(\bar{x}, \bar{y}) = (0, 0)$. By how much have you moved away from the road surface by this point?

1.1.1 MATRIX NOTATION

In the case of Cartesian coordinates, we can utilize the machinery of *linear algebra* to work with vectors in a straightforward manner. For example, we can group the components of a vector $\mathbf{v}$ together according to

$$\mathbf{v} = v^x \hat{\mathbf{e}}_x + v^y \hat{\mathbf{e}}_y + v^z \hat{\mathbf{e}}_z \rightarrow V = \begin{bmatrix} v^x \\ v^y \\ v^z \end{bmatrix}_{\text{Cartesian}}, \tag{1.17}$$

to form a *column vector*. Note that the expression on the left-hand side (LHS) is a true *vector* in the coordinate-independent sense, whereas the final term on the right-hand side (RHS) expression after the arrow is 'merely' a grouping together of the components of $\mathbf{v}$ *in a specific basis* and thus a more limited object[4]. I explicitly indicated the coordinate system (*Cartesian*) here, but this I will often omit for brevity when it is obvious from context. If the entries are grouped horizontally rather than vertically, we have a row vector

$$V^T = \begin{bmatrix} v^x & v^y & v^z \end{bmatrix}, \tag{1.18}$$

obtained by transposing V.

---

[4]In linear algebra the notation using straight brackets is common. In this choice, we follow e.g. [22].

Vector equations are a convenient means to concisely encode multiple scalar equations, e.g.

$$3\mathbf{v} + 2\mathbf{u} = 0 \rightarrow \begin{cases} 3v^x + 2u^x = 0 \\ 3v^y + 2u^y = 0 \\ 3v^z + 2u^z = 0 \end{cases} \rightarrow \begin{bmatrix} 3v^x \\ 3v^y \\ 3v^z \end{bmatrix} + \begin{bmatrix} 2u^x \\ 2u^y \\ 2u^z \end{bmatrix} = \begin{bmatrix} 0 \\ 0 \\ 0 \end{bmatrix} \tag{1.19}$$

The inner product between two vectors results in a scalar that gives the projected length of one onto the other, multiplied by the other's length, i.e. $\mathbf{v} \cdot \mathbf{w} = vw\cos\theta$ with v and w the scalar lengths of the vectors and θ the angle between the vectors. Between general and basic-specific linear algebra notation of the inner product, we have

$$\mathbf{v} \cdot \mathbf{w} = V^T W = \begin{bmatrix} v^x & v^y & v^z \end{bmatrix} \begin{bmatrix} w^x \\ w^y \\ w^z \end{bmatrix} = v^x w^x + v^y w^y + v^z w^z. \tag{1.20}$$

In matrix notation the inner product dot is typically omitted, and the inner product is implied by the juxtaposition of a vector and its transposed version. Although in matrix notation we have made a specific choice of basis (Cartesian in the case above), the inner product is a scalar quantity independent of this basis, and the equal signs throughout this expression remain appropriate.

Outside of matrix notation *the absence of a dot between vectors means that no inner product is to be taken between them*. In fluid dynamics, in particular, we will soon encounter equations like

$$\frac{\partial \rho \mathbf{v}}{\partial t} + \nabla \cdot (\rho \mathbf{v} \mathbf{v}) + \nabla p = 0, \tag{1.21}$$

(which we will at some point recognise as expressing conservation of momentum density). The most straightforward way of reading this expression is to think once again in terms of the total information it is meant to encode, as the following example makes clear:

$$2\mathbf{v}\mathbf{w} + \mathbf{v}\mathbf{v} = 0 \rightarrow \begin{cases} 2v^x\mathbf{w} + v^x\mathbf{v} = 0 \rightarrow \begin{cases} 2v^xw^x + v^xv^x = 0 \\ 2v^xw^y + v^xv^y = 0 \\ 2v^xw^z + v^xv^z = 0 \end{cases} \\ 2v^y\mathbf{w} + v^y\mathbf{v} = 0 \rightarrow \begin{cases} 2v^yw^x + v^yv^x = 0 \\ 2v^yw^y + v^yv^y = 0 \\ 2v^yw^z + v^yv^z = 0 \end{cases} \\ 2v^z\mathbf{w} + v^z\mathbf{v} = 0 \rightarrow \begin{cases} 2v^zw^x + v^zv^x = 0 \\ 2v^zw^y + v^zv^y = 0 \\ 2v^zw^z + v^zv^z = 0 \end{cases} \end{cases} \tag{1.22}$$

The same nine expressions can also compactly be written using matrix notation:

$$2\begin{bmatrix} v^xw^x & v^xw^y & v^xw^z \\ v^yw^x & v^yw^y & v^yw^z \\ v^zw^x & v^zw^y & v^zw^z \end{bmatrix} + \begin{bmatrix} v^xv^x & v^xv^y & v^xv^z \\ v^yv^x & v^yv^y & v^yv^z \\ v^zv^x & v^zv^y & v^zv^z \end{bmatrix} = \begin{bmatrix} 0 & 0 & 0 \\ 0 & 0 & 0 \\ 0 & 0 & 0 \end{bmatrix}. \tag{1.23}$$

Both the expression in the form of a series of equations and in matrix notation implied a choice of basis (Cartesian). That we can use matrix notation here is illustrative of the fact that we can choose to interpret the combination **vw** as a *tensor* and write $\mathbf{v} \otimes \mathbf{w}$, but we will not pursue tensor algebra in much depth[5]. Typically, the only tensors we will need we can represent in the form of vectors (a *rank-1 tensor*) and matrices (a *rank-2 tensor*).

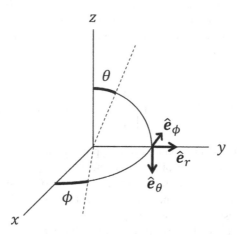

Figure 1.3 The orientation of spherical coordinate unit vectors depends on the point in the grid and is ill-defined in the origin.

1.1.2 LIMITATIONS OF MATRIX NOTATION

While it is somewhat possible to push the matrix notation approach beyond Cartesian coordinates, it is not long before we run out of steam. Problem 1.3 explores the impact of using matrix notation in a basis with basis vectors that are not normalized and/or orthogonal. Even for coordinate systems that are both normalized and orthogonal, things are bound to go wrong when working in a coordinate system that is position-dependent, such as spherical coordinates (see Figure 1.3), if one does not tread very carefully. Remember that a vector does not intrinsically have a location in space (it merely has a length and a direction, and if it is used to describe some quantity at a certain point in space, this is an extra piece of contextual information).

Consider, for example[6], the inner product of two vectors **v** and **w** at the origin as shown in the left image of Figure 1.4, expressed in Cartesian coordinates according

---

[5]Also, do not confuse tensor product $\otimes$ with cross product $\times$. The two are unrelated. Note that the term 'outer product' refers to the tensor product rather than the cross product, something which can confuse e.g. native speakers of German or Dutch (who might be tempted to directly translate 'äusseres Produkt' and 'uitwendig product' respectively).

[6]See e.g. [16] for a very similar discussion of these limitations, electrodynamics being another branch of physics where confusion about vectors is not unheard of among students.

to

$$\mathbf{v} \cdot \mathbf{w} = \begin{bmatrix} 0 & 1 & 0 \end{bmatrix}_{\substack{\text{Cartesian} \\ \text{at } 0,0,0}} \begin{bmatrix} 0 \\ -1 \\ 0 \end{bmatrix}_{\substack{\text{Cartesian} \\ \text{at } 0,0,0}} = -1. \qquad (1.24)$$

The label indicating that these vectors are expressed in a basis defined at the origin is pointless for the case of Cartesian coordinates, since these are independent of position. But in spherical coordinates (r, θ, ϕ), this inner product cannot even be meaningfully expressed for vectors at the origin,

$$\mathbf{v} \cdot \mathbf{w} = \begin{bmatrix} 1 & ? & ? \end{bmatrix}_{\substack{\text{Spherical} \\ \text{at } 0,?,?}} \begin{bmatrix} 1 \\ ? \\ ? \end{bmatrix}_{\substack{\text{Spherical} \\ \text{at } 0,?,?}} = ?, \qquad (1.25)$$

because the origin is a singular point where the angles θ and ϕ are undefined[7], and it is unclear which values to set for the distances along $\hat{\mathbf{e}}_\theta$ and $\hat{\mathbf{e}}_\phi$. Note how it might be tempting to pick both lengths equal to 0 and end up erroneously with an inner product 1 instead of -1, even though the scalar inner product is supposedly coordinate-independent. Evaluating this same inner product elsewhere on the grid— which makes no difference for coordinate-independent Cartesian unit vectors, but means avoiding a singularity in spherical coordinates—works just fine as far as the inner product is concerned (right image of Figure 1.4):

$$\mathbf{v} \cdot \mathbf{w} = \begin{bmatrix} 0 & 1 & 0 \end{bmatrix}_{\substack{\text{Cartesian} \\ \text{at } 0,1,0}} \begin{bmatrix} 0 \\ -1 \\ 0 \end{bmatrix}_{\substack{\text{Cartesian} \\ \text{at } 0,1,0}} = -1, \qquad (1.26)$$

and

$$\mathbf{v} \cdot \mathbf{w} = \begin{bmatrix} 1 & 0 & 0 \end{bmatrix}_{\substack{\text{Spherical} \\ \text{at } 1, \frac{\pi}{2}, \frac{\pi}{2}}} \begin{bmatrix} -1 \\ 0 \\ 0 \end{bmatrix}_{\substack{\text{Spherical} \\ \text{at } 1, \frac{\pi}{2}, \frac{\pi}{2}}} = -1. \qquad (1.27)$$

However, the bottom line remains that one should be extremely careful when attempting to extend matrix notation beyond Cartesian coordinates.

PROBLEM 1.3
Matrix notation in an unsuitable coordinate system

In this problem, we explore what happens if matrix notation is erroneously applied without regard for the length and orthogonality of a coordinate system.

i) Consider a coordinate system vw in two dimensions with basis vectors $\mathbf{e}_v = 2\hat{\mathbf{e}}_x$ and $\mathbf{e}_w = \hat{\mathbf{e}}_y$. Express the vectors $\mathbf{s} = a\hat{\mathbf{e}}_x + b\hat{\mathbf{e}}_y$ and $\mathbf{t} = c\hat{\mathbf{e}}_x + d\hat{\mathbf{e}}_y$ in this basis and demonstrate that a wrong outcome results for the dot product between the two when computed in matrix-notation in coordinate system vw without regard for the length of its basis vectors.

---

[7] Remember, its the base of the arrow that matters here, not the tip.

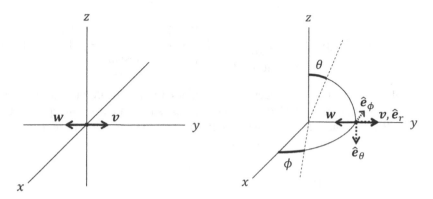

Figure 1.4 Because spherical coordinates are ill-defined at the origin, vectors **v** and **w** can directly be seen to lie parallel to $\hat{\mathbf{e}}_r$ in the right figure but cannot be expressed as such in the left figure.

ii) What if the two vectors **s** and **t** from (i) are added in the same matrix notation?

iii) Now consider a coordinate system vw with basis vectors that are of unit length but not orthogonal:

$$\hat{\mathbf{e}}_v = \frac{\hat{\mathbf{e}}_x - \hat{\mathbf{e}}_y}{\sqrt{2}}, \qquad \hat{\mathbf{e}}_w = \hat{\mathbf{e}}_y.$$

Demonstrate that a wrong outcome results when the dot product of **s** and **t** (as defined previously) is computed in matrix notation in this coordinate system vw without regard for the orthogonality of its basis vectors.

iv) Will the outcome also be wrong when the inner product is computed using matrix notation in a coordinate system that is merely rotated relative to an initial Cartesian basis? That is, when we take

$$\hat{\mathbf{e}}_v = \cos\phi\,\hat{\mathbf{e}}_x + \sin\phi\,\hat{\mathbf{e}}_y,$$
$$\hat{\mathbf{e}}_w = -\sin\phi\,\hat{\mathbf{e}}_x + \cos\phi\,\hat{\mathbf{e}}_y.$$

1.1.3   THE CROSS-PRODUCT AND COORDINATE TRANSFORMATIONS

The familiar cross-product is defined in Cartesian coordinates according to

$$A \times B \rightarrow \begin{bmatrix} a^x \\ a^y \\ a^z \end{bmatrix} \times \begin{bmatrix} b^x \\ b^y \\ b^z \end{bmatrix} = \begin{bmatrix} a^y b^z - a^z b^y \\ -a^x b^z + a^z b^x \\ a^x b^y - a^y b^x \end{bmatrix}. \tag{1.28}$$

The cross product of two vectors produces a vector that is perpendicular to both, e.g.

$$\begin{bmatrix} 1 \\ 0 \\ 0 \end{bmatrix} \times \begin{bmatrix} 0 \\ 1 \\ 0 \end{bmatrix} = \begin{bmatrix} 0 \\ 0 \\ 1 \end{bmatrix}, \tag{1.29}$$

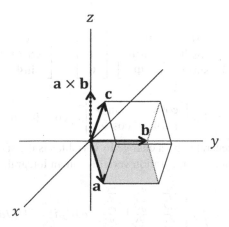

Figure 1.5 Using the cross and dot product to span a volume, where $V = (\mathbf{a} \times \mathbf{b}) \cdot \mathbf{c}$. Here, vectors $\mathbf{a}$ and $\mathbf{b}$ lie in the xy-plane, so their cross-product runs along z. Vector $\mathbf{c}$ is tilted at an angle relative to the z-axis.

etc. Unlike the dot product, it is anti-symmetric under swapping of the vectors. Like the dot product, it is proportional to the length of the input vectors, albeit that the dot product obeys $\mathbf{a} \cdot \mathbf{b} = \cos\theta \, |\mathbf{a}| \, |\mathbf{b}|$ for an angle θ between the vectors, whereas the cross product produces a vector with length $\sin\theta \, |\mathbf{a}| \, |\mathbf{b}|$.

These dependencies on $\cos\theta$ and $\sin\theta$ have as consequence that the volume spanned by three vectors can neatly be expressed using a dot and a cross product according to

$$V = \mathrm{abs}\,(\mathbf{a} \times \mathbf{b}) \cdot \mathbf{c}, \tag{1.30}$$

using 'abs' to denote the absolute value. This is illustrated in Figure 1.5. The (scalar) volume V is independent of the overall orientation of the set $\mathbf{a}, \mathbf{b}, \mathbf{c}$ but is impacted by their lengths and their relative angles (with a fully perpendicular set maximising V). The above can also be written (in Cartesian coordinates) as

$$V = \mathrm{abs}\,|\mathrm{A}|, \qquad \mathrm{A} = \begin{bmatrix} a^x & b^x & c^x \\ a^y & b^y & c^y \\ a^z & b^z & c^z \end{bmatrix}, \tag{1.31}$$

i.e. the determinant of A is a measure of the volume spanned by $\mathbf{a}, \mathbf{b}, \mathbf{c}$. The matrix A applied to vectors $\hat{\mathbf{e}}_x, \hat{\mathbf{e}}_y, \hat{\mathbf{e}}_z$, results in

$$\mathrm{A} \cdot \hat{\mathbf{e}}_x = \mathbf{a}, \qquad \mathrm{A} \cdot \hat{\mathbf{e}}_y = \mathbf{b}, \qquad \mathrm{A} \cdot \hat{\mathbf{e}}_z = \mathbf{c}, \tag{1.32}$$

which therefore amounts to a basis transformation that includes a volume scaling by a factor V. As an example, the matrix corresponding to a rotation around the x-axis for example will have $V = 1$, because it does not alter any of the lengths of the basis vectors:

$$A \cdot \hat{e}_y \rightarrow \begin{bmatrix} 1 & 0 & 0 \\ 0 & \cos\Phi & -\sin\Phi \\ 0 & \sin\Phi & \cos\Phi \end{bmatrix} \begin{bmatrix} 0 \\ 1 \\ 0 \end{bmatrix} = \begin{bmatrix} 0 \\ \cos\Phi \\ \sin\Phi \end{bmatrix}, \text{etc.,} \quad (1.33)$$

with

$$V = |A| = 1 \times \begin{vmatrix} \cos\Phi & -\sin\Phi \\ \sin\Phi & \cos\Phi \end{vmatrix} = \cos^2\Phi + \sin^2\Phi = 1. \quad (1.34)$$

This volume V generated by a change in variables is the same scaling factor that appears when changing the integration variables in an integral:

$$\int f dx' dy' dz' = \int f \begin{vmatrix} \frac{\partial x'}{\partial x} & \frac{\partial x'}{\partial y} & \frac{\partial x'}{\partial z} \\ \frac{\partial y'}{\partial x} & \frac{\partial y'}{\partial y} & \frac{\partial y'}{\partial z} \\ \frac{\partial z'}{\partial x} & \frac{\partial z'}{\partial y} & \frac{\partial z'}{\partial z} \end{vmatrix} dx dy dz = \int f J dx dy dz, \quad (1.35)$$

where J is the *Jacobian* of the coordinate transformation and provides the volume scaling that goes with the coordinate transformation. As a practical example, consider going from Cartesian to spherical coordinates. The coordinates relate via

$$x = r \sin\theta \cos\phi, \qquad y = r \sin\theta \sin\phi, \qquad z = r \cos\theta, \quad (1.36)$$

which leads to a Jacobian of the form

$$J = \begin{vmatrix} \frac{\partial x}{\partial r} & \frac{\partial x}{\partial \theta} & \frac{\partial x}{\partial \phi} \\ \frac{\partial y}{\partial r} & \frac{\partial y}{\partial \theta} & \frac{\partial y}{\partial \phi} \\ \frac{\partial z}{\partial r} & \frac{\partial z}{\partial \theta} & \frac{\partial z}{\partial \phi} \end{vmatrix} = \begin{vmatrix} \sin\theta\cos\phi & r\cos\theta\cos\phi & -r\sin\theta\sin\phi \\ \sin\theta\sin\phi & r\cos\theta\sin\phi & r\sin\theta\cos\phi \\ \cos\theta & -r\sin\theta & 0 \end{vmatrix} = r^2 \sin\theta. \quad (1.37)$$

PROBLEM 1.4
Transformations to spherical and polar coordinates

Show that the determinant as shown for the transformation to spherical coordinates indeed yields $r^2 \sin\theta$. Also show that $J = r$ for a two-dimensional integral transformed from Cartesian to polar coordinates.

1.1.4 GRADIENT, DIVERGENCE AND CURL

The ∇ ('nabla' or 'del') operator that we already briefly encountered in Equation 1.21 is a well-known vector operator and can be expressed in matrix notation as

$$\nabla = \begin{bmatrix} \frac{\partial}{\partial x} \\ \frac{\partial}{\partial y} \\ \frac{\partial}{\partial z} \end{bmatrix} \Rightarrow \nabla f = \begin{bmatrix} \frac{\partial f}{\partial x} \\ \frac{\partial f}{\partial y} \\ \frac{\partial f}{\partial z} \end{bmatrix}, \quad (1.38)$$

with ∇f being the *gradient* of a scalar function $f(\mathbf{r})$. The function f can also be interpreted as a *scalar field*, associating a scalar value with every position $\mathbf{r}$.

A vector field associates both a value and direction, i.e. a vector, with every position **r**. Its *divergence* is given by

$$\nabla \cdot \mathbf{v} = \begin{bmatrix} \frac{\partial}{\partial x} & \frac{\partial}{\partial y} & \frac{\partial}{\partial z} \end{bmatrix} \begin{bmatrix} v^x \\ v^y \\ v^z \end{bmatrix} = \frac{\partial v^x}{\partial x} + \frac{\partial v^y}{\partial y} + \frac{\partial v^z}{\partial z}. \tag{1.39}$$

The *curl* of a vector field is given by

$$\nabla \times V = \begin{bmatrix} \frac{\partial}{\partial x} \\ \frac{\partial}{\partial y} \\ \frac{\partial}{\partial z} \end{bmatrix} \times \begin{bmatrix} v^x \\ v^y \\ v^z \end{bmatrix} = \begin{bmatrix} \frac{\partial v^z}{\partial y} - \frac{\partial v^y}{\partial z} \\ -\frac{\partial v^z}{\partial x} + \frac{\partial v^x}{\partial z} \\ \frac{\partial v^y}{\partial x} - \frac{\partial v^x}{\partial y} \end{bmatrix}. \tag{1.40}$$

Let us look at a few example vector fields to illustrate divergence and curl. First, consider a vector field of increasing strength in the x-direction, $\mathbf{v} = x\hat{\mathbf{e}}_x$. This field has a divergence $\nabla \cdot \mathbf{v} = 1$. The curl $\nabla \times \mathbf{v} = 0$ (looking ahead to a fluid dynamics interpretation: there are no eddies in this stream, but the stream flows steadily faster with distance along x).

Second, consider a spherically symmetric vector field $\mathbf{v} = v^r(r)\hat{\mathbf{e}}_r$. For this field, the divergence is given by

$$\nabla \cdot \mathbf{v} = \frac{2v^r}{r} + \frac{\partial v^r}{\partial r}. \tag{1.41}$$

In the case of constant v^r, or v^r increasing less steeply than linear with r, this divergence becomes smaller as one moves further outwards. In the limiting case of large r, the divergence is effectively zero as the vector field divergence is diluted over an increasing spherical surface. The vector field is without curl, $\nabla \times \mathbf{v} = 0$.

Third, consider a vector field generated by the ϕ basis vector of spherical coordinates (see Figure 1.1), given by $\mathbf{v} = \hat{\mathbf{e}}_\phi$. This vector field has no divergence, $\nabla \cdot \mathbf{v} = 0$, but it appears to show an obvious curl. Indeed, the non-zero curl is given by

$$\nabla \times V = \begin{bmatrix} 0 \\ 0 \\ \frac{1}{\sqrt{x^2+y^2}} \end{bmatrix}. \tag{1.42}$$

Note how the cross-product picks out the axis along which the curl occurs, in this case, the z-axis. In a manner similar to the behaviour at increasing distance from the origin of the spherically symmetric field from the previous example, the impact of the curl in this field is diluted with increasing distance from the z-axis.

1.1.5 GRADIENT, DIVERGENCE, CURL AND INTEGRATION

For a function $f(x)$ of one coordinate in a one-dimensional space, the fundamental theorem of calculus states that

$$\int_a^b \frac{df(x)}{dx} d = f(b) - f(a). \tag{1.43}$$

This can be readily understood when building up the integral as a Riemann sum and using the definition of the derivative:

$$
\begin{aligned}
\int_{x_0}^{x_N} \frac{df(x)}{dx} dx &= \lim_{h \to 0} \sum_{i=0}^{N} \left. \frac{df(x)}{dx} \right|_{x=x_i} h \\
&= \lim_{h \to 0} \sum_{i=0}^{N} \frac{f(x_i + \frac{h}{2}) - f(x_i - \frac{h}{2})}{h} h, \\
&= \lim_{h \to 0} \left(f(x_N + \frac{h}{2}) - f(x_0 - \frac{h}{2}) \right) \\
&= f(x_N) - f(x_0).
\end{aligned} \tag{1.44}
$$

All the internal terms cancel each other, and we are left with the values on the boundaries of the domain. This notion generalizes to integrals along a line in a higher-dimensional space,

$$
\int_{\mathbf{a}}^{\mathbf{b}} \nabla f(\mathbf{x}) d\mathbf{x} = f(\mathbf{b}) - f(\mathbf{a}). \tag{1.45}
$$

The notion of internal terms cancelling carries over to Gauss' theorem (also known as the divergence theorem:

$$
\int_{volume} (\nabla \cdot \mathbf{v}) \, dV = \oint_{surface} \mathbf{v} \cdot d\mathbf{S}. \tag{1.46}
$$

Instead of having two individual endpoints, the boundary is now defined by an oriented surface encompassing the volume of integration. The surface element $d\mathbf{S}$ is oriented perpendicularly outward from the volume. The divergence theorem also works with tensors T rather than vectors $\mathbf{v}$,

$$
\int_{volume} (\nabla \cdot T) \, dV = \oint_{surface} T \cdot d\mathbf{S}. \tag{1.47}
$$

An analogous theorem for the cross product is Stokes' theorem,

$$
\int_{surface} (\nabla \times \mathbf{v}) \cdot d\mathbf{S} = \oint_{perimeter} \mathbf{v} \cdot d\mathbf{l}, \tag{1.48}
$$

linking the curl of a vector on a surface to its properties at the perimeter of the surface. Internal curl contributions cancel in a manner similar to the cancellation of internal divergence and gradient contributions. We will return to the implications of Stokes' theorem for the curl of a fluid flow in Section 6.6.

PROBLEM 1.5
Spherically symmetric vector field

Consider a spherically symmetric vector field $\mathbf{v} = v^r \hat{\mathbf{e}}_r$. Use the *divergence theorem* to quickly show that

$$
\nabla \cdot \mathbf{v} = \frac{2v^r}{r} + \frac{\partial v^r}{\partial r}.
$$

1.1.6 DIVERGENCE, CURL AND CURVILINEAR COORDINATES

When acting on scalars, the ∇-operator just follows the product rule, e.g.:

$$\nabla(abc) = ab\nabla c + ac\nabla b + bc\nabla a. \tag{1.49}$$

But the gradient of a scalar does suffer complications deriving from the nature of the coordinate system:

$$\nabla a = \frac{\partial a}{\partial x}\hat{\mathbf{e}}_x + \frac{\partial a}{\partial y}\hat{\mathbf{e}}_y + \frac{\partial a}{\partial z}\hat{\mathbf{e}}_z, \tag{1.50}$$

in e.g. spherical coordinates takes the form

$$\nabla a = \frac{\partial a}{\partial r}\hat{\mathbf{e}}_r + \frac{1}{r}\frac{\partial a}{\partial \theta}\hat{\mathbf{e}}_\theta + \frac{1}{r\sin\theta}\frac{\partial a}{\partial \phi}\hat{\mathbf{e}}_\phi. \tag{1.51}$$

In principle, this equation can be derived using a brute force approach (a straightforward but lengthy process). This involves substituting

$$\hat{\mathbf{e}}_x = \sin\theta\cos\phi\,\hat{\mathbf{e}}_r + \cos\theta\cos\phi\,\hat{\mathbf{e}}_\theta - \sin\phi\,\hat{\mathbf{e}}_\phi,$$
$$\hat{\mathbf{e}}_y = \sin\theta\sin\phi\,\hat{\mathbf{e}}_r + \cos\theta\sin\phi\,\hat{\mathbf{e}}_\theta + \cos\phi\,\hat{\mathbf{e}}_\phi,$$
$$\hat{\mathbf{e}}_z = \cos\theta\,\hat{\mathbf{e}}_r - \sin\theta\,\hat{\mathbf{e}}_\theta, \tag{1.52}$$

for the unit basis vectors. Furthermore, the product rule

$$\frac{\partial}{\partial x} = \frac{\partial r}{\partial x}\frac{\partial}{\partial r} + \frac{\partial \theta}{\partial x}\frac{\partial}{\partial \theta} + \frac{\partial \phi}{\partial x}\frac{\partial}{\partial \phi}, \tag{1.53}$$

along with its counterparts in the y- and z-directions can be applied to the partial derivatives, while making use of

$$x = r\sin\theta\cos\phi,$$
$$y = r\sin\theta\sin\phi,$$
$$z = r\cos\theta, \tag{1.54}$$

and

$$r = \sqrt{x^2 + y^2 + z^2},$$

$$\theta = \arccos\left(\frac{z}{\sqrt{x^2 + y^2 + z^2}}\right),$$

$$\phi = \begin{cases} \arctan\left(\frac{y}{x}\right), & x > 0,\, y \geq 0, \\ \frac{\pi}{2} - \arctan\left(\frac{x}{y}\right), & |x| > 0,\, y > 0, \\ \pi + \arctan\left(\frac{y}{x}\right), & x < 0, \\ \frac{3\pi}{2} - \arctan\left(\frac{x}{y}\right), & y < 0, \end{cases} \tag{1.55}$$

to deal with partial derivatives of the type $\partial\theta/\partial x$, etc. (the various expressions above for ϕ across the different quadrants of the xy-plane are chosen to overlap, and they all lead to the same outcome).

Likewise, the action of the ∇-operator on a vector is still straightforward in Cartesian coordinates,

$$\nabla \cdot \mathbf{v} = \nabla \cdot (v^x \hat{\mathbf{e}}_x + v^y \hat{\mathbf{e}}_y + v^z \hat{\mathbf{e}}_z) = \hat{\mathbf{e}}_x \cdot \nabla v^x + \hat{\mathbf{e}}_y \cdot \nabla v^y + \hat{\mathbf{e}}_z \cdot \nabla v^z, \tag{1.56}$$

but less obvious in, e.g. spherical coordinates since the coordinate vectors themselves now become dependent on position:

$$\nabla \cdot \mathbf{v} = \nabla \cdot (v^r \hat{\mathbf{e}}_r + v^\theta \hat{\mathbf{e}}_\theta + v^\phi \hat{\mathbf{e}}_\phi) = \hat{\mathbf{e}}_r \cdot \nabla v^r + v^r \nabla \cdot \hat{\mathbf{e}}_r + \dots \tag{1.57}$$

In theory, this again can be sorted out using a brute force approach using Equations 1.9–1.11 and the sets 1.52, 1.54 and 1.55. In practice, we will take the expressions for gradient, divergence and curl in curvilinear coordinates as a given and not bother too much with rederiving them (using either brute force or more elegant methods). To complete the example above, the divergence in spherical coordinates is given by

$$\nabla \cdot \mathbf{v} = \frac{1}{r^2}\frac{\partial}{\partial r}\left(r^2 v^r\right) + \frac{1}{r\sin\theta}\frac{\partial}{\partial\theta}\left(\sin\theta v^\theta\right) + \frac{1}{r\sin\theta}\frac{\partial}{\partial\phi}\left(v^\phi\right). \tag{1.58}$$

This expression can also be found in Appendix B, along with other expressions for gradient, divergence and curl in curvilinear coordinates. Appendix C provides the means to derive such expressions directly.

PROBLEM 1.6
Gradient and divergence in plane polar coordinates

Polar coordinates (r, φ) in two dimensions obey

$$x = r\cos\varphi, \qquad y = r\sin\varphi,$$

$$\hat{\mathbf{e}}_x = \cos\varphi\hat{\mathbf{e}}_r - \sin\varphi\hat{\mathbf{e}}_\varphi, \qquad \hat{\mathbf{e}}_y = \sin\varphi\hat{\mathbf{e}}_r + \cos\varphi\hat{\mathbf{e}}_\varphi$$

$$r = \left(x^2 + y^2\right)^{\frac{1}{2}}, \qquad \varphi = \arctan\left(\frac{y}{x}\right), \text{ assuming } x, y > 0.$$

For the purpose of this problem, it is sufficient to limit ourselves to the quadrant where $x, y > 0$ indeed applies. Consider a scalar a and a vector $\mathbf{v} = v^r\hat{\mathbf{e}}_r + v^\varphi\hat{\mathbf{e}}_\varphi$. Using the 'brute force' approach, demonstrate that in this coordinate system we have

$$\nabla a = \frac{\partial a}{\partial r}\hat{\mathbf{e}}_r + \frac{1}{r}\frac{\partial a}{\partial\varphi}\hat{\mathbf{e}}_\varphi,$$

and

$$\nabla \cdot \mathbf{v} = \frac{\partial v^r}{\partial r} + \frac{v^r}{r} + \frac{1}{r}\frac{\partial v^\varphi}{\partial\varphi}.$$

1.2 INDEX NOTATION

In practice, vector identities including product rules will be provided where needed, e.g.

$$\nabla \cdot (\mathbf{a} \times \mathbf{b}) = \mathbf{b} \cdot (\nabla \times \mathbf{a}) - \mathbf{a} \cdot (\nabla \times \mathbf{b}), \qquad (1.59)$$

$$\nabla \cdot (f\mathbf{a}) = f\nabla \cdot \mathbf{a} + \mathbf{a} \cdot \nabla f, \qquad (1.60)$$

$$\nabla (\mathbf{a} \cdot \mathbf{b}) = \mathbf{a} \times (\nabla \times \mathbf{b}) + \mathbf{b} \times (\nabla \times \mathbf{a}) + (\mathbf{a} \cdot \nabla)\mathbf{b} + (\mathbf{b} \cdot \nabla)\mathbf{a}. \qquad (1.61)$$

Various vector identities are provided for convenience in Appendix B.

To avoid at least some of the complexity of working with vectors, we introduce index notation. Among other things, index notation will help us to quickly understand why, for example

$$\nabla \cdot (\mathbf{ab}) = (\nabla \cdot \mathbf{a})\mathbf{b} + \mathbf{a} \cdot \nabla \mathbf{b}, \qquad (1.62)$$

and

$$\nabla \cdot (\mathbf{ab}) \neq (\nabla \cdot \mathbf{a})\mathbf{b} + \mathbf{a}(\nabla \cdot \mathbf{b}). \qquad (1.63)$$

Eventually, index notation will provide an extremely concise notation for (special) relativistic fluid dynamics.

For index notation we use numbers rather than labels, i.e. instead of x, y, z or r, θ, ϕ we use x^1, x^2, x^3 or y^1, y^2, y^3, and instead of v^x, v^y, v^z we use v^1, v^2, v^3 etc. If we use a roman letter, e.g. i, this means that we have not yet specified a particular entry (unless the letter we use is one that we already assigned a special meaning, such as x), and x^i can be taken to refer to x^1, x^2 or x^3. Using this notation, a vector can be expressed as

$$\mathbf{v} = v^1 \hat{\mathbf{e}}_1 + v^2 \hat{\mathbf{e}}_2 + v^3 \hat{\mathbf{e}}_3 = \sum_i v^i \hat{\mathbf{e}}_i, \qquad (1.64)$$

when expressed in the basis consisting of the set of basis vectors $\{\hat{\mathbf{e}}_i\}$. The components v^i of a vector are called *contravariant components*.

In Cartesian coordinates, the basis $\{\hat{\mathbf{e}}_i\}$ can be conflated with its *dual* basis $\{\hat{\mathbf{e}}^i\}$ (a concept that we have not yet defined), so if we continue for now under an implicit assumption that we are working with Cartesian coordinates, we can postpone thinking about the difference between the two bases and proceed as if the only difference between them is the placement of the suffix. We will return to this issue below but keep matters simple at first. This means that we take both

$$v^i = v_i \text{ (Cartesian)}, \qquad \hat{\mathbf{e}}^i = \hat{\mathbf{e}}_i \text{ (Cartesian)}. \qquad (1.65)$$

Using this notation, a vector can be expressed alternatively as

$$\mathbf{v} = v_1 \hat{\mathbf{e}}^1 + v_2 \hat{\mathbf{e}}^2 + v_3 \hat{\mathbf{e}}^3 = \sum_i v_i \hat{\mathbf{e}}^i. \qquad (1.66)$$

We have

$$\mathbf{v} = \sum_i v_i \hat{\mathbf{e}}^i = \sum_i v_i \hat{\mathbf{e}}_i = \sum_i v^i \hat{\mathbf{e}}^i = \sum_i v^i \hat{\mathbf{e}}_i \text{ (Cartesian)}. \qquad (1.67)$$

We further introduce a strictly *typographical* convention, the *Einstein summation convention*, where we take a summation to be implied once an index occurs both as sub- and superscript:

$$\mathbf{v} = v_i \hat{\mathbf{e}}^i = v^i \hat{\mathbf{e}}_i = \sum_i v^i \hat{\mathbf{e}}_i. \tag{1.68}$$

This does not alter our earlier statement that in Cartesian coordinates, we make no difference in interpretation between sub- and superscripts. However, it does mean that it remains true that

$$\mathbf{v} \neq v^i \hat{\mathbf{e}}^i \text{ (in general).} \tag{1.69}$$

This is because here the RHS either refers to (if expressed in Cartesian coordinates) $v^x \hat{\mathbf{e}}^x$, $v^y \hat{\mathbf{e}}^y$ or $v^z \hat{\mathbf{e}}^z$. But not to the sum, since we never established any typographical convention about special treatment where a superscript is not matched with a subscript index.

Any index that disappears upon summation is called a *dummy index*. You can always replace a pair of dummy indices with a different symbol and reuse the same symbol elsewhere in the equation, e.g.

$$v^i w_i + 3 v^i w_i = v^i w_i + 3 v^j w_j. \tag{1.70}$$

On the other hand,

$$v^i + 3 w^i \neq v^i + 3 w^j \text{ (in general),} \tag{1.71}$$

since the LHS is a shorthand for three options $i = 1, 2, 3$ yet to be specified, whereas for the RHS we have nine combinations drawn from $i = 1, 2, 3$ *and* $j = 1, 2, 3$. The expression would therefore only hold true for special cases of $\mathbf{v}$ and $\mathbf{w}$ (i.e. where $w^1 = w^2 = w^3$). The conventions for index notation carry over to objects with multiple indices:

$$A^{ii} = \begin{cases} A^{xx} \\ A^{yy} \\ A^{zz} \end{cases}, \qquad A^i_{\ i} = A^x_{\ x} + A^y_{\ y} + A^z_{\ z}. \tag{1.72}$$

This brings us to the Kronecker delta δ_{ij}, defined according to:

$$\begin{cases} \delta_{ij} = 1, \text{ if } i = j \\ \delta_{ij} = 0, \text{ if } i \neq j \end{cases}, \qquad \delta_{ij} = \delta^i_j = \delta^{ij}. \tag{1.73}$$

In Cartesian coordinates, the Kronecker delta encodes the unit length and orthogonality of the basis vectors in the inner product:

$$\mathbf{v} \cdot \mathbf{w} = \left(v^i \hat{\mathbf{e}}_i \right) \cdot \left(w^j \hat{\mathbf{e}}_j \right) = v^i w^j \hat{\mathbf{e}}_i \cdot \hat{\mathbf{e}}_j = v^i w^j \delta_{ij} \text{ (Cartesian).} \tag{1.74}$$

When represented in matrix notation, the Kronecker delta would take the form of a unit matrix $\mathbb{1}$. In Cartesian coordinates, we have straightforward access to matrix notation in order to write the inner product of two vectors $\mathbf{v}$ and $\mathbf{w}$ in various ways:

$$v^i \delta_{ij} w^j = \begin{bmatrix} v^x & v^y & v^z \end{bmatrix} \begin{bmatrix} 1 & 0 & 0 \\ 0 & 1 & 0 \\ 0 & 0 & 1 \end{bmatrix} \begin{bmatrix} w^x \\ w^y \\ w^z \end{bmatrix} = v^x w^x + v^y w^y + v^z w^z. \tag{1.75}$$

The Kronecker delta can be used to 'raise or lower' an index in Cartesian coordinates, since $v^i \left(\delta_{ij} w^j \right) = v^i w_i$ in this case.

We can extend the Einstein summation convention to cover derivatives as well. For this typographical purpose, in case an index is used beneath a dividing line, the interpretation is flipped between sub- and superscript:

$$\sum_i \frac{\partial}{\partial x^i} v^i = \frac{\partial}{\partial x^i} v^i \tag{1.76}$$

Index notation allows us to work with vector entries directly, which in practice resemble scalars in their behaviour instead of vectors. The value of a vector or matrix entry still depends on a choice of coordinate system, but once this choice has been made the entry will act as a scalar. At this point, we can address the issue raised at the beginning of this section, and see how $\nabla \cdot (\mathbf{ab})$ expands. In index notation and with the dot product expressed in a Cartesian basis, we have

$$\frac{\partial}{\partial x^i} \left(a^i b^j \right) = \left(\frac{\partial}{\partial x^i} a^i \right) b^j + a^i \frac{\partial}{\partial x^i} b^j, \tag{1.77}$$

which follows directly from applying the usual product rule for scalars. But even though we temporarily assumed a Cartesian geometry that dictated the particulars of this expression, this statement remains applicable to any arbitrary vectors $\mathbf{a}$, $\mathbf{b}$, because we have not yet put any constraints on their actual values. We can therefore go back from Equation 1.77 to vector notation, and obtain

$$\nabla \cdot (\mathbf{ab}) = (\nabla \cdot \mathbf{a}) \, \mathbf{b} + \mathbf{a} \cdot \nabla \mathbf{b}, \tag{1.78}$$

confirming indeed Equations 1.62 and 1.63 from the beginning of this section.

1.2.1 THE CROSS PRODUCT AND INDEX NOTATION

In case you are wondering whether the cross product too can be represented using index notation, the answer is yes, although we will avoid writing expressions involving cross products in index notation throughout this book (so this section may safely be skipped). The cross product involves three vectors, however, and therefore requires us to introduce a rank-3 tensor ε^{ijk} with *three* indices rather than two (e.g. rank-2 tensor δ^{ij}; vectors are rank-1 tensors). If $\mathbf{c}$ is the vector produced by the cross product between vectors $\mathbf{a}$ and $\mathbf{b}$, we have

$$c^i = \varepsilon^{ijk} a_j b_k. \tag{1.79}$$

Remember that at this point, we have made no distinction in meaning between subscript and superscript, which we carry over to tensors of any rank, such that $\varepsilon^{ijk} = \varepsilon^i{}_j{}^k$ etc., although the horizontal position of the index while remain important in what follows. Furthermore, the repeated indices at different height placements imply that the Einstein summation convention needs to be applied twice, summing over j and k. In order for these summations to yield entries c^i that match those of the cross product, we need to define the entries ε^{ijk} such that they are equal to $+1$ whenever ijk is an *even* permutation of 123, equal to -1 for *odd* permutations of 123 and equal to *zero* if any two indices are the same. As examples, $\varepsilon^{132} = -1$, $\varepsilon^{312} = 1$, $\varepsilon^{122} = 0$, and so forth. That this indeed leads to correct result can easily be verified by hand by writing out the summations. The tensor ε^{ijk} is also known as the *permutation symbol*. It can also be verified by hand that it obeys

$$\varepsilon^{ijk}\varepsilon_i^{\ lm} = \delta_l^j \delta_m^k - \delta_m^j \delta_l^k. \tag{1.80}$$

1.2.2 FOUR-VECTORS, INDICES AND (SQUARE) BRACKETS

Although further details will be provided in Chapter 8 on relativistic fluid dynamics, we introduce the notion of *four-vectors* already here for completeness. In relativity, space and time are merged together in a single four-dimensional spacetime and a timelike component is added to vector quantities in spacetime. For example, the spacetime coordinates of a single event can be given by (ct, x, y, z) (where light speed c is usually included in front of time t). We extend the index notation approach by a convention that *Greek* indices will cover 0, 1, 2, 3, unlike Roman indices that only cover space entries 1, 2, 3. So in our example x^μ refers to either $x^0 = ct$ or any of the entries x^i. The Einstein summation convention carries over as expected, e.g. $x_\mu y^\mu = x_0 y^0 + x_1 y^1 + x_2 y^2 + x_3 y^3$.

It will remain notationally convenient to limit the use of **boldface** to threedimensional vectors though, and like other text books we will on various occasions refer to 'x^μ' as a four-vector without adding the qualifier that it actually represents the contravariant components of said four-vector in a particular basis.

We can connect index notation to matrix notation as introduced previously as well. By bracketing an object with multiple entries using square brackets we can connect four-vectors (and regular vectors) to matrix notation. The resulting object is then to be considered a vector or matrix in the matrix-notation sense, i.e., for a vector **x**, we have a corresponding matrix-notation grouping of entries

$$[\mathbf{x}] \equiv [x^i] \equiv \begin{bmatrix} x^1 \\ x^2 \\ x^3 \end{bmatrix}. \tag{1.81}$$

These entries are part of a larger four-dimensional object

$$[x^\mu] \equiv \begin{bmatrix} x^0 \\ x^1 \\ x^2 \\ x^3 \end{bmatrix} \equiv \begin{bmatrix} x^0 \\ \mathbf{x} \end{bmatrix}. \tag{1.82}$$

PROBLEM 1.7
Vector notation conventions

For the following, we use three vectors **u**, **v** and **w**, defined in matrix notation via

$$U = \begin{bmatrix} u^1 \\ u^2 \\ u^3 \end{bmatrix}, \qquad V = \begin{bmatrix} v^1 \\ v^2 \\ v^3 \end{bmatrix}, \qquad W = \begin{bmatrix} w^1 \\ w^2 \\ w^3 \end{bmatrix}.$$

The vectors have lengths u, v and w respectively. Below follows a list of vector expressions. Determine whether they make sense by writing the LHS and RHS in Cartesian matrix notation, note where you think they are mismatched. Be careful: sometimes the LHS or RHS is just gibberish in the context of the notational conventions we have established. You may use that in Cartesian coordinates $v^i = v_i$, if you wish to use the summation convention for concise notation.

a) $\nabla \cdot (\mathbf{uv}) = 3\mathbf{w}$.   b) $\left[u^i v^j \right] = \left[u_i v_j \right]$.

c) $\mathbf{uv} \cdot \mathbf{w} + 4\mathbf{w} = \mathbf{u} \cdot \mathbf{v} \cdot \mathbf{w}$.   d) $4u^i = 2w^j$.

e) $\left[u^i u^i \right] = \left[\nabla^j v^j \right]$.   f) $\left[u^i u_i \right] = \mathbf{vw}$.

g) $\left[u^i u_j \right] = \mathbf{v} \cdot \mathbf{w}$.   h) $v \nabla \cdot \mathbf{w} = \nabla \cdot (\mathbf{uv})$.

i) $u^i v_i v^j = 3u^j$.   j) $u^i v^j u_i = 3u^j$.

k) $\left[u^i v^j u_i \right] = \mathbf{w}$.   l) $\mathbf{u} \cdot \mathbf{v} = 10$.

m) $\left[\left(u^i - v^i \right) w^j \right] = \mathbb{1}$.   n) $\left(u^i - 3v^i \right) v^j = 1 - 3$.

o) $\mathbf{v}^2 = v^2$.   p) $\left[u^i u^j \right] \cdot \mathbf{v} = \mathbf{w} \cdot \mathbb{1}$.

q) $\left[u^i u^j \right] \left[u^i u^j \right] = \mathbb{1} \mathbf{v}^2$.   r) $\mathbf{v} |\mathbf{w}| = \mathbf{u} \cdot \mathbb{1}$.

1.3   COVARIANT COMPONENTS AND THE DUAL BASIS

While for Cartesian coordinates we have not drawn a real distinction between $\mathbf{e}_i$ and $\mathbf{e}^i$, this distinction does exist in general and manifests itself when working with coordinate systems with a set of basis vectors that are no longer orthogonal or of length unity. Here we go briefly into some of the details, but because we tend to use normalized basis vectors even for curvilinear coordinates such as spherical or cylindrical coordinates, this section is mostly relevant to relativistic fluid dynamics (Chapter 8) where the time dimension of spacetime adds an unavoidable clear distinction between normalized basis four-vectors and their *dual basis* even in flat space. As long as it is kept in mind that spherical and cylindrical bases are position-dependent, the preceding discussion of vector algebra should be sufficient for understanding non-relativistic fluid dynamics (note that the default curvilinear coordinate systems are orthogonal), and this section may be skipped safely.

We have already defined the natural basis vectors $\{\mathbf{e}_i\}$ at some position $\mathbf{r}$ as the set of vectors spanning the dimensions of a local *tangent space*, and the entries v^i as the components of a vector $\mathbf{v}$ expressed in this basis. We now need to figure out what $\mathbf{e}^i$ and v_i mean, and how they relate to the natural basis. Recall that for the inner product between two vectors we have

$$\mathbf{v} \cdot \mathbf{w} = \left(v^i \mathbf{e}_i \right) \cdot \left(w^j \mathbf{e}_j \right) = v^i w^j \left(\mathbf{e}_i \cdot \mathbf{e}_j \right). \tag{1.83}$$

Because Cartesian basis vectors are orthogonal and of unit length, for Cartesian coordinates this reduces further to

$$v^i w^j \mathbf{e}_i \cdot \mathbf{e}_j = v^i w^j \delta_{ij} = v^i w_i \text{ (Cartesian)}. \tag{1.84}$$

We know that for Cartesian coordinates w_i is the same as w^i, and since $\left(\delta_{ij}w^j\right)$ is an object with a single remaining non-dummy subscript index i we might as well identify this object with w_i, like we did above. We now move into new territory by generalizing this approach to cases where the inner product between two basis vectors $\mathbf{e}_i$, $\mathbf{e}_i$, is *not* necessarily equal to δ_{ij}:

$$v^i w^j \mathbf{e}_i \cdot \mathbf{e}_j \equiv v^i w^j g_{ij} = v^i w_i, \qquad w_i \equiv g_{ij} w^j. \tag{1.85}$$

This defines a *metric tensor* g_{ij}. One of its uses appears to be that it can help us translate a contravariant component into a covariant component. It stands to reason that this process can be inverted, and we might as well define another metric tensor g^{ij} (which can also be considered as being merely a different manifestation of the same underlying metric) that makes this happen: $v^i = g^{ij}v_j$. We then have

$$v^i = g^{ij}v_j = g^{ij}g_{jk}v^k = \delta^i_k v^k, \tag{1.86}$$

which tells us that g_{ij} and g^{ij} are each other's inverse. So if we can construct a tensor $g_{ij} = \mathbf{e}_i \cdot \mathbf{e}_j$, represent this in matrix-notation as a matrix $G \equiv [g_{ij}]$, we can simply invert this matrix in order to obtain $\left[g^{ij}\right] = G^{-1}$.

To make the above more concrete, for the natural basis in spherical coordinates this implies

$$[g_{ij}] = G = \begin{bmatrix} 1 & 0 & 0 \\ 0 & r^2 & 0 \\ 0 & 0 & r^2\sin^2\theta \end{bmatrix}_{r=r,\theta=\theta,\phi=\phi}, \tag{1.87}$$

where I have added a label to the matrix to indicate that this is not only a matrix that is expressed in terms of spherical *coordinates* r, θ, ϕ but also[8] in a spherical coordinate basis at a *particular position* r, θ, ϕ (see also our previous discussion of Section 1.1.2). The inverse can be obtained quickly and obeys

$$\left[g^{ij}\right] = G^{-1} = \begin{bmatrix} 1 & 0 & 0 \\ 0 & r^{-2} & 0 \\ 0 & 0 & r^{-2}\sin^{-2}\theta \end{bmatrix}_{r=r,\theta=\theta,\phi=\phi}. \tag{1.88}$$

So how should g_{ij} be interpreted? Based on its definition as summarizing the possible permutations of $\mathbf{e}_i \cdot \mathbf{e}_j$, we know that it provides a full characterization of the basis. It also features when co- and contravariant vector components are combined in a summation (a *contraction*):

$$v^i w_i = v^i g_{ij} w^j = V^T G W, \qquad V = \left[v^i\right], W = \left[w^i\right]. \tag{1.89}$$

This shows how the metric can be used to generalize the inner product. On the one hand, the summation rule stays what it is, and $v^i w_i = v^1 w_1 + v^2 w_2 + v^3 w_3$. On the other hand, $v^i \neq v_i$ in general, so we do not run into the issues identified in

---

[8]At the risk of being accused of splitting hairs, the issue is that strictly speaking the same set of symbols r, θ, ϕ has been overloaded with two meanings.

Section 1.1.2 when working with basis vectors in a manner that failed to account for cases were these were not of length unity or mutually orthogonal. For the natural basis of spherical coordinates, for example, expression 1.89 leads to $v^1 v^1 + r^2 v^2 w^2 + r^2 \sin^2 \theta v^3 w^3$ (do not confuse the squares and the superscripts!), which is properly consistent with the outcome that is obtained by simply summing over the components directly in a Cartesian basis.

There remains the issue that for any representation in a particular basis, manipulations such as those of Equation 1.89 only work out correctly if the same basis is applied consistently to all terms. Which means one cannot mix spherical coordinates basis vectors from different positions in space, for example. This might seem overly limiting at first, but in fact it does leave us with arguably the most important role of g_{ij} intact: to account for the *local* metric properties of a coordinate system. Consider an infinitesimal displacement $dW = (dR, d\theta, d\phi)$ in natural spherical coordinates. When contracted with itself, we obtain

$$
\begin{aligned}
(dW)^2 &= (dW)^i (dW)_i \\
&= (dr)^2 + r^2 (d\theta)^2 + r^2 \sin^2 \theta (d\phi)^2 \\
&= (dx)^2 + (dy)^2 + (dz)^2 .
\end{aligned}
\tag{1.90}
$$

The last step was added to emphasize once again the coordinate-independent nature of this scalar quantity. It should also help remind you of the well-known Pythagoras' theorem (i.e. for a displacement in two dimensions, we have $|dW| = \sqrt{(dx)^2 + (dy)^2}$). Indeed, the *line element* $g_{ij} dx^i dx^j$ is the generalization of this theorem to arbitrary coordinate systems, and g_{ij} can be used to work out what actual distance corresponds to a displacement dW.

That leaves only the interpretation of the dual basis vectors $\{\mathbf{e}^i\}$. The space spanned by the dual base is *not* the same as that spanned by the natural basis. Instead, it describes an alternate tangent space to the natural basis. It is merely in Euclidean space that we can conflate all three spaces: the Euclidean space of interest and its mappings using a natural and dual basis. But even when considered to be existing in strictly conceptually distinct tangent spaces, there remains a direct connection between a basis vector $\mathbf{e}_i$ and its dual counterpart, and dual vectors can be considered as *operators*, or functions taking vector arguments:

$$
\mathbf{e}^i (\mathbf{e}_j) \equiv g^{ik} \mathbf{e}_k \cdot \mathbf{e}_j = g^{ik} g_{kj} = \delta^i_j .
\tag{1.91}
$$

It follows directly from the above that for any dual vector $\mathbf{v}^\star$ and vector $\mathbf{w}$ we have

$$
\mathbf{v}^\star (\mathbf{w}) = v_i \mathbf{e}^i \left(w^j \mathbf{e}_j \right) = v_i w^j g^{ik} \mathbf{e}_k \cdot \mathbf{e}_j = v_i w^j \delta^i_j = v_i w^i = v^i w_i .
\tag{1.92}
$$

In other words, it is this juxtaposition of a vector and a dual vector that generalizes the inner product of Euclidean space to arbitrary geometry.

1.4 PARTIAL AND FULL DIFFERENTIATION

The conservation laws of fluid dynamics are often derived by considering the behaviour of a fluid parcel. This is defined as a small blob of fluid containing a fixed number of constituent particles, or total amount of mass, and we trace its motion within the flow. Expressing conservation laws for this parcel will amount to formulating the conservation laws for fluid dynamics. We will see this in Chapter 3, but for now let us start at the very basics and remind ourselves of the product rule, applied to some function $F(x(t), y(t), t)$:

$$\frac{dF(x(t), y(t), t)}{dt} = \frac{\partial F}{\partial t} + \frac{dx}{dt}\frac{\partial F}{\partial x} + \frac{dy}{dt}\frac{\partial F}{\partial y}, \tag{1.93}$$

Here the full derivative takes into consideration everything that changes while moving from t to $t + dt$. The partial derivatives are taken while keeping the other variables constant. Note that there is an agreement here about which set of variables we are using, implied by what we included in the parentheses following F on the LHS of the equation. In particular, we appear to have agreed to use x, y, t and not r, θ, t (i.e. Cartesian, not polar, if we interpret these as coordinates), nor even some weird hybrid x, θ, t. If we also limit ourselves to considering a single trajectory $x(t), y(t), t$ instead of the full 2+1 dimensional (2 space, one time) volume, we additionally have a specific path along which the full derivative is taken. We are taking a *co-moving* derivative.

To illustrate that, let us specify a fixed path in the case of two coordinates x, y, ignoring time dependence for this example. Say we have a function $F(x, y) \equiv x^2 + y^2$, describing some characteristic of a particle confined to move back and forth only along a straight line $y = x\cos\theta$ (i.e. our path). For the total change in F that comes with a displacement in the x-direction, we obtain

$$\frac{dF(x, y)}{dx} = \frac{dx}{dx}\frac{\partial F}{\partial x} + \frac{dy}{dx}\frac{\partial F}{\partial y} = \frac{\partial F}{\partial x} + \frac{dy}{dx}\frac{\partial F}{\partial y}, \tag{1.94}$$

which evaluates to

$$\frac{dF(x, y)}{dx} = 2x + 2y\cos\theta, \tag{1.95}$$

when using $dy/dx = \cos\theta$ for our chosen path. Had we instead first incorporated our path into the equation for F, i.e. $F(x) = x^2 + x^2\cos^2\theta$, we would have been left with a function of a single variable and no distinction between full and partial derivative:

$$\frac{dF(x)}{dx} = \frac{\partial F(x)}{\partial x} = 2x + 2(x\cos\theta)\cos\theta, \tag{1.96}$$

which results in the same. Equation 1.93 demonstrates the same principle, where with our notation we have suggested a path tracing out a trajectory in time.

In fluid dynamics, there tends to be an obvious candidate for the path that is followed when a full derivative is taken: the path carved out by the local fluid flow velocity $\mathbf{v}_{\text{flow}}$ or just $\mathbf{v}$ when dropping the *flow* label (see Figure 1.6). Some textbooks use a capital D notation for this case:

$$\frac{D\rho}{Dt} \equiv \frac{\partial \rho}{\partial t} + \frac{dx}{dt} \cdot \frac{\partial}{\partial \mathbf{x}}\rho, \quad \frac{dx}{dt} \equiv \mathbf{v}_{\text{flow}}, \tag{1.97}$$

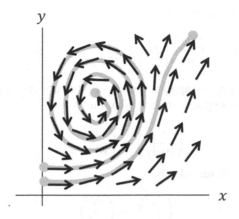

Figure 1.6 The local fluid velocity as a vector field. An object moving with flow lines will have its path in space dictated by the flow velocity. The grey curves show two trajectories with different starting points. We assume that the velocity field remains unchanged.

as opposed to the more generic

$$\frac{d\rho}{dt} \equiv \frac{\partial\rho}{\partial t} + \frac{d\mathbf{x}}{dt} \cdot \frac{\partial}{\partial\mathbf{x}}\rho, \quad \frac{d\mathbf{x}}{dt} \equiv \mathbf{v}_{\text{path}}, \tag{1.98}$$

that leaves the path to be specified. In this book, we will stick with lower case d and assume that the path is that of the flow unless explicitly specified otherwise. The derivative along a flow line is also called the *co-moving* or *Lagrangian* derivative. For notational convenience, we will use an overhead dot on occasion to denote the co-moving time derivative, that is

$$\dot{X} \equiv \frac{dX}{dt}, \tag{1.99}$$

and similar for the co-moving time derivative of vectors etc.

PROBLEM 1.8
The product rule

Let us build a little on the preceding example by adding a time dependency. For the same function $F(x,y,t) = x^2 + y^2$, let us require $x = -\frac{1}{2}gt^2$ (with g some constant), in addition to $y = x\cos\theta$. Please calculate $\frac{dF}{dt}$ twice, once using the product rule and once by first eliminating x and y. Show that the results are identical.

1.4.1 EXACT DIFFERENTIALS

The local physics of fluids is to a large extent dictated by thermodynamics, and we will borrow various concepts and methods from this branch of physics. An exact differential is a differential that obeys

$$df = a\,dx + b\,dy = \left(\frac{\partial f}{\partial x}\right)_y dx + \left(\frac{\partial f}{\partial y}\right)_x dy = \frac{\partial f}{\partial x}dx + \frac{\partial f}{\partial y}dy, \qquad (1.100)$$

where in the middle step we apply a notation commonly encountered in thermodynamics but in the last step left implicit which the other variable(s) were to be kept constant. Since

$$\left(\frac{\partial^2 f}{\partial x \partial y}\right) = \left(\frac{\partial^2 f}{\partial y \partial x}\right), \qquad (1.101)$$

it follows that exact differentials obey

$$\left(\frac{\partial a}{\partial y}\right)_x = \left(\frac{\partial b}{\partial x}\right)_y \qquad (1.102)$$

There are plenty of differentials to be found in thermodynamics that we will need to make use of on occasion in fluid dynamics as well (see Appendix A for a summary of relevant thermodynamics). For reversible processes, the famous first law of thermodynamics states that

$$d\varepsilon = T\,ds - p\,d\mathcal{V}, \qquad (1.103)$$

rewritten for *specific* internal energy ε. *Specific* quantities are per unit mass: $\varepsilon \equiv E_{int}/M$ where E_{int} and M respectively the total internal energy and mass of the system under consideration (maybe a fluid parcel, maybe a larger volume). Similarly, s is the specific entropy $s \equiv S/M$ and $\mathcal{V} = V/M$ the specific volume. Temperature and density are denoted T and p respectively.

Since the first law of thermodynamics is an exact differential, it follows that

$$\left(\frac{\partial T}{\partial \mathcal{V}}\right)_s = -\left(\frac{\partial p}{\partial s}\right)_{\mathcal{V}}. \qquad (1.104)$$

There are many more thermodynamic identities that can be derived when expressions for enthalpy, Helmholtz free energy and Gibbs free energy are considered in addition to the first law. These are collected in Appendix A.

PROBLEM 1.9
Mathematical identities involving partial differentiation

We have three interdependent variables s, T and p. Derive the following mathematical identity (known as the cyclic chain rule, triple product rule or Euler's chain rule),

$$\left(\frac{\partial T}{\partial p}\right)_s = -\frac{(\partial s/\partial p)_T}{(\partial s/\partial T)_p},$$

starting from an expression for the differential ds.

1.4.2 INEXACT DIFFERENTIALS

Not all differential forms are necessarily exact. Consider a function

$$f(x,y) = x^2 y + x. \tag{1.105}$$

Its corresponding exact differential form is given by

$$df = \left(\frac{\partial f}{\partial x}\right)_y dx + \left(\frac{\partial f}{\partial y}\right)_x dy = (2xy+1)dx + x^2 dy. \tag{1.106}$$

The integral of df along any closed path should yield zero. If this were not the case, $f(x,y)$ would not be uniquely specified. This generalizes the one-dimensional case where for example a function $f(x)$ obeys

$$\int_{x_0}^{x_1} df = f(x_1) - f(x_0) = 0, \text{ if } x_0 = x_1. \tag{1.107}$$

We can illustrate this for the path (x,y) that goes $(0,0) \rightarrow (1,0) \rightarrow (1,1) \rightarrow (0,1) \rightarrow (0,0)$:

$$\oint df = \int_{x=0}^{1} (2x \cdot 0 + 1)dx \Big|_{\text{at } y=0} + \int_{y=0}^{1} 1^2 dy \Big|_{\text{at } x=1}$$

$$+ \int_{x=1}^{0} (2x \cdot 1 + 1)dx \Big|_{\text{at } y=1} + \int_{y=1}^{0} 0^2 dy \Big|_{\text{at } x=0} \tag{1.108}$$

$$= [x]_0^1 + [y]_0^1 + [x^2+x]_1^0 + 0 \tag{1.109}$$

$$= 1 - 0 + 1 - 0 + 0 - (1+1) = 0. \tag{1.110}$$

Now define a dg and dh according to $df = dg + dh$, $dg \equiv (2xy+1)dx$ and $dh \equiv x^2 dy$. Generally speaking such redefinitions do not lead to exact differentials, and we can for example easily demonstrate that dg is an inexact differential using an integration along the same path as previously (but now in the absence of a dy term, so all displacement in the y-direction will have zero impact):

$$\oint dg = \int_{x=0}^{1} (2x \cdot 0 + 1)dx + 0 + \int_{x=1}^{0} (2x \cdot 1 + 1)dx + 0 = -1 \neq 0. \tag{1.111}$$

The differential form $dg = (2x+1)dx + 0dy$ also fails to obey Equation 1.102:

$$\left(\frac{\partial(2xy+1)}{\partial x}\right)_y = 2y \neq \left(\frac{\partial 0}{\partial y}\right)_x. \tag{1.112}$$

The core issue is that dg has been defined from a differential, and not from g, which allows for cases where $g(x,y)$ cannot be specified uniquely. Thermodynamics contains classic examples of exact and inexact differentials. Instead of Equation 1.103, we could have written

$$d\varepsilon = dq + dw, \qquad dq \equiv Tds, \qquad dw \equiv -pd\mathcal{V}. \tag{1.113}$$

Here specific heat change dq and specific work change dw are both inexact differentials, that have been defined through their differentials rather than by differentiating a uniquely determined function. Inexact differentials are sometimes denoted as $đq$ and $đw$. The total amount of heat (or work) you end up with depends on the path taken.

2 The Conservation Laws of Fluid Dynamics

There are four fundamental states of matter in the universe, solids, gases, liquids and plasmas, plus a plethora of more exotic states that only exist at the intersections of the four and/or at very specific extreme physical conditions (such as extreme cold or pressure). Solids excepted, the remaining three fundamental states describe *continuum* states of matter that can flow and deform freely and that, up to a point, can be split up in arbitrary smaller subdivisions that maintain the same physical properties. The behaviour of these three states—gases, liquids and plasmas—are governed by the laws of *fluid dynamics*. Viscosity will play a larger role for liquids than it does for gases, and plasmas are unique in that they contain many freely moving charged particles and therefore respond directly to electromagnetic forces. But all types of fluids share a common framework in that they obey a series of *conservation laws* that are cast in a form capable of dealing with the continuum nature of the fluid. It is the assumption that the fluid under consideration is something that can be subdivided arbitrarily that makes fluid dynamics into a *macroscopic* description of matter, as opposed to a *microscopic* theory that explicitly describes individual particles.

This book is about formulating these conservation laws and exploring their consequences when they are adapted to various situations. When applied to liquids, this is the subject of *hydrodynamics*, although the terminology is not strict, and the prefix *hydro-* is admittedly sometimes also applied to fluids other than liquids. When adapted to plasmas, this is the subject of *magnetohydrodynamics*, abbreviated 'MHD', which connects to the rich and complicated field of *plasma physics*. The term MHD coincidentally is already an example of a loose use of the hydro- prefix.

Many authors build up piecewise to a full set of coupled conservation laws, but we take a slightly different approach here. Instead, we immediately move to formulate the complete set of partial differential equations describing conservation of mass, energy and momentum density in the simplest case (no viscosity, no MHD, no relativistic fluid motion), introducing as few novel concepts as possible. This is the topic of the current chapter. *Euler's equations*, as the set of conservation laws are called in the case of frictionless gases, will provide a guiding framework for the rest of the book. Subsequent chapters will incorporate additional physics and discuss in more depth relevant theoretical concepts (such as fluid parcels) that help us understand and explore the implications of the laws of hydrodynamics introduced here.

DOI: 10.1201/9781003095088-2

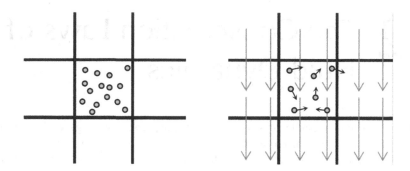

Figure 2.1 A virtual division of space into cells. Left image illustrates that the virtual fluid cells contain macroscopic particle numbers (where *macroscopic* is typically orders of magnitude larger than the fifteen depicted particles). The right image illustrates that long-range forces such as gravity may act throughout the cell volumes, while the individual constituent particles of the fluid exhibit random motion.

2.1 A DERIVATION OF THE CONSERVATION LAWS

2.1.1 MASS CONSERVATION: THE CONTINUITY EQUATION

We start by mentally subdividing the fluid continuum into cells, or zones (as illustrated in Figure 2.1). The total mass M in a given cell of volume V_{fixed} is given by integrating over mass density $\rho(\mathbf{r},t)$,

$$M = \int_{V_{fixed}} \rho \, dV_{fixed}. \tag{2.1}$$

Here we use the subscript *fixed* to emphasize that the virtual cells remain fixed in space. The number of particles in a single cell is assumed to be so large that even at scales smaller than the cell scale the massive number of constituent particles effectively blurs into a continuous mass density ρ (we discuss the nature of fluids in more depth in Section 2.4). If we integrate out the spatial dimensions as above in order to cover the entire cell, the change in time of this total mass is given by

$$\frac{dM(t)}{dt} = \frac{\partial M(t)}{\partial t} = \frac{\partial}{\partial t} \int_{V_{fixed}} \rho \, dV_{fixed} = \int_{V_{fixed}} \left(\frac{\partial \rho}{\partial t} \right) dV_{fixed}. \tag{2.2}$$

Since M is a function of time only, there is no distinction between the full and partial derivative. Because we keep our grid fixed in space, we do not need to worry about any cell displacement occurring while we take the time derivative (a displacement would have introduced a time-dependency of the spatial coordinates, since the integration domain is tied to the position the cell occupies in space). The partial derivative can therefore be taken inside the integral without any trouble.

But if we do use a fixed grid, individual constituent particles of the fluid may well end up outside of the grid cell under consideration due to their individual random motions (first, it is after all a fluid and not a solid, so the particles are not bound to particular positions; second, the boundary is only an imaginary one). We therefore need to account for a possible loss through the grid cell boundaries, which yields an alternative expression for $\partial M / \partial t$ where we introduce a *flow velocity* $\mathbf{v}$ and a *particle density flux* $\rho \mathbf{v}$:

$$\frac{\partial M}{\partial t} = -\oint_{A_{fixed}} \rho \mathbf{v} \cdot d\mathbf{S}_{fixed} = -\int_{V_{fixed}} \nabla \cdot (\rho \mathbf{v}) \, dV_{fixed}. \tag{2.3}$$

Here we applied the divergence theorem in the second step. The minus sign is to account for the fact that the surface orientation vector tends to be defined to point *outwards* from the volume and velocity flow along this direction should lead to a *loss* in mass. The inner product between flow velocity and surface normal expresses that we only care about what actually leaves the grid cell, not about what merely moves along the surface. Equating the two expressions for the time change in total cell mass and realizing that they must hold for arbitrary (fictional) cell volume, we can conclude that:

$$\int_{V_{fixed}} \left(\frac{\partial \rho}{\partial t} + \nabla \cdot (\rho \mathbf{v}) \right) dV = 0 \Rightarrow \frac{\partial \rho}{\partial t} + \nabla \cdot (\rho \mathbf{v}) = 0, \tag{2.4}$$

This is the *continuity equation*, expressing conservation of mass[1]. It tells us that a local change in density must be accompanied by a net difference between the amount of density flux arriving and the amount of flux departing (i.e. a divergence in flux) to compensate for this change. Because if this were not the case, some of our fluid would be disappearing into (or appearing out of) thin air.

Note that the continuity equation is *not* stating that the full derivative of density is zero, as a comparison between application of the chain rule and the continuity equation demonstrates:

$$\frac{d\rho}{dt} = \frac{\partial \rho}{\partial t} + \frac{d\mathbf{x}}{dt} \cdot \nabla \rho = -\rho \nabla \cdot \mathbf{v}. \tag{2.5}$$

Fluids for which the additional condition $d\rho / dt = 0$ holds, are labelled *incompressible*.

2.1.2 SNEAK PEEK AT FULL SET OF EULER'S EQUATIONS

Let us jump ahead for a second and deliver the conclusion of this section first. The full set of Euler's equations, grouped together, is as follows. First,

$$\frac{\partial \rho}{\partial t} + \nabla \cdot (\rho \mathbf{v}) = 0, \tag{2.6}$$

---

[1]One can argue that we did not really derive this expression as much as assert a definition of 'mass flux' consistent with this expression. That is a fair point, but we will rederive mass conservation twice more, in Sections 2.4.1 and 3.1, in ways that do not require a prior definition of mass flux as $\rho \mathbf{v}$.

i.e. mass conservation, which we have seen already. Second, there is an analogous equation for momentum density $\rho\mathbf{v}$:

$$\frac{\partial\rho\mathbf{v}}{\partial t} + \nabla\cdot(\rho\mathbf{v}\mathbf{v} + p\mathbb{1}) = \mathbf{f}. \tag{2.7}$$

where p is pressure and f an external force density acting on the fluid. $\mathbb{1}$ is the identity tensor, represented by an identity matrix when the expression is cast in matrix-notation form. Since this is a vector equation, it actually represents as many equations as we have dimensions in our system. To clarify the vector notation here once more: in Cartesian coordinates matrix notation, the momentum flux term would look like

$$
\begin{aligned}
[\nabla\cdot(\rho\mathbf{v}\mathbf{v})] &= \begin{bmatrix} \frac{\partial}{\partial x} & \frac{\partial}{\partial y} & \frac{\partial}{\partial x} \end{bmatrix} \begin{bmatrix} \rho v_x v_x & \rho v_x v_y & \rho v_x v_z \\ \rho v_y v_x & \rho v_y v_y & \rho v_y v_z \\ \rho v_z v_x & \rho v_z v_y & \rho v_z v_z \end{bmatrix} \\
&= \begin{bmatrix} \frac{\partial}{\partial x}(\rho v_x v_x) + \frac{\partial}{\partial y}(\rho v_y v_x) + \frac{\partial}{\partial z}(\rho v_z v_x) \\ \frac{\partial}{\partial x}(\rho v_x v_y) + \frac{\partial}{\partial y}(\rho v_y v_y) + \frac{\partial}{\partial z}(\rho v_z v_y) \\ \frac{\partial}{\partial x}(\rho v_x v_z) + \frac{\partial}{\partial y}(\rho v_y v_z) + \frac{\partial}{\partial z}(\rho v_z v_z) \end{bmatrix}^T.
\end{aligned} \tag{2.8}
$$

No inner or outer product acts between the two $\mathbf{v}$'s. The ∇ operator acts on $\rho\mathbf{v}\mathbf{v}$ entirely, but is contracted only with the first $\mathbf{v}$, as shown above[2].

Third, we have a conservation law for the total energy density of the fluid $\mathscr{E} = e + \frac{1}{2}\rho v^2$ (i.e. internal energy density plus kinetic energy density). The conservation law is given by

$$\frac{\partial\mathscr{E}}{\partial t} + \nabla\cdot([\mathscr{E} + p]\mathbf{v}) = \rho\dot{q} + \mathbf{v}\cdot\mathbf{f}, \tag{2.9}$$

where $\dot{q}$ is a term accounting for any additional local deposition (or loss, if negative) of heat within the fluid.

The set of Equations 2.4, 2.7 and 2.9 contain all the *macrophysics* of an *inviscid* fluid, i.e., a fluid where friction and viscosity are absent. Completely neglecting friction and viscosity is perfectly acceptable in many astrophysical settings where the fluid can indeed be approximated as an *ideal gas*. The *source terms* q and f are either specified externally or kept at zero. We therefore have $n + 2$ equations (n the number of dimensions), and $n + 3$ unknowns (ρ, $\mathscr{E}$, n unknowns from $\mathbf{v}$, and p). In fluid dynamics, we assume that p can be expressed as a local function $p(\rho, e)$ in a manner that does not violate the fundamental continuum assumption. The exact function for p expresses how the nature of the fluid continuum is ultimately dictated by the *microphysics* of the constituent particles and their interactions. We will return to this point in Section 2.4. Together with this description for p, we then have a closed set of equations which we can solve completely (analytically, if possible, or on a computer) once the initial state of the fluid is specified.

---

[2]The result of this operation is a row vector in matrix notation, so if Equation 2.7 were written out in matrix notation a transposition needs to be applied to align the various terms again. Index notation benefits here from its explicit use of subscripts and superscripts.

If the velocity v is constant throughout the fluid, such that the partial derivative with respect to time in Equations 2.4, 2.7 and 2.9 is zero, then the flow is a *steady* flow. In the more limiting case where the velocity is actually zero, hydrodynamics reduces to *hydrostatics*. In cases where it (approximately) holds, the assumption of steady flow or even hydrostatic equilibrium can be very helpful in providing a simplified but useful description of a system. We will discuss hydrostatics in more detail in Chapter 4.

2.1.3   DERIVATION OF THE MOMENTUM EQUATION

Let us now derive Equation 2.7. Consider classical mechanics as a guide, where total momentum $\mathbf{P}$ and total force $\mathbf{F}$ relate according to $\frac{d\mathbf{P}}{dt} = \mathbf{F}$, i.e., the time derivative of momentum is equal to a force. Considering once more a fixed fictional grid, we again will be able to compare two expressions for the same. On the one hand, we have,

$$\frac{\partial \mathbf{P}}{\partial t} = \frac{\partial}{\partial t} \int_{V_{fixed}} \rho \mathbf{v} dV_{fixed} = \int_{V_{fixed}} \frac{\partial (\rho \mathbf{v})}{\partial t} dV_{fixed}. \tag{2.10}$$

On the other hand, the momentum within a volume changes under the influence of forces and due to momentum flux through the volume boundaries. Forces can act throughout the volume or between fluid elements. Dubbing the latter volume and surface forces respectively, we have

$$\frac{\partial \mathbf{P}}{\partial t} = - \oint_{A_{fixed}} (\rho \mathbf{v}) \mathbf{v} \cdot d\mathbf{S}_{fixed} + \mathbf{F}_{volume} + \mathbf{F}_{surface}. \tag{2.11}$$

In this case, we have a flux of a vector quantity $\rho \mathbf{v}$ rather than a scalar quantity ρ, which explains the repeated $\mathbf{v}$ in the equations.

For the forces acting on the entire volume we can define a force density $\mathbf{f}$ (force per unit volume). For the surface forces we can define a surface force density $\mathbf{t}$ (force per unit area), leading to

$$\int_{V_{fixed}} \frac{\partial (\rho \mathbf{v})}{\partial t} dV_{fixed} = - \int_{V_{fixed}} \nabla \cdot (\rho \mathbf{v} \mathbf{v}) dV_{fixed} + \int_{V_{fixed}} \mathbf{f} dV_{fixed}$$
$$+ \oint_{A_{fixed}} \mathbf{t} dS_{fixed}. \tag{2.12}$$

Note that the surface force density integral is not in terms of directed surface area elements $d\mathbf{S} \equiv \hat{\mathbf{s}} dS$ (where $\hat{\mathbf{s}}$ the outward normal on the surface, see Figure 2.2) but in terms of scalar surface area size elements dS. The directional information for the surface force is contained in $\mathbf{t}(\mathbf{x}, \hat{\mathbf{s}})$, which is itself a function of position $\mathbf{x}$ and surface orientation $\hat{\mathbf{s}}$. This allows us to take into account surface forces that are not perpendicular to the surface (i.e. cases where $\hat{\mathbf{s}}$ and $\mathbf{t}$ are not aligned, as we will find to be the case when treating viscosity in Chapter 9).

But what we would like to do is to apply the (tensor-analog) of the divergence theorem like we did when deriving the continuity equation, in order to move from a

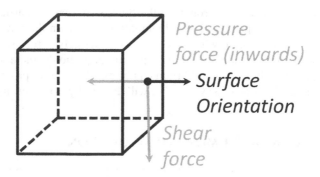

Figure 2.2 Virtual grid cells can experience surface forces that can be split into shear forces and perpendicular forces. Pressure forces are perpendicular.

surface integral to a volume integral. For this purpose, we introduce the *stress tensor* T, via

$$\mathbf{t} = \mathbf{T} \cdot \hat{\mathbf{s}} \Rightarrow \mathbf{t}dS = \mathbf{T} \cdot d\mathbf{S}. \tag{2.13}$$

For an ideal gas, where there is no viscosity and no net forces tangential to the surface are allowed to exist, the stress tensor T (in matrix notation T) is given by

$$\mathbf{T} = -p\mathbb{1} \rightarrow T = \begin{bmatrix} -p & 0 & 0 \\ 0 & -p & 0 \\ 0 & 0 & -p \end{bmatrix}, \tag{2.14}$$

Because we are considering forces acting *on* the grid cell, we needed a minus sign when expressing T in terms of pressure p, such that pressure acts in the direction opposite to the normal vectors pointing outwards from the fluid surface. We can now apply the *divergence theorem*, i.e.

$$\int_{A_{fixed}} \mathbf{t}dS = \int_{A_{fixed}} \mathbf{T} \cdot d\mathbf{S} = \int_{V_{fixed}} (\nabla \cdot \mathbf{T}) dV = -\int_{V_{fixed}} (\nabla \cdot p\mathbb{1}) dV \tag{2.15}$$

Putting the pieces together, we end up with the momentum equation Equation 2.7 after realizing once more that the integral expression should hold for arbitrary cell size. In index notation[3], the momentum equation reads

$$\frac{\partial}{\partial t} \left(\rho v^i \right) + \sum_j \frac{\partial}{\partial x^j} \left(\rho v^j v^i + p \delta^{ji} \right) = f^i. \tag{2.16}$$

With help of the continuity equation, the momentum equation may be rewritten as a velocity equation

$$\frac{\partial \mathbf{v}}{\partial t} + \mathbf{v} \cdot \nabla \mathbf{v} = -\frac{1}{\rho} \nabla \cdot p\mathbb{1} + \frac{\mathbf{f}}{\rho}, \tag{2.17}$$

---

[3]To ease the reader into this notation, we will leave the summation signs explicit in this chapter, even though Einstein's summation convention means they can be safely omitted.

which is also often called the 'momentum' equation although it strictly speaking is more of a 'velocity' equation. Although the inviscid fluid conservation laws collectively are referred to as Euler's equations, this one is also known as *the* Euler equation. In index notation, it reads

$$\frac{\partial v^i}{\partial t} + \sum_j v^j \frac{\partial}{\partial x^j} v^i = -\sum_j \frac{1}{\rho} \frac{\partial}{\partial x^j} p \delta^{ji} + \frac{f^i}{\rho} \tag{2.18}$$

PROBLEM 2.1
The velocity equation

Starting from the momentum equation Equation 2.7, derive the velocity equation Equation 2.17 by making use of the product rule and the continuity equation Equation 2.4. If you find the vector nature of the equations cumbersome, you can also start from

$$\frac{\partial \rho v^i}{\partial t} + \sum_j \frac{\partial}{\partial x^j} \left(\rho v^j v^i + p \delta^{ji} \right) = f^i, \tag{2.19}$$

and derive

$$\frac{\partial v^i}{\partial t} + \sum_j v^j \frac{\partial}{\partial x^j} v^i = \sum_j -\frac{1}{\rho} \frac{\partial}{\partial x^j} p \delta^{ji} + \frac{f^i}{\rho},$$

which has the advantage of being able to work with all the terms as if they were scalars.

2.1.4   DERIVATION OF THE ENERGY EQUATION

The total energy density of the fluid $\mathscr{E}$ is given by the sum of internal energy density e and kinetic energy density $\frac{1}{2}\rho v^2$. Let us start by formulating an expression for the evolution of the internal energy density, making use of the first law of thermodynamics to do so (the relevant thermodynamics is briefly reviewed in appendix A). Now, the first law of thermodynamics is typically formulated for a closed system with a fixed number of particles, so our previous virtual grid cell is less than ideal as a starting point given that it does not conserve particle number (or mass, which is equivalent in this case). Instead, we therefore consider the internal energy E_{int} of a total mass M of particles. Dividing the two[4], we obtain *specific* internal energy $\varepsilon = E_{int}/M$. Because we are not considering a fixed region of space, but instead a collection of particles, it is important to keep in mind that $\varepsilon(\mathbf{x},t)$ is a function of space as well as time (we will expand upon this approach later in Chapter 3, where we discuss fluid parcels and Lagrangian fluid dynamics).

---

[4]In practice we will conflate ε defined in this manner with $\varepsilon \equiv e/\rho$. This means that we are strictly speaking confusing ε with its average value weighed by mass across our virtual cell, i.e., $\langle \varepsilon \rangle = \int \rho \varepsilon dV / \int \rho dV$. This is only a minor sin, assuming we have been considering a *very small* cell, since in that case $\langle \varepsilon \rangle \approx \rho \varepsilon \Delta V / \rho \Delta V = \varepsilon$. The same applies to other specific quantities introduced in this section.

For a system in thermal equilibrium, the first law of thermodynamics reads

$$d\varepsilon = dq - pd\mathscr{V},\tag{2.20}$$

in terms of specific heat q and specific volume $\mathscr{V} = V/M = \rho^{-1}$. For reversible processes we could slightly rewrite this expression using $dq = Tds$, with specific entropy $s = S/M$. However, making this substitution introduces the notions of thermal equilibrium and thermodynamic temperature T into our system. Since the fluid is not required to be in thermodynamic equilibrium, we will work with q for now. For a change over a time interval dt, we obtain

$$\frac{d\varepsilon}{dt} = \dot{q} + p\rho^{-2}\frac{d\rho}{dt},$$

$$\Rightarrow \quad \rho\frac{d\varepsilon}{dt} = \rho\dot{q} + p\rho^{-1}\frac{d\rho}{dt},$$

$$\Rightarrow \quad \frac{d(\rho\varepsilon)}{dt} - \varepsilon\frac{d\rho}{dt} = \rho\dot{q} + p\rho^{-1}\frac{d\rho}{dt}.\tag{2.21}$$

In terms of notation, $\dot{q}$ is equivalent to dq/dt. From this point onwards, all it takes is a consistent application Equation 2.5, which tells us that $d\rho/dt = -\rho\nabla\cdot\mathbf{v}$, to obtain the result

$$\frac{\partial}{\partial t}(\rho\varepsilon) + \nabla\cdot((\rho\varepsilon + p)\mathbf{v}) = \mathbf{v}\cdot\nabla p + \rho\dot{q},\tag{2.22}$$

This expression for the internal energy density $e \equiv \rho\varepsilon$ has the same features on the LHS as the other conservation laws, in particular, a combination of a local time derivative and a flux term. It also includes a RHS term $\rho\dot{q}$ which reflects a gain in internal energy due to heating, or loss due to e.g. radiation. The other RHS term is an interesting one that we will see appearing for kinetic energy density as well, but with an opposing sign. This means that it disappears in the total energy equation and therefore apparently signifies a mechanism to shift energy between internal and kinetic.

To complete a derivation of the equation of conservation of total energy, we need to consider kinetic energy in addition to internal energy. To derive the kinetic energy density $\frac{1}{2}\rho v^2$ equation, with v the absolute value of velocity defined by $v^2 = \mathbf{v}\cdot\mathbf{v}$, we can start from the velocity equation multiplied by $\mathbf{v}$,

$$\mathbf{v}\cdot\frac{\partial\mathbf{v}}{\partial t} + \mathbf{v}\cdot(\mathbf{v}\cdot\nabla\mathbf{v}) = -\frac{\mathbf{v}}{\rho}\cdot\nabla p + \mathbf{v}\cdot\frac{\mathbf{f}}{\rho},\tag{2.23}$$

(note that for the second term, the first inner product contracts the two outer $\mathbf{v}$'s, as index notation would make clear), and work our way to

$$\frac{\partial}{\partial t}\left(\frac{1}{2}\rho v^2\right) + \nabla\cdot\left(\frac{1}{2}\rho\mathbf{v}v^2\right) + \mathbf{v}\cdot\nabla\cdot(p\mathbb{1}) = \mathbf{v}\cdot\mathbf{f},\tag{2.24}$$

by adding $\frac{1}{2}v^2$ times the continuity equation. Equations 2.22 and 2.24 can now be summed to obtain the energy conservation equation 2.9.

PROBLEM 2.2
Derivation of the energy equation

Derive Equation 2.24 from 2.23. Add Equation 2.24 and 2.22 to arrive at Equation 2.9. If the vector algebra is confusing, you can work in index notation instead and derive

$$\frac{\partial \mathscr{E}}{\partial t} + \sum_i \frac{\partial}{\partial x^i} \left([\mathscr{E} + p] v^i \right) = \rho \dot{q} + \sum_i v_i f^i, \tag{2.25}$$

PROBLEM 2.3
Spherically symmetric fluid dynamics

Write down expressions for the equations of fluid dynamics (Equations 2.4, 2.7 and 2.9) in spherical coordinates. You may assume spherical symmetry, which means that you can skip many of the terms. You do not need to derive anything, just look up the appropriate expressions for gradients and divergences in spherical coordinates.

2.2 FLUID DYNAMICS AND THE CONTINUUM APPROXIMATION

The discussion from Section 2.1 on Euler's equations uses a key assumption in fluid dynamics, which is that we can safely treat the fluid as a continuum. We have assumed that we had a substance that we could arbitrarily divide into smaller and smaller pieces (our virtual grid cells) while the essential properties of these cells remained unchanged. In this sense, fluid dynamics is *scale free*, and applies unchanged across a vast range of scales from the laboratory scale to the extragalactic scale and beyond. It is noteworthy that, absent external forces, no constants of nature appear in Euler's equations to set natural length, time or energy scales. At no point did we introduce analogues to Planck's constant h, light speed c Newton's gravitational constant G or anything else that would set an actual scale.

Nevertheless, the fluid continuum approximation will inevitably break down at some length scale, so even if we did not introduce a length scale, we did imply that we were on the safe side of a size scale where the individual constituent particles of the fluid effectively begin to blur together for the purpose of our analysis. It is not necessarily obvious for a given (astro-)physical problem that this assumption is applicable. Additionally, we have assumed that fluid cells interact with their direct neighbours only (through our use of a local divergence). In particular, this is why we have mentioned gravity thus far as an externally imposed force due to its long range.

For a continuum approximation to be meaningful, we need to have a sufficiently large number of particles and particle interactions per grid cell to indeed be able to move away from individual particles to a bulk approximation. Furthermore, the grid cells should be sufficiently small to allow for a large number of them throughout the fluid under consideration.

Take for example water, a liquid. A litre of water weighs about a kilogram by definition. The diameter l_D of a water molecule is roughly 3×10^{-8} cm. The weight of a water molecule H_2O is 18.02 atomic mass units (amu), which translates to

1.661×10^{-24} g and therefore 6×10^{26} water molecules per litre. The number of particles is therefore definitely large enough for a litre of water to qualify as a fluid.

What about the number of interactions between water molecules? The specific volume $\mathscr{V}$ of 1 litre per kilogram implies 1.661×10^{-24} cm^3 per molecule. The length scale $l_{\mathscr{V}}$ associated with this volume is $(1.66 \times 10^{-24})^{1/3} \approx 1.2 \times 10^{-8}$ cm. Since $l_{\mathscr{V}} < l_D$, there will also definitely be more than enough interactions between water molecules per reasonable unit time. As a matter of fact, for the specific case of water the answer to this question was particularly obvious: ice floats, which means that the molecules in the solid state of water that are bound into place by molecular bonds are actually on average further apart than in the liquid state.

2.2.1 THE CONTINUUM APPROXIMATION AND THE MEAN FREE PATH

As illustrated above for water, in liquids the interaction distance between constituent particles (i.e. molecules, atoms, etc.) is of the same order as the effective size of the elements. For this reason, liquids are mostly well approximated as being *incompressible*, obeying $d\rho/dt = 0$. The short interaction distance means that the granularity of a liquid will begin to diminish in importance already at distance scales of the order of a few times this distance. But for more dilute gases and plasmas this will no longer be the case. Here the average distance a constituent particle can travel on average before experiencing an interaction with its environment, i.e. its *mean free path* λ, can well be larger than the particle size scale by orders of magnitude. In order for the fluid continuum approximation to be valid, λ will need to be significantly smaller than the distance scales in the system under consideration. So much so, that the system can in theory be subdivided into a large number of virtual cells where *each cell* size is still far larger than λ. This condition is parametrized by the *Knudsen number* Kn, defined as the ratio between mean free path λ and a relevant length scale L:

$$\mathrm{Kn} \equiv \frac{\lambda}{L}. \tag{2.26}$$

Once $\mathrm{Kn} \gtrsim 10^{-2}$, the continuum flow approximation will start to become compromised, and systems with large particle numbers but large Knudsen numbers are the subject of statistical mechanics rather than fluid mechanics. The Knudsen number is an example of a *similarity parameter*, a dimensionless parameter that marks a boundary between two different regimes of physical behaviour. Over the course of the book we will encounter a number of such parameters, each triggered by the introduction of some scale into the system that breaks the scale-invariance of the basic Euler equations.

Certainly in astrophysics, with its staggeringly vast range of physical scales from the densest neutron stars to the dilute reaches of empty space or even the universe as a whole, it is by no means obvious whether the condition $\mathrm{Kn} \ll 1$ is fulfilled for a given system. Nor is it always the same physical process which dictates λ.

For neutral gases, the mean free path obeys

$$\lambda = \frac{1}{n\sigma}, \tag{2.27}$$

with n a number density and σ a cross section for binary interaction. In the absence of interactions with a reach extending significantly beyond the scale of the particles, this cross section is determined by the effective area of the particles (think colliding billiard balls), and typically $\sigma \sim l_D^2 \sim 10^{-15}$ cm^2.

As an example, H I clouds consisting mostly of neutral hydrogen form a key part of the *interstellar medium* (ISM), and have number densities on the order of 1 cm^{-3}, more dilute than any laboratory-obtained vacuum on earth. However, since their sizes are immense, $L \sim 10^{19} - 10^{20}$ cm, we still have $\lambda \ll L$ and the fluid approximation is appropriate.

Plasmas, by contrast, are subject to long-range electromagnetic forces, which can end up sustaining the fluid nature of the plasma. It is therefore not unusual for the mean free path for binary interactions to become comparable to or even exceed the scale of the system without this jeopardizing the continuum approximation. We will return to this in Chapter 12.

Finally, while it is certainly beyond the scope of this book, it is worthwhile to point out that there are systems other than physical fluids and plasmas that can be described with some degree of success through an application of continuum conservation laws derived from fluid dynamics. A fascinating example is traffic flow, and the treatment of traffic flow as a continuum dynamics problem has a venerable history dating back to 1955[5].

2.3 ENERGY TRANSPORT IN ATMOSPHERES AND OTHER FLUIDS

There are various means by which energy can be transported throughout a fluid, be it fluids on earth, our atmosphere or the interior of a star. The conservation law for energy has been introduced so far for inviscid fluids including a potential heating term $\rho \dot{q}$ and a flux term for the energy carried by fluid parcels (Equation 2.9):

$$\frac{\partial \mathscr{E}}{\partial t} + \nabla \cdot ([\mathscr{E} + p] \mathbf{v}) = \rho \dot{q} + \mathbf{v} \cdot \mathbf{f}. \tag{2.28}$$

The term $\rho \dot{q}$ can act as a heat loss term, for example through radioactive decay, but can also be part of a more detailed modelling of the fluid that accounts for the impact of radiation (*radiation hydrodynamics*). Radiation, emitted at one part of the flow and intercepted elsewhere within the flow, is one of the means by which energy can be redistributed. This type of long-range energy redistribution does not need to take place uniquely through photons. Examples of other particles capable of transporting energy across long distances, and even of exerting a non-negligible pressure, are cosmic rays and neutrinos. Cosmic rays can be considered to be massive particles (protons and atomic nuclei) that have managed to escape the (virtual) confines of their fluid parcel, and the collective pressure from cosmic rays is an actual component of galaxy models. Neutrinos are nearly massless and carry no electric charge, so

---

[5]Two early key papers [26, 44] are titled *shock waves on the highway* and *On kinematic waves II. A theory of traffic flow on long crowded roads*, respectively. Describing traffic flow using a continuum approximation is still an active area of research.

physical conditions need to be very extreme for them to play a role in fluid behaviour. Nevertheless, neutrino pressure and energy transfer is a key component of supernova explosion models.

Apart from radiation and similar long-range processes, the energy can be transported directly by the fluid parcels through an energy flux term $\mathscr{E}\mathbf{v}$. This is most obvious in the case of a clear bulk flow velocity. But even in an ostensibly hydrostatic fluid, it is not a given that individual fluid parcels remain neatly in place indefinitely. Minimal displacements will always occur, and the fluid may not actually be stable under such disruptions. This will lead to *convection*, which we will discuss in our chapter on fluid instabilities (Chapter 10).

The last option to highlight here is that of *conduction*. While it does not really play a big role in astrophysics, it is of course well-known in earth-based fluid dynamics. In *conduction*, heat is transported between adjacent fluid parcels by direct interactions between their respective constituent particles through collisions and through the diffusion of particles that mix in with neighbouring regions. It is not included in Equation 2.28, but since this describes a local process we can do so using a flux divergence similar to $\mathscr{E}\mathbf{v}$. We consider *heat flux* q, leading to an expression of the form:

$$\frac{\partial \mathscr{E}}{\partial t} + \nabla \cdot ((\mathscr{E} + p)\mathbf{v}) + \nabla \cdot q = \mathbf{v} \cdot \mathbf{f}. \tag{2.29}$$

(Do not confuse the heat flux vector q with scalar local heat production $\dot{q}$). Of course we can also include both q and $\dot{q}$ at the same time, as these describe different physical mechanisms that might both be active (e.g heat conduction and radioactive decay). Thanks to Fourier's law of heat conduction we have an experimental law that provides us with an expression for q as a function of the gradient in temperature:

$$q = -\mathscr{K}\nabla T, \tag{2.30}$$

where $\mathscr{K}$ is the coefficient of thermal conductivity. Dimensional analysis reveals the dimension of $\mathscr{K}$ to be [erg] $[K^{-1}]$ $[cm^{-1}]$ $[s^{-1}]$ in terms of cgs units (or [W]$[K^{-1}]$ $[m^{-1}]$ in SI units including unit of power 'Watt' through W), which helps us to interpret $\mathscr{K}$: it characterizes the thermal energy throughput velocity ([erg] [cm] $[s^{-1}]$) per unit area ($[cm^{-2}]$), including a conversion factor between temperature in Kelvin and energy in erg. q has the dimension of an energy density flux, [erg] $[cm^{-3}]$ [cm] $[s^{-1}]$. The sign in Fourier's law reflects that heat flows from hot to cold temperature. In a hydrostatic fluid of constant density, substituting Fourier's law eventually yields a *heat equation* of characteristic form.

$$\rho T \frac{ds}{dt} = \mathscr{K}\nabla^2 T \Rightarrow \frac{\partial T}{\partial t} = \frac{\mathscr{K}}{c_V \rho}\nabla^2 T. \tag{2.31}$$

The heat equation is a famous example of a *parabolic partial differential equation* and a clean-cut example of *diffusion*, and as such has been widely studied in pure and applied mathematics.

PROBLEM 2.4
The heat equation

Derive heat equation 2.31. You may start from a version of Equation 2.22 with the local heat source replaced by Fourier's law,

$$\frac{\partial}{\partial t}(\rho\varepsilon) + \nabla\cdot((\rho\varepsilon + p)\mathbf{v}) = \mathbf{v}\cdot\nabla p + \mathscr{K}\nabla^2 T,$$

and backtrack from there to include the relevant thermodynamic relations (appendix A).

2.4 BOLTZMANN'S EQUATION AND FLUID DYNAMICS

2.4.1 THE CONTINUITY EQUATION

In Section 2.1 we presented $n+2$ equations for fluid dynamics (where n is the number of dimensions in the system, typically 3), but with $n+3$ unknowns. The $n+2$ equations could be formulated by only looking at the conservation laws affecting the fluid on a *macroscopic* level, and by merely assuming the fluid to be some sort of continuum with *microscopic* properties still to be determined. To close our set of equations and make them solvable, we need an extra equation, and this equation will express the impact of the underlying micro-physics on the macro-physical (continuum) scale—typically it will be an expression relating pressure p to the fluid quantities ρ and ε. Such an equation is called an *equation of state* ('EOS'). We will now proceed to rederive Euler's equations starting at the microphysical scale, given certain assumptions about the nature and interactions of the constituent particles of

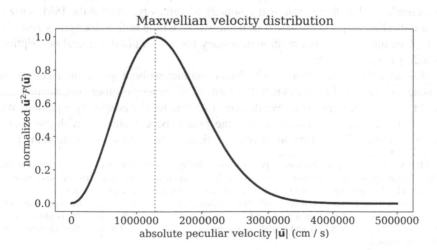

Figure 2.3 Maxwellian distribution of peculiar velocities $\tilde{\mathbf{u}} \equiv \mathbf{u} - \mathbf{v}$ for the ISM, taking $n = 1$ cm^{-3}, $m = m_p$ (the proton mass), and $T = 10^4$ K. The peak velocity per particle is over 10^6 cm s^{-1}.

the fluid, an approach which will also provide us with a suggestion for an equation of state under these conditions.

This takes us to statistical physics and thermodynamic ensembles, for it is typically not feasible to describe the trajectories separately for all particles typically contained in a fluid element (on the order of e.g. 10^{23}). We denote the distribution of point particles in a volume element in *phase space*[6] $d\mathbf{x}d\mathbf{u}$, where $\mathbf{u}$ is now the velocity of an individual particle, by $\mathscr{F}(\mathbf{x},\mathbf{u},t)$. In the case of point particles, there are no further degrees of freedom to consider that make up phase space. The total number of particles in the system N now follows by definition from integrating over the phase space of particles,

$$N \equiv \int \mathscr{F}(\mathbf{x},\mathbf{u},t)d\mathbf{x}d\mathbf{u}. \tag{2.32}$$

If we do not integrate over space but use this approach to define local quantities such as mass density and momentum density, we can write these as[7]

$$\rho \equiv \int m\mathscr{F}(\mathbf{x},\mathbf{u},t)d\mathbf{u}, \qquad \rho\mathbf{v} \equiv \int m\mathbf{u}\mathscr{F}(\mathbf{x},\mathbf{u},t)d\mathbf{u}. \tag{2.33}$$

As an example of a distribution function, the *Maxwellian* velocity distribution for a population of particles in thermal equilibrium is

$$\mathscr{F}(\mathbf{x},\mathbf{u},t)d\mathbf{u} = n(\mathbf{x},t)\left(\frac{m}{2\pi k_B T(\mathbf{x},t)}\right)^{\frac{3}{2}} \exp\left(-\frac{m(\mathbf{u}-\mathbf{v})^2}{2k_B T(\mathbf{x},t)}\right)d\mathbf{u}, \tag{2.34}$$

which is set up to be consistent with Equations 2.33 in that it indeed produces ρ and $\rho\mathbf{v}$ when integrated over the velocity dimensions of phase space. Figure 2.3 provides an example of this distribution using numerical values typical for the ISM, shown for a collection of particles in a coordinate frame where bulk velocity v is zero. In fact, $\mathscr{F}$ contains all the information necessary for a full fluid dynamical description including equation of state.

To derive the conservation laws of fluid dynamics without specifying an equation of state, one does not need to know the form of $\mathscr{F}$ exactly—otherwise, Section 2.1 would not have been possible. We demonstrate this for the continuity equation. We have $\mathscr{F}d\mathbf{u}d\mathbf{x} = dN$ or, in other words, the phase space multiplied with the distribution function will yield the number of particles in that phase space element. This

---

[6]Here we use phase space in terms of position and velocity. Alternatively, we could use position and momentum $d\mathbf{x}d\mathbf{p}$, which would allow us to work with massless particles that can have various momenta but cannot be differentiated through velocity (i.e. they all move with c, like a *photon gas*).

[7]Of course a single fluid cell will actually not be infinitesimally small, so some coarse-graining from the microscopic scale of individual particles to the macroscopic scale of fluid elements is glossed over here. That is, we conflate $\mathbf{x}_{coarse}$ and $\mathbf{x}$ while assuming the spatial scale h of a single fluid element to be extremely small, i.e.

$$\rho(x_{coarse},y_{coarse},z_{coarse}) = \frac{\int_{x_{coarse}-h/2}^{x_{coarse}+h/2} \int_{y_{coarse}-h/2}^{y_{coarse}+h/2} \int_{z_{coarse}-h/2}^{z_{coarse}+h/2} \rho(x,y,z)dxdydz}{\int_{x_{coarse}-h/2}^{x_{coarse}+h/2} \int_{y_{coarse}-h/2}^{y_{coarse}+h/2} \int_{z_{coarse}-h/2}^{z_{coarse}+h/2} dxdydz}$$

is a countable quantity (for the very patient, that is, because the continuum approximation also assumes very large particle numbers within each element of interest). So absent any drastic external interference, such as collisional interactions with particles from an entirely different range of phase space, this number should remain unchanged while this collection of dN particles evolves towards a different corner in phase space. The distribution function obeys

$$\frac{d\mathscr{F}}{dt} = \frac{\partial \mathscr{F}}{\partial t}\bigg|_{collision},$$

(2.35)

where the RHS would have been zero if it had not been for the potential occurrence of abrupt changes due to collisions between particles of drastically different velocities. If this expression holds while $dN(t + \delta t) = dN(t)$ also holds, the implication is that the phase space element $d\mathbf{u}d\mathbf{x}$ is invariant as well. This is indeed the case, and can be shown using *Liouville's theorem* from statistical and Hamiltonian mechanics. The proof lies outside the scope of this book, although later in Chapter 8 we will confirm that the invariance of phase space also holds in special relativity.

If we apply the product rule to the time derivative of $\mathscr{F}$, we obtain (turning to index notation for a bit):

$$\frac{d\mathscr{F}}{dt} = \frac{\partial \mathscr{F}}{\partial t} + \sum_i \frac{dx^i}{dt}\frac{\partial \mathscr{F}}{\partial x^i} + \sum_i \frac{du^i}{dt}\frac{\partial \mathscr{F}}{\partial u^i}.$$

(2.36)

Note how $\mathscr{F}(\mathbf{x}, \mathbf{u}, t)$ is also a function of particle velocity, as reflected in the rightmost term above.

The time derivative of particle positions at a given velocity-point in phase space will be that velocity. The time change in velocity will be due to the action of a force $\mathbf{F}$, so for particles with mass m we have

$$\frac{\partial \mathscr{F}}{\partial t} + \sum_i u^i \frac{\partial \mathscr{F}}{\partial x^i} + \sum_i \frac{F^i}{m}\frac{\partial \mathscr{F}}{\partial u^i} = \frac{\partial \mathscr{F}}{\partial t}\bigg|_{collision}.$$

(2.37)

The above are iterations of *Boltzmann's transport equation*. Recall our discussion of differentiation from Chapter 1: there is always an unspoken assumption about which variables are kept constant (alternatively, we explicitly denote what is happening, as on the RHS, where we mark that for this differentiation everything but the collisional processes are kept constant). For example, in Cartesian coordinates, partial derivatives with respect to x will not affect t, y, z, u^x, u^y or u^z of our phase space.

Note that the term on the RHS and the third term on the LHS both amount to a redistribution over different velocities, the RHS version in a more violent manner than the third term on the LHS. Therefore, if we integrate over all possible velocities (and assuming no particles with infinite velocity in any direction), these integrals should yield zero. We are left with

$$\int \frac{\partial \mathscr{F}}{\partial t}d\mathbf{u} + \sum_i \int u^i \frac{\partial \mathscr{F}}{\partial x^i}d\mathbf{u} = 0,$$

(2.38)

which we can systematically manipulate in order to arrive at the equation of continuity, via

$$m \int \frac{\partial \mathscr{F}}{\partial t} d\mathbf{u} + m \sum_i \int u^i \frac{\partial \mathscr{F}}{\partial x^i} d\mathbf{u} = 0$$

$$\Rightarrow \quad \frac{\partial}{\partial t} \int m\mathscr{F} d\mathbf{u} + \sum_i \frac{\partial}{\partial x^i} \int mu^i \mathscr{F} d\mathbf{u} = 0$$

$$\Rightarrow \quad \frac{\partial \rho}{\partial t} + \sum_i \frac{\partial}{\partial x^i} \left(\rho v^i \right) = 0. \tag{2.39}$$

We were able to move the partial derivatives with respect to $\mathbf{x}$ and t safely outside of the integral, since $\mathbf{x}$, t and $\mathbf{u}$ are independent parameters.

In a way, our approach is in the inverse direction from that of Section 2.1, where we started big and subsequently refined to arbitrary sized elements. Instead of *introducing* a coordinate $\mathbf{x}$ to a system (i.e. to the virtual fluid cell, or the collection of particles of fixed mass M) that was a function only of time t, we now *integrate out* a phase space coordinate $\mathbf{u}$.

2.4.2 THE MOMENTUM EQUATION

Derivations of the equations for momentum and energy follow a similar pattern. Let us do momentum as well (leaving energy as an exercise), in order to be able to interpret pressure in terms of the distribution function $\mathscr{F}$. The second of Equations 2.33 identifies $\mathbf{v}$ as the ensemble-average velocity of the distribution, and we can subtract this from the individual particle velocities $\mathbf{u}$ to define the remaining random *peculiar* velocity $\tilde{\mathbf{u}} \equiv \mathbf{u} - \mathbf{v}$. According to this definition, we have

$$\int \tilde{\mathbf{u}} \mathscr{F} d\tilde{\mathbf{u}} = 0. \tag{2.40}$$

Keeping Equation 2.40 in mind for use during our derivation, we can again use Boltzmann's equation as a starting point. For simplicity, let us assume that the collision term is zero and no external forces are present. Starting from Equation 2.37, we get

$$\frac{\partial \mathscr{F}}{\partial t} + \sum_i u^i \frac{\partial \mathscr{F}}{\partial x^i} + \sum_i \frac{F^i}{m} \frac{\partial \mathscr{F}}{\partial u^i} = \left. \frac{\partial \mathscr{F}}{\partial t} \right|_{collision}$$

$$\Rightarrow \quad \frac{\partial \mathscr{F}}{\partial t} + \sum_i u^i \frac{\partial \mathscr{F}}{\partial x^i} = 0$$

$$\Rightarrow \quad mu^i \frac{\partial \mathscr{F}}{\partial t} + \sum_j mu^i u^j \frac{\partial \mathscr{F}}{\partial x^j} = 0$$

$$\Rightarrow \quad \frac{\partial}{\partial t} \int mu^i \mathscr{F} d\mathbf{u} + \sum_j \frac{\partial}{\partial x^j} \int mu^i u^j \mathscr{F} d\mathbf{u} = 0$$

$$\Rightarrow \quad \frac{\partial}{\partial t}(\rho v^i) + \sum_j \frac{\partial}{\partial x^j} \int m\left(\tilde{u}^i + v^i\right)\left(\tilde{u}^j + v^j\right)\mathscr{F}d\mathbf{u} = 0$$

$$\Rightarrow \quad \frac{\partial}{\partial t}(\rho v^i) + \sum_j \frac{\partial}{\partial x^j} \int m\left(\tilde{u}^i\tilde{u}^j + v^i v^j\right)\mathscr{F}d\mathbf{u} = 0$$

$$\Rightarrow \quad \frac{\partial}{\partial t}(\rho v^i) + \sum_j \frac{\partial}{\partial x^j}(\rho v^i v^j) + \sum_j \frac{\partial}{\partial x^j} \int m\tilde{u}^i\tilde{u}^j\mathscr{F}d\mathbf{u} = 0. \tag{2.41}$$

In this derivation we were able to jettison terms combining $\mathbf{v}$ and $\mathbf{u}$, since in these cases $\mathbf{v}$ is just a pre-factor to an integral of the sort shown in Equation 2.40. The result of our derivation becomes an expression of conservation of momentum once we identify the pressure term from a comparison to Equation 2.7:

$$p\delta^{ij} \equiv \int m\tilde{u}^i\tilde{u}^j\mathscr{F}d\mathbf{u}$$

$$\Rightarrow \quad \sum_i p\delta^i_i = \sum_i \int m\tilde{u}^i\tilde{u}_i\mathscr{F}d\mathbf{u}$$

$$\Rightarrow \quad 3p = \int m\tilde{u}^2\mathscr{F}d\mathbf{u}. \tag{2.42}$$

(If we interpret δ^{ij} and $\tilde{u}^i\tilde{u}^j$ as matrices $[\delta^{ij}]$, $[\tilde{u}^i\tilde{u}^j]$, this is akin to taking the *trace* on the left and right, which for the identity matrix is equal to number of dimensions n). In terms of physical interpretation, pressure is represented on an individual particle level by a flux of peculiar momentum $(m\tilde{u})\tilde{u}$ through local fluid parcel boundaries, carried by the peculiar velocities.

PROBLEM 2.5
Boltzmann's equation and energy

Derive conservation of energy Equation 2.9 in the absence of external heating $(\dot{q} = 0)$, starting from Boltmann's equation Equation 2.37. For simplicity, assume that the collision term on the RHS of Equation 2.37 is zero (although this is not actually necessary to arrive at the correct result). Identify $\int \frac{1}{2}m\tilde{u}^2\mathscr{F}d\mathbf{u}$ as internal energy $\rho\varepsilon$ (does this make physical sense?). Furthermore, for an ideal gas, we can assume no conduction heat flux, meaning that $\int \frac{1}{2}m\tilde{u}^2.\tilde{\mathbf{u}}\mathscr{F}d\mathbf{u} = 0$

2.4.3 EQUATIONS OF STATE

Nowhere in deriving Euler's equations from the Boltzmann transport equation did we need to know the form of $\mathscr{F}$. However, in order to derive an actual equation of state that closes the system of equations, we *do* need to know something about the nature of the constituent particles. In a fluid of point particles with no remaining internal degrees of freedom (which could be present as e.g. vibrational modes in molecules), the kinetic energy of the individual particles adds up completely to the total of the fluid energy, both kinetic and internal. The fluid bulk kinetic energy is set by $\mathbf{v}$, the fluid internal energy by $\tilde{\mathbf{u}}$. For example, for a Maxwellian distribution of velocities, this implies:

$$\rho \varepsilon(\mathbf{x}, t) = \frac{1}{2} m \int \tilde{u}^2 \mathscr{F}(\mathbf{x}, \mathbf{u}, t) d\mathbf{u} \Rightarrow \varepsilon = \frac{3}{2} \frac{k_B}{m} T, \qquad (2.43)$$

which also serves as a definition of the *kinetic* temperature. For the equilibrium case obeying a Maxwellian velocity distribution, this temperature is identical to the *absolute* temperature from thermodynamics.

PROBLEM 2.6
Equilibrium specific internal energy

Perform the integral from Equation 2.43. If you have not done this integral before, note that the integral over velocity phase space is best done in 'spherical' velocity coordinates, where $\tilde{u}$ plays the role of radial variable and the isotropy of the Maxwellian distribution implies that the integral over the angular variables reduces to 4π. In other words, $d\tilde{u} \rightarrow 4\pi \tilde{u}^2 d\tilde{u}$. You may use the standard result that $\int_0^\infty x^4 \exp[-x^2] dx = 3\sqrt{\pi}/8$.

From the previous section we already know that

$$p = \frac{1}{3} \int m \tilde{u}^2 \mathscr{F} d\mathbf{u}. \qquad (2.44)$$

It therefore follows that for a gas of non-relativistic point particles with no additional internal degrees of freedom,

$$p = \frac{2}{3} \rho \varepsilon \Rightarrow p = (\hat{\gamma}_{ad} - 1) \rho \varepsilon, \ \hat{\gamma}_{ad} = \frac{5}{3}, \qquad (2.45)$$

introducing a constant $\hat{\gamma}_{ad}$, the *adiabatic exponent*. Equation 2.45 with $\hat{\gamma}_{ad} = 5/3$ is an example of an EOS, specifically the EOS for a *perfect* monatomic gas. If we wanted, we could use Equation 2.45 to eliminate p from the fluid dynamical equations. For this EOS, we did not require the gas to be in an equilibrium state leading to the Maxwellian velocity distribution; it is sufficient to use that, at a microscopic level, the fluid consists of non-relativistic point particles in three dimensions without a preferred direction in the frame of the bulk flow (the latter to ensure consistency with our definition of pressure, which does not distinguish between directions).

We will revisit the adiabatic exponent and the closely related *polytropic exponent* further in Section 3.4, where we will introduce *polytropic* fluids.

PROBLEM 2.7
A single-temperature gas with multiple particle species

For this exercise, you will need: speed of light $c = 2.99792458 \times 10^{10}$ cm, Boltzmann's constant $k_B = 1.380658 \times 10^{-16}$ [erg][K^{-1}], proton mass $m_p = 1.6726231 \times 10^{-24}$ g, electron mass $m_e = 9.1093897 \times 10^{-28}$ g. We consider a stellar atmosphere consisting of a charge-neutral gas of free protons and electrons. Both particle species share the same temperature T and the gases are in thermal equilibrium. Assume the temperature $T = 10^4$ K, and $n_e = n_p = 10^{14}$ cm^{-3} for the electron (e) and proton (p) number densities.

i) For two ideal gases ($\hat{\gamma} = 5/3$) sharing the same temperature, we have

$$\sum_s n_s = n_e + n_p = n, \qquad \sum_s p_s k_B T = p k_B T,$$

where s sums over the species (two, in this case, protons and electrons). Show that the internal energy densities are the same for protons and electrons (i.e. there is *equipartition* of energy), by computing this value in erg for both species.

ii) Re-express the previous result in terms of the energy per particle both for electrons and protons in units of their respective $m_e c^2$, $m_p c^2$.

iii) Provide a brief physical interpretation of the previous result.

iv) Compute the *average* velocities of an electron in the fluid frame and of a proton, that is $\langle \tilde{u} \rangle$ where $\tilde{u} = |\tilde{\mathbf{u}}|$ the magnitude of the peculiar velocity. You may use the standard integral result $\int_0^\infty x^3 \exp[-x^2] dx = 1/2$.

v) At what temperature does the assumption of non-relativistic electrons break down?

PROBLEM 2.8
Frame-invariance of Euler's equations

i) Consider Euler's equations in one dimension without external forces or heating:

$$\frac{\partial \rho}{\partial t} + \frac{\partial}{\partial x}(\rho v) = 0,$$

$$\frac{\partial \rho v}{\partial t} + \frac{\partial}{\partial x}\left(\rho v^2 + p\right) = 0,$$

$$\frac{\partial \mathscr{E}}{\partial t} + \frac{\partial}{\partial x}(\mathscr{E}v + pv) = 0.$$

Show that Euler's equations are *invariant* under Galilean frame transformations where two reference frames differ by a fixed velocity v_S. The transformation rules are $x = x' + v_S t'$ and $t = t'$. Note that the partial derivative with respect to time is *different* between the two frames. The generalization to three dimensions and the inclusion of external forces and heating (both of which unaffected by the frame transformation) is straightforward. You can take p, ρ and e to be frame-independent, but $\mathscr{E}$ includes a bulk kinetic energy term that is not.

ii) A fluid flow is steady if the partial derivatives with respect to time are all equal to zero (the flow velocity can still be non-zero). Show how steady flow in one frame will not be steady flow in other frames in general, even for fixed frame velocity transformations.

3 Lagrangian Fluid Dynamics

3.1 FLUID PARCELS

In Section 2.1 we used the *Eulerian* approach to deriving the continuity equation, which means that we took a fixed (virtual) grid in space as our starting point and explicitly accounted for flux terms through fixed boundaries. By contrast, we used a *Lagrangian* approach to derive the expression for the internal energy density. Instead of considering a fixed region in space and keeping track of what flowed in and out across a set of virtual boundaries, we considered a free-floating 'closed system' of particles. Although we did not dwell on it any further other than by switching from partial derivative $\partial/\partial t$ to a full derivative d/dt, what we did there was a first implicit introduction of the concept of a *fluid parcel*. Here we define this concept more carefully as we rederive the continuity equation.

A fluid parcel is defined to encompass a fixed number of constituent particles of the fluid, and therefore has a fixed mass M:

$$\frac{dM}{dt} = 0. \tag{3.1}$$

Figure 3.1 A fluid parcel is defined to group a fixed mass, and will therefore enclose a fixed (macroscopic) set of fluid constituent particles.

Recall (Section 1.4) that we defined $d\mathbf{x}/dt \equiv \mathbf{v}_{flow}$ by default, i.e. as the co-moving derivative, unless specified otherwise. This we apply to the fluid parcel as well: a fluid parcel follows the flow velocity $\mathbf{v}$ from Euler's equation, stretching and shrinking in order to contain the same amount of mass over time (see Figure 3.1). The total mass in the parcel is given by

$$M \equiv \int_V \rho \, dV, \tag{3.2}$$

where V is now notably no longer a fixed volume in space but defined by the fluid parcel. For the sake of notational simplicity, we will assume for now an extremely small fluid parcel of mass ΔM for which we can approximate $\Delta M \approx \rho \Delta V$ (i.e. the integral can be approximated by the central value of the integrand times the integration domain). We will find that the generalization to actual integration is straightforward.

DOI: 10.1201/9781003095088-3

Re-deriving the continuity equation from mass conservation in the fluid parcel, we have

$$\frac{d}{dt}\Delta M = \frac{d\rho}{dt}\Delta V + \rho\frac{d\Delta V}{dt} = 0 \Rightarrow \left(\frac{\partial\rho}{\partial t} + \mathbf{v}\cdot\nabla\rho\right)\Delta V + \rho\frac{d\Delta V}{dt} = 0 \quad (3.3)$$

New here is the time derivative of the volume element, which we did not need to worry about in the case of a fixed grid. So what is this time derivative? Given that we already know the form of the continuity equation, we can 'cheat' and conclude based on Equation 3.3 that $d\Delta V/dt = (\nabla\cdot\mathbf{v})\Delta V$. This makes intuitive sense as well: the time derivative of ΔV is dictated by the stretching of the volume element, which in turn is a result of a divergence in velocities across the parcel.

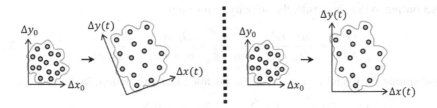

Figure 3.2 The connection between a fluid parcel moving and coordinate transformations. The coordinate systems at each point in time are defined ad-hoc for the purpose of evaluating the fluid parcel volume. For the sake of simplicity, we can limit ourselves to transformations without rotation.

Let us now proceed in the proper way and actually rederive the continuity equation strictly from a Lagrangian approach, which at this point means deriving $d\Delta V/dt = (\nabla\cdot\mathbf{v})\Delta V$. We can express $\Delta V(t)$ in terms of a baseline size at a fixed time t_0, taking $\Delta V(t) \equiv J(t)\Delta V_0$. Here, J is the factor by which the size of the fluid parcel grows or shrinks over time. The use of the symbol J as a factor to multiply the volume element $\Delta V_0 = \Delta x_0\Delta y_0\Delta z_0$ and the implied connection to coordinate transformations is intentional. J is the Jacobian corresponding to the transformation from x_0, y_0, z_0 to $x(t), y(t), z(t)$:

$$J(t) = \begin{vmatrix} \frac{\partial x(t)}{\partial x_0} & \frac{\partial x(t)}{\partial y_0} & \frac{\partial x(t)}{\partial z_0} \\ \frac{\partial y(t)}{\partial x_0} & \frac{\partial y(t)}{\partial y_0} & \frac{\partial y(t)}{\partial z_0} \\ \frac{\partial z(t)}{\partial x_0} & \frac{\partial z(t)}{\partial y_0} & \frac{\partial z(t)}{\partial z_0} \end{vmatrix}. \quad (3.4)$$

However, as illustrated in Figure 3.2, the orientations of these ad-hoc coordinate systems are arbitrary, so we might as well pick an orthogonal pair without a relative rotation, simplifying $J(t)$ to

$$J(t) = \begin{vmatrix} \frac{\partial x(t)}{\partial x_0} & 0 & 0 \\ 0 & \frac{\partial y(t)}{\partial y_0} & 0 \\ 0 & 0 & \frac{\partial z(t)}{\partial z_0} \end{vmatrix} = \frac{\partial x(t)}{\partial x_0}\frac{\partial y(t)}{\partial y_0}\frac{\partial z(t)}{\partial z_0}. \quad (3.5)$$

At this point, $d\left(\Delta V\right)/dt = \left(dJ(t)/dt\right)\Delta V_0$ becomes an exercise in applying the product rule:

$$
\begin{aligned}
\frac{dJ}{dt} &= \frac{d}{dt}\left(\frac{\partial x(t)}{\partial x_0}\right)\frac{\partial y(t)}{\partial y_0}\frac{\partial z(t)}{\partial z_0} + \frac{\partial x(t)}{\partial x_0}\frac{d}{dt}\left(\frac{\partial y(t)}{\partial y_0}\right)\frac{\partial z(t)}{\partial z_0} + \\
&\quad \frac{\partial x(t)}{\partial x_0}\frac{\partial y(t)}{\partial y_0}\frac{d}{dt}\left(\frac{\partial z(t)}{\partial z_0}\right), \\
&= \frac{\partial v^x(t)}{\partial x_0}\frac{\partial y(t)}{\partial y_0}\frac{\partial z(t)}{\partial z_0} + \frac{\partial x(t)}{\partial x_0}\frac{\partial v^y(t)}{\partial y_0}\frac{\partial z(t)}{\partial z_0} + \\
&\quad \frac{\partial x(t)}{\partial x_0}\frac{\partial y(t)}{\partial y_0}\frac{\partial v^z(t)}{\partial z_0}.
\end{aligned}
\tag{3.6}
$$

According to the chain rule, the velocity terms obey

$$
\frac{\partial v^x}{\partial x_0} = \frac{\partial x}{\partial x_0}\frac{\partial v^x}{\partial x} + \frac{\partial y}{\partial x_0}\frac{\partial v^x}{\partial y} + \frac{\partial z}{\partial x_0}\frac{\partial v^x}{\partial z} = \frac{\partial x}{\partial x_0}\frac{\partial v^x}{\partial x},
\tag{3.7}
$$

and similar for v^y and v^z, relying on the assumed lack of rotation. In conclusion,

$$
\frac{dJ}{dt} = \left(\frac{\partial v^x}{\partial x} + \frac{\partial v^y}{\partial y} + \frac{\partial v^z}{\partial z}\right)\left(\frac{\partial x}{\partial x_0}\frac{\partial y}{\partial y_0}\frac{\partial z}{\partial z_0}\right) = (\nabla \cdot \mathbf{v})J(t).
\tag{3.8}
$$

Given that

$$
\frac{d}{dt}\Delta M = \left(\frac{\partial \rho}{\partial t} + \nabla \cdot (\rho \mathbf{v})\right)\Delta V = 0,
\tag{3.9}
$$

needs to hold for arbitrary ΔV, we have therefore at this point re-derived the continuity equation. In the integral version, we similarly obtain:

$$
\begin{aligned}
\frac{dM}{dt} = 0 \quad &\Rightarrow \quad \frac{d}{dt}\int_{V_{moving}} \rho dV_{moving} = 0 \\
&\Rightarrow \quad \frac{d}{dt}\int_{V_0} \rho J(t)dV_0 = 0, \\
&\Rightarrow \quad \int_{V_0} \frac{d}{dt}\left(\rho J(t)\right)dV_0 = 0,
\end{aligned}
\tag{3.10}
$$

which has to hold for arbitrary integration domain.[1]

---

[1] It is also possible to derive Equation 3.8 without resorting to the assumptions that no rotation takes place and that the coordinate vectors remain perpendicular when going from one coordinate system to the next. For completeness we provide this derivation here, using index notation and the permutation symbol from Section 1.2.1. We move from a set of coordinates x_0^1, x_0^2, x_0^3 to a set x^1, x^2, x^3. In terms of the permutation symbol, the Jacobian can be written as

$$
J = \varepsilon^{ijk}\frac{\partial x^1}{\partial x_0^i}\frac{\partial x^2}{\partial x_0^j}\frac{\partial x^3}{\partial x_0^k},
$$

with the summation convention taking care of the indices and the permutation symbol ensuring that what results is indeed the determinant (this is easily checked by writing out the terms). Differentiating J with

3.2 DERIVATION OF CONSERVATION LAWS IN A LAGRANGIAN APPROACH

In order to derive the momentum equation in a Lagrangian approach, consider

$$\frac{d}{dt} \int_V \rho \mathbf{v} \, dV = \int_V \mathbf{f} \, dV + \int_A \mathbf{t} \, dS, \tag{3.12}$$

where S denotes the moving surface of the fluid parcel. From here, the derivation proceeds analogous to the argument from Section 2.1.3, except that we need to account for the changing fluid element as per the discussion above.

A Lagrangian derivation of the energy equation would not be different from the derivation that we have already provided in Section 2.1.4, since the first law of thermodynamics $d\varepsilon = -p \, d\mathcal{V} + dq$ already contains a term $d\mathcal{V}$ explicitly accounting for the change in size of a volume element. The kinetic energy can again be derived from earlier results for momentum and density, like we did in the Eulerian case.

It is sometimes useful to write Euler's equations

$$\frac{\partial \rho}{\partial t} + \nabla \cdot (\rho \mathbf{v}) = 0, \tag{3.13}$$

$$\frac{\partial \rho \mathbf{v}}{\partial t} + \nabla \cdot (\rho \mathbf{v}\mathbf{v} + p\mathbb{1}) = \mathbf{f}, \tag{3.14}$$

$$\frac{\partial \mathcal{E}}{\partial t} + \nabla \cdot ([\mathcal{E} + p]\mathbf{v}) = \rho \dot{q} + \mathbf{v} \cdot \mathbf{f}, \tag{3.15}$$

not in Eulerian form as above, but explicitly in Lagrangian form:

$$\frac{d\rho}{dt} + \rho \nabla \cdot \mathbf{v} = 0, \tag{3.16}$$

$$\frac{d\mathbf{v}}{dt} = -\frac{\nabla p}{\rho} + \frac{\mathbf{f}}{\rho}, \tag{3.17}$$

$$\frac{d\mathcal{E}}{dt} + \mathbf{v} \cdot \nabla p + (\mathcal{E} + p)\nabla \cdot \mathbf{v} = \rho \dot{q} + \mathbf{v} \cdot \mathbf{f}. \tag{3.18}$$

Whereas the Eulerian form is written with a fixed grid in mind, the Lagrangian form suggests *Lagrangian* coordinates where the grid moves with the fluid elements.

---

respect to time now leads to

$$\frac{dJ}{dt} = \varepsilon^{ijk} \frac{\partial v^1}{\partial x_0^i} \frac{\partial x^2}{\partial x_0^j} \frac{\partial x^3}{\partial x_0^k} + \varepsilon^{ijk} \frac{\partial x^1}{\partial x_0^i} \frac{\partial v^2}{\partial x_0^j} \frac{\partial x^3}{\partial x_0^k} + \varepsilon^{ijk} \frac{\partial x^1}{\partial x_0^i} \frac{\partial x^2}{\partial x_0^j} \frac{\partial 3^3}{\partial x_0^k}.$$

The chain rule applied to the velocity divergence gives

$$\frac{\partial v^1}{\partial x_0^i} = \frac{\partial v^1}{\partial x^m} \frac{\partial x^m}{\partial x_0^i}, \tag{3.11}$$

and similar for v^2 and v^3. It can then be demonstrated by writing out all terms that it again follows that $dJ/dt = (\nabla \cdot \mathbf{v}) J$, using the properties of the permutation symbol (in particular, when it equals zero).

These different ways of thinking about the grid suggest different algorithmic approaches to numerically solving the equations of fluid dynamics in computer simulations and either approach has its pros and cons. We will briefly return to this in Chapter 13.

The difference between the Eulerian and Lagrangian formulation might strike you as superficial, and in a way that is true. After all, these are ultimately the exact same equations, albeit written slightly differently. But the way the equations are expressed is a guide to solving them, as the following example demonstrates.

3.2.1   EXAMPLE: STELLAR WIND

Suppose we have a steady *stellar wind*, where a star has been losing mass at a steady rate $\dot{M}$, leading to a steady outflow profile with fixed radial velocity v. (For the sun, we have approximately v = 500 km s^{-1}, and $\dot{M}_\odot = 3 \times 10^{-14} M_\odot$ yr^{-1}. By contrast, a massive Wolf-Rayet star ($M \sim 25 M_\odot$) can have $v = 10^3$ km s^{-1} and $\dot{M} = 10^{-5} M_\odot$ yr^{-1}).

In a Eulerian framework, we would be considering the equation

$$\frac{\partial \rho}{\partial t} + \frac{1}{r^2}\frac{\partial}{\partial r}\left(r^2 \rho v\right) = 0, \tag{3.19}$$

which tells us that $\rho = Cr^{-2}$ if the flow is steady (i.e. the partial derivative with respect to time is zero) at fixed v. The constant of proportionality C follows from accounting for all mass streaming out over a time t:

$$\int \dot{M} dt = M = \int_0^{vt} 4\pi r^2 \rho \, dr \Rightarrow M = 4\pi C v t \Rightarrow C = \frac{\dot{M}}{4\pi v}. \tag{3.20}$$

The density at a fixed position r is therefore given by $\rho = \dot{M}/\left(4\pi v r^2\right)$, as long as we are within the sphere of influence of the star over its lifetime (the 'stellar wind bubble').

Working in a Lagrangian framework instead, we could consider

$$\frac{d\rho}{dt} = -\frac{\rho}{r^2}\frac{\partial v r^2}{\partial r} = -\frac{\rho}{r^2}2vr = -\frac{2\rho}{t}. \tag{3.21}$$

Now r is no longer a fixed grid coordinate but instead the changing position of a fluid parcel $r(t) = vt$. We can solve this further according to

$$\frac{d\rho}{\rho} = -2\frac{dt}{t} \Rightarrow \ln\left(\frac{\rho}{\rho_0}\right) = -2\ln\left(\frac{t}{t_0}\right) \Rightarrow \frac{\rho}{\rho_0} = \left(\frac{t}{t_0}\right)^{-2}. \tag{3.22}$$

This tells us that a fluid parcel starting at time t_0 with density ρ_0 will decrease in density quadratically over time as it advects away from the star. Its mass will be spread out over a surface that increases quadratically in distance and time.

PROBLEM 3.1
Evolution of a shocked fluid parcel

Consider a strong spherically symmetric explosion depositing a large amount of energy in a small volume at the origin and giving rise to a hydrodynamical blast wave expanding radially in the environment surrounding the origin. This could be a powerful bomb on earth, or a star going supernova. The blast wave itself will outrun the gas that it shocks, rapidly overtaking fluid parcels at increasing radii. The position R of the blast wave obeys $R = Ct^{2/3}$, where C is some constant setting the scale of the explosion. The velocity profile of the shocked fluid at radii r between 0 and R obeys

$$v = \frac{1}{2}\frac{r}{t}.$$

It is often of interest to compute for a given fluid parcel the time since shock passage (for example when it radiates for a finite amount of time after being excited by a shock crossing). In this exercise, we compute a measure for this time.

i) Show how the information on fluid flow velocity and blast wave position given above confirms that the shock indeed outruns the fluid.
ii) A fluid parcel shocked by passage of the blast wave will fall progressively further behind the shock front. From considering the evolution dr_{parcel}/dt of position r_{parcel} of a parcel, show that

$$\frac{r_{parcel}}{R(t_0)} = \left(\frac{t}{t_0}\right)^{\frac{1}{2}},$$

where t_0 the time when the parcel was initially overtaken by the shock.
iii) Looking at the blast wave at a given time t, when it has reached a radius $R(t)$, we wish to know how the crossing times t_0 are distributed within the shocked fluid. Show that

$$t_0 = t\left(\frac{r}{R}\right)^6.$$

iv) Starting from the Lagrangian form of the continuity equation,

$$\frac{d\rho}{dt} + \rho\nabla\cdot\mathbf{v} = 0,$$

show that the shocked fluid parcels become more dilute following shock passage according to

$$\frac{\rho}{\rho_0} = \left(\frac{t}{t_0}\right)^{-\frac{3}{2}},$$

where ρ_0 the parcel density at the moment of shock passage.

3.3   MASS COORDINATES

Even though in the previous section we moved to a Lagrangian formalism that was intended to shift the emphasis from grid coordinates towards fluid parcel trajectories, in practice we still expressed the conservation laws in terms of spatial coordinates (in the terms involving the ∇ operator). However, it can sometimes be practical to take this emphasis one step further, and shift to coordinates that are fully Lagrangian. Let us illustrate this for *mass coordinates* in spherical symmetry, which are often deployed in stellar structure models.

We define the cumulative mass m up to a radius r according to

$$m = 4\pi \int_0^r (r')^2 \rho(r') dr'. \tag{3.23}$$

We are going to use this 'coordinate' m to replace r. One of the advantages it has when modelling stellar structure is that its range (from 0 to total stellar mass M) remains relatively constant over the lifetime of a star, whereas stars might vary far more strongly in radius across over smaller time scales. In terms of the old coordinates r, t, a change dm in mass is given by

$$dm = \left(\frac{\partial m}{\partial r}\right)_t dr + \left(\frac{\partial m}{\partial t}\right)_r dt, \tag{3.24}$$

invoking the notation first introduced in Section 1.4.1 to keep track of which variables are kept constant when partial derivatives are taken. On the RHS, the first term tracks the mass contribution that results from adding a shell of width dr (at any fixed time t). This term therefore will be equal to $4\pi r^2 \rho dr$ (i.e. the integrand of the integral in Equation 3.23, which is solved at a fixed point in time). The second term on the RHS side tracks the flow of mass over a time interval dt passing through a fixed radius r. This term will thus be a flux term, with an added minus sign to account for the relative direction of flux versus surface orientations, $-4\pi r^2 \rho v dt$. Because the surface is oriented outwards, any flux along this direction should be lost to the volume. Combined, we get

$$dm = 4\pi r^2 \rho dr - 4\pi r^2 \rho v dt. \tag{3.25}$$

The mass differential dm is an *exact* differential (referring again to terminology introduced in Section 1.4.1), and therefore obeys

$$\frac{\partial}{\partial t}\frac{\partial m}{\partial r} = \frac{\partial}{\partial r}\frac{\partial m}{\partial t} \Rightarrow 4\pi r^2 \frac{\partial \rho}{\partial t} = -4\pi \frac{\partial}{\partial r}\left(r^2 \rho v\right). \tag{3.26}$$

In this expression, we can recognize the continuity equation in spherical coordinates (just shift the r^2 factor to the RHS and divide out the 4π). In spherically symmetric Lagrangian mass coordinates the role of the continuity equation can be served by an expression that traces the radial position of a mass shell:

$$\frac{\partial r(m,t)}{\partial m} = \frac{1}{4\pi r^2 \rho}, \tag{3.27}$$

obtaining by inverting our previous expression for the change in mass over radius.

A second stellar structure equation can be obtained when starting from the velocity equation (using Lagrangian version Equation 3.17):

$$\frac{dv}{dt} = -\frac{1}{\rho}\left(\frac{\partial p}{\partial r}\right)_t + \frac{f}{\rho} = -\frac{1}{\rho}\left(\frac{\partial p}{\partial r}\right)_t - \frac{Gm}{r^2}. \qquad (3.28)$$

Here we used that the gravity experienced by a shell is provided by the mass enclosed by that shell. Moving from radius to mass coordinate, and writing shell velocity explicitly as the full time derivative of shell radius, we get

$$\frac{d^2r}{dt^2} = -\frac{1}{\rho}\left(\frac{\partial m}{\partial r}\right)_t\left(\frac{\partial p}{\partial m}\right)_t - \frac{Gm}{r^2} = -4\pi r^2\left(\frac{\partial p}{\partial m}\right)_t - \frac{Gm}{r^2}. \qquad (3.29)$$

The full derivative we have taken to be defined along the fluid parcel trajectory by default, and this case is no different. This points to a strength of mass coordinates, because moving along with the shell is equivalent to staying at fixed m position (the shell might move further out or closer in, but it will not 'leap-frog' over one of its neighbouring shells in spherical symmetry). In mass coordinates, therefore, $d/dt \rightarrow \partial/\partial t$, so we can obtain the following result:

$$\frac{\partial p(m,t)}{\partial m} = -\frac{1}{4\pi r^2}\frac{\partial^2 r(m,t)}{\partial t^2} - \frac{Gm}{4\pi r^4}. \qquad (3.30)$$

Equations 3.27 and 3.30 are to be contrasted to their Eulerian counterparts (Equation 3.13 and 3.14, expressed in spherical symmetry, see Appendix C.2). They have fewer terms than their counterparts and track the physical changes in a fluid parcel (or mass shell, in this case) rather than local conditions on the grid.

To compute the structure of a star under the assumption of spherical symmetry, Equations 3.27 and 3.30 need to be paired with a Lagrangian equation for energy and as many equations of state as needed to close the system (depending on which additional variables are introduced in the energy equation). It is often assumed that the star is in hydrostatic equilibrium, at which point Equation 3.30 simplifies to the following expression for pressure:

$$\frac{\partial p(m,t)}{\partial m} = -\frac{Gm}{4\pi r^4}. \qquad (3.31)$$

Here we have used mass coordinates mostly as an example to emphasize how a Lagrangian approach works. But we will revisit mass coordinates in the next chapter, where we discuss hydrostatics in more detail (in particular, in Section 4.4).

3.4 POLYTROPIC PROCESSES AND THERMODYNAMICS

When following fluid parcels it is typically possible to make some simplifying assumption about the evolution of the fluid along these trajectories, relative to the behaviour of the fluid across space more generally. Consider a fluid parcel with initial pressure p_0 and mass density ρ_0. As the fluid parcel follows its flow trajectory, the pressure $p(t)$ and mass density $\rho(t)$ will vary according to its equation of state and interactions with its fluid environment. In a *polytropic* process, the relation

$$\frac{p}{p_0} = \left(\frac{\rho}{\rho_0}\right)^{\hat{\gamma}}, \tag{3.32}$$

holds, where $\hat{\gamma}$ is known as the *polytropic exponent*. Both ideal and non-ideal gases can undergo polytropic processes. Equation 3.32 can be considered purely along fluid parcel trajectories without assuming the constant of proportionality between p and $\rho^{\hat{\gamma}}$ is the same along each direction of coordinate change. If a more strict equation of state

$$p = K\rho^{\hat{\gamma}}, \tag{3.33}$$

applies throughout the fluid, with K a constant of proportionality, the fluid is termed a *polytropic fluid* or, equivalently, a *polytrope*.

A first, simple example of a polytropic process is that of isothermal flow in an ideal gas (again, recall that appendix A recaps thermodynamics and thermodynamical concepts). Under the condition of constant temperature, we have

$$p = \frac{\rho k_B T}{\mu m} \Rightarrow p \propto \rho^1. \tag{3.34}$$

In this expression of the ideal gas law, μ is the mean molecular weight and m the atomic mass unit. Because constant T is merely a number that can be absorbed in the constant K, an isothermal ideal gas is therefore a polytrope with $\hat{\gamma} = 1$.

As a second example, consider adiabatic flow in a perfect gas. In the cases of isentropic and adiabatic flow, $d\varepsilon = dq - pd\mathcal{V}$ reduces to $d\varepsilon = -pd\mathcal{V}$. Manipulating this expression, substituting the ideal gas law and making use of the perfect gas relation between $c_\mathcal{V}$ and c_p from appendix A we move along the following derivation:

$$d\varepsilon + pd\mathcal{V} = 0$$

$$\Rightarrow \quad \left(\frac{d\varepsilon}{dT}\right)_\mathcal{V} dT + pd\rho^{-1} = 0$$

$$\Rightarrow \quad c_\mathcal{V} dT - \frac{p}{\rho^2}d\rho = 0$$

$$\Rightarrow \quad c_\mathcal{V} dT - \frac{k_B T}{\mu m}\frac{1}{\rho}d\rho = 0$$

$$\Rightarrow \quad c_\mathcal{V}\frac{dT}{T} = (c_p - c_\mathcal{V})\frac{d\rho}{\rho}. \tag{3.35}$$

The last equation can be solved along a flow trajectory. A perfect gas has constant specific heat capacity, so we get

$$\frac{T}{T_0} = \left(\frac{\rho}{\rho_0}\right)^{\frac{c_p - c_\mathcal{V}}{c_\mathcal{V}}}. \tag{3.36}$$

We therefore have a means of expressing temperature changes purely in terms of changes in density. Combining this with the ideal gas law, we end up with an expression for polytropic flow:

$$\frac{p}{p_0} = \frac{\rho}{\rho_0}\frac{T}{T_0} = \left(\frac{\rho}{\rho_0}\right)^{\frac{c_p}{c_V}}.$$ (3.37)

In this case, the polytropic exponent is equal to the ratio of heat capacities, $\hat{\gamma} = c_p/c_V$.

For an adiabatic process we have an *adiabatic* exponent:

$$\hat{\gamma}_{ad} \equiv \left(\frac{\log p}{\log \rho}\right)_q.$$ (3.38)

In this example, $\hat{\gamma}_{ad}$ is equal to $\hat{\gamma}$. But, in general, polytropic processes are not required to be adiabatic.

For processes that are both adiabatic and isentropic, we have $c_p/c_V = \hat{\gamma} = \hat{\gamma}_{ad}$. For a perfect gas we have $c_p = c_V + k_B/(\mu m)$, which for such processes implies that

$$(\hat{\gamma}_{ad} - 1) = \frac{k_B}{\mu m c_V}.$$ (3.39)

Combined with the ideal gas law and $\varepsilon = c_V T$, this reveals an equation of state that can be written in a form we have encountered before (see Equation 2.45 of Section 2.4.3):

$$p = \rho \frac{k_B T}{\mu m} \Rightarrow p = (\hat{\gamma}_{ad} - 1)\rho \varepsilon.$$ (3.40)

We found $\hat{\gamma}_{ad} = 5/3$ for the case of a monatomic perfect gas, but nowhere in the current derivations did we require the gas to be monatomic. As long as the conditions for a perfect gas hold (small particles, no long-range interactions or correlations between particles), a range of $\hat{\gamma}_{ad}$ values are allowed for. For example, at ordinary temperatures in earth's lower atmosphere, a diatomic gas such as air (dominated by N_2) has $\hat{\gamma}_{ad} = 7/5$.

Note that Equation 3.40 does not follow from polytropic flow by default, and isothermal flow with $\hat{\gamma} = 1$ serves as a counter example. In this case, $c_p/c_V \neq \hat{\gamma}$ (instead, a nonzero dq gets dragged along during the derivation starting at Equation 3.35, and Equation 3.36 ends up with a multiplicative factor on the RHS accounting for an integrated change over entropy differential $ds = dq/T$). As a consequence, Equation 3.39 and Equation 3.40 do not follow. This is why when introducing Equation 3.40 first as Equation 2.45 we then referred to $\hat{\gamma}_{ad}$ as the *adiabatic* exponent rather than the *polytropic* exponent. Physically, to maintain an isothermal process typically requires the presence of a thermalizing heat bath (in astrophysics, a coupling to an external radiation field can play this role), at which point we start to move away from the notion of an isolated perfect gas[2].

3.4.1   THE POLYTROPIC EXPONENT OF A GAS DOMINATED BY RADIATION PRESSURE

It is not uncommon in astrophysics to encounter gases that produce copious amounts of radiation that is not able to escape its local environment, instead ending up dominating the local energy and pressure budget of the system of fluid plus radiation. An obvious example of this would be the interiors of stars. If the majority of the energy of the system indeed resides in the photons, and it is the radiation pressure that provides the dominant contribution to the overall pressure, such a system can actually be described effectively as a polytropic gas[3] with a polytropic exponent obeying $\hat{\gamma} = 4/3$, rather than $5/3$ for classical massive point particles. In exceptional cases, such as supernova collapse, a similar scenario can play out with neutrinos, particles which closely approach key characteristics of photons on account of their near light-speed velocity and near-zero mass, and the same polytropic exponent applies. Let us therefore derive this polytropic exponent.

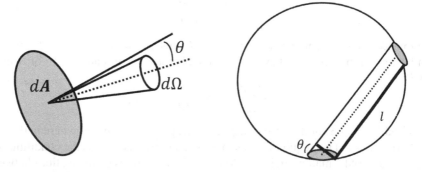

Figure 3.3   Left figure shows a ray of intensity I passing through a surface element $d\mathrm{A}$ in the direction covered by solid angle element $d\Omega$. Right figure shows the 'tube' of volume $dV = l\cos\theta d\mathrm{A}$ carved out by a ray crossing a volume V.

To do so, we briefly introduce the concept of *intensity*, dipping our toes into the area of physics known as *radiative transfer*. Intensity I is usually defined implictly,

---

[2]There is also the well-known scenario in thermodynamics of 'Joule expansion', or free expansion, of an ideal gas into a vacuum. Here an ideal gas is thermally isolated in half of a container while the other half remains empty. The barrier between both halves is then removed and the gas is allowed to expand and fill the whole container, remaining thermally isolated. But even though initial state and end state temperature are identical in this setup, once the barrier is removed, pressure and volume become ill-defined *while the process plays out* [2]. In astrophysical scenarios, sudden expansions of gases into a near vacuum are not uncommon (e.g. supernova and gamma-ray burst blast waves), but are usually modelled as adiabatic, not isothermal processes.

[3]We are at the moment considering adiabatic processes by default, and can therefore conflate $\hat{\gamma}_{ad}$ and $\hat{\gamma}$.

by considering an amount of energy dE contained in a beam of light of intensity I over a time interval dt. The beam passes through a surface element dA into a solid angle $d\Omega$, and the relevant expression is

$$dE = I \cos \theta dt dA d\Omega. \tag{3.41}$$

Here we added a factor $\cos \theta$ to account only for the energy that actually passes *through* the surface element. The angle θ is the angle between the normal to the surface element and the direction of the beam heading towards $d\Omega$ (see Figure 3.3, left image). For example, if $\theta = \pi$, then $\cos \theta$ is zero and $dE = 0$ reflects that no energy passes through. Equation 3.41 implies that the dimensions of intensity are ergs s^{-1} cm^{-2} ster^{-1}. The rays of light defined in this manner are known as 'pencil beams', and for a given ray I will not change over distance unless absorbed or scattered[4].

We can use the concept of intensity to obtain expressions for both the pressure p and the energy density e of the radiation field in order to verify whether indeed $p = (\hat{\gamma} - 1) e = e/3$ applies. By definition, the radiation energy density in a local volume V consists of the total energy of all the rays passing through this volume in all possible directions. Note how it is perfectly fine for multiple rays to simultaneously pass through a point in opposing directions. Also, given the local entrapment of the radiation within a fluid, we have no reason to believe the rays to look any different from one direction to the other: the radiation field can reasonably be considered *isotropic*. No matter which direction $d\Omega = \sin \theta d\theta d\phi$ is considered, there will always be an outgoing ray with intensity I.

Assume each ray to traverse a distance l through our local volume V of interest, moving with light speed c. The time spent within the volume for each ray will then be equal to $dt = l/c$. If a ray enters the volume through a surface element dA on its boundary, it will carve out a little 'tube' of volume $dV = l \cos \theta dA$ within the volume V. The $\cos \theta$ factor here is used to map the area of the surface element onto the diameter of the tube, which will, in general, not be normal to the surface of the volume (see right image of Figure 3.3). Eliminating l in favour of cdt, we therefore have

$$dV = \cos \theta c dt dA. \tag{3.42}$$

Looking at the energy of a given ray as set by Equation 3.41 per volume carved out by the ray, $de = dE/dV$, it follows that

$$de = I d\Omega/c, \tag{3.43}$$

in a given solid angle direction $d\Omega$. Integrating over all solid angles, and recalling that the radiation field can be considered isotropic, we therefore have

$$e = \int de = \int \frac{I}{c} d\Omega = \frac{4\pi I}{c}. \tag{3.44}$$

---

[4]That for example the earth would receive less energy per unit time from the sun if it were further away is not due to I diminishing over distance, but rather on account of a more distant earth covering a smaller solid angle element $d\Omega$ on the sky when seen from the surface of the sun.

To obtain an expression for the pressure, we begin by considering the momentum $|\mathbf{p}|$ contained in the ray. For photons, the amount of momentum is given by

$$d|\mathbf{p}| = \frac{dE}{c} = \frac{I}{c}\cos\theta dt dA d\Omega, \tag{3.45}$$

owing to the straightforward relation between photon energy and momentum. Previously (Section 2.4.2), we formulated the pressure in terms of a net momentum transfer across fluid parcel boundaries, carried by individual constituent particles. If the photons making up the radiation field dominate this process, we can use the momentum equation above with one small tweak. Because we consider pressure, which acts perpendicularly on the surface element, it is the *perpendicular* component $\mathbf{p}_\perp$ of the above that is relevant:

$$d|\mathbf{p}_\perp| = \frac{I}{c}\cos^2\theta dt dA d\Omega. \tag{3.46}$$

Note the extra factor $\cos\theta$ in this expression relative to the expression for $d|\mathbf{p}|$. To obtain the local pressure p, defined as momentum transferred per unit time dt per unit of local surface area dA for incoming rays of each direction, we compute

$$p \equiv \int \frac{d|\mathbf{p}_\perp|}{dt dA} = \int \frac{I}{c}\cos^2\theta d\Omega = \frac{4\pi I}{3c}. \tag{3.47}$$

Comparing Equation 3.47 to Equation 3.44, this indeed confirms that for a gas dominated by radiation pressure and energetics, $p = e/3$ holds and $\hat{\gamma} = 4/3$ is applicable. We will refer to this result on a few occasions later on, when discussing the hydrostatics of stellar structure (Chapter 4) and when expanding our conservations laws to cover *relativistic* fluids (Chapter 8), fluids for which the peculiar velocities of the massive particles that make up the fluid achieve velocities close to the speed of light.

4 Hydrostatics, Atmospheres and Stellar Structure

In *hydrostatics*, the velocity of the fluid is zero. This is a more restrictive condition than *steady flow*, for which the velocity field is merely time-independent. Meaningful models for atmospheres and stellar structure, in particular, can be devised assuming an approximately hydrostatic fluid. To illustrate this, we discuss a range of models of varying complexity.

4.1 HYDROSTATICS OF A PLANE-PARALLEL ATMOSPHERE

In hydrostatic equilibrium v is zero. The only part of the three Euler equations that does not collapse to a triviality under this assumption (assuming $\dot{q} = 0$), is

$$\nabla p = \mathbf{f}. \tag{4.1}$$

Let us consider an atmosphere (stellar or planetary). If we are considering a thin layer sufficiently far from the centre of mass of the star or planet, we can use the plane-parallel approximation (i.e. forget about curvature) and ignore any increase in total mass throughout the atmosphere. In other words:

$$\frac{dp}{dz} = -\rho g, \tag{4.2}$$

with g the magnitude of the gravitational acceleration around the radii under consideration (note that, being a magnitude, g is always positive). We can use the ideal gas law that connects pressure and temperature for N particles in a volume V to set a relation between pressure and density,

$$\rho = \frac{\mu m}{k_B T} p. \tag{4.3}$$

Here μ is the mean molecular weight, m the atomic mass unit and k_B Boltzmann's constant. The value of μ is not necessarily constant throughout an atmosphere. For now, we take μ and T constant (i.e. we assume an isothermal atmosphere).

$$\frac{dp}{dz} = -\left(\frac{\mu m g}{k_B T}\right) p = -\frac{p}{H}, \tag{4.4}$$

where H the *pressure scale-height*. Solving the system, we find

$$p = p_0 e^{-(z-z_0)/H}, \qquad \rho = \rho_0 e^{-(z-z_0)/H}, \tag{4.5}$$

where z_0, p_0 and ρ_0 the values at the base of the atmosphere. This solution illustrates how the scale height provides a representative length scale across which the

DOI: 10.1201/9781003095088-4

atmosphere properties change significantly. For the earth's atmosphere, H is about 10 km.

In modelling stellar structure and hydrostatic equilibrium, total column mass m is typically used instead of distance (see also Section 3.3), i.e. $dm = -\rho dz$ (also integrating downward rather than upward). Independent of whether the atmosphere is isothermal (i.e. fixed T), this reduces the problem to $dp/dm = g$ and implies $p = p_0 + gm$.

PROBLEM 4.1
Hydrostatic atmospheres

For this exercise you will need to following constants: Boltzmann's constant $k_B = 1.380658 \times 10^{-16}$ erg K^{-1}, the atomic mass unit $m = 1.6605 \times 10^{-24}$ g, the solar radius $R_\odot = 6.9599 \times 10^{10}$ cm, the solar mass $M_\odot = 1.98892 \times 10^{33}$ g, the gravitational constant $G = 6.6743 \times 10^{-8}$ cm^3 g^{-1} s^{-2}, the earth radius $R_\oplus = 6.378 \times 10^8$ cm and the earth mass $M_\oplus = 5.974 \times 10^{27}$. We will calculate characterizations of the sun and earth atmospheres, assuming that the ideal gas law to be valid:

$$p\mathscr{V} = \frac{k_B T}{\mu m},$$

where the symbols have their usual meaning.

i) In addition to the ideal gas assumption, we further assume that pressure and density are related according to

$$p = K\rho^{1+\frac{1}{n}},$$

where K and n two constants. If $n \to \infty$, what does this imply for the temperature profiles in the atmospheres?

ii) We start with the case of the sun's photosphere. You can assume a planar atmosphere and constant T. For the purpose of computing the effect of gravity, you may further assume that the atmosphere has a total mass content that is negligible relative to the mass of the sun and an equally negligible radial extent relative to the radius of the sun (which also justifies the planar assumption). Starting from momentum conservation in the hydrostatic case,

$$\nabla p = \mathbf{f},$$

show that the pressure profile obeys

$$p = p_0 \exp\left[-(z - z_0)\frac{\mu m g}{k_B T}\right],$$

where g gravitational acceleration.

iii) Identify a measure of 'scale height' in this expression for pressure and interpret the meaning of a 'scale height' in an atmosphere.

iv) Compute the scale heights for the solar photosphere at temperature $T_\odot = 5770$ K and the earth at $T_\oplus = 300$ K. The solar photosphere is highly ionized and has a mean molecular weight $\mu = 0.6$, while the earth atmosphere consists largely of dinitrogen (N_2) and has a mean molecular weight $\mu = 28$. The scale heights for sun and earth should come out as $H = 2.9 \times 10^7$ cm and 9.0×10^5 cm respectively.

v) Mauna Kea in Hawaii hosts a number of the world's most powerful telescopes requiring astronomers to work at an altitude of 4.2 km. What is the implication for the atmospheric *density* at this elevation relative to sea level?

4.2 AN ISOTHERMAL SLAB MODEL

The hydrostatic plane-parallel atmosphere model discussed above involves a number of simplifying assumptions that will not be generally applicable. It assumes the atmosphere to be isothermal (1), plane-parallel (2), with negligible gravitational pull of its own (3) and obeying a single equation of state (4; both on account of being isothermal and consisting of the same particles throughout). Devising a stellar structure model will quickly prompt us to drop at least assumptions 1–3, even while retaining the still massively simplifying assumption (4), and we will return to this below in Section 4.3. Meanwhile, let us relax only assumption (3), as a stepping stone to stellar structure modelling and to construct a slightly more sophisticated atmospheric profile model (the 'isothermal slab').

4.2.1 SOME QUICK NOTES ON GRAVITY

At this point, it is helpful to quickly cover some features of Newtonian gravity that we will need when calculating the density profile of an atmosphere exerting gravitational pull. The gravitational force density on a fluid at position $\mathbf{r}$ due to a mass M at $\mathbf{r}'$ obeys

$$\mathbf{f}(\mathbf{r}) = \rho \mathbf{g} = -\frac{GM\rho}{(\mathbf{r}-\mathbf{r}')^2} \frac{\mathbf{r}-\mathbf{r}'}{|\mathbf{r}-\mathbf{r}'|}, \qquad (4.6)$$

where G the gravitational constant and $\mathbf{g}$ the local gravitational acceleration. In the simple case where M is located in the origin, we have

$$\mathbf{f} = -\frac{GM\rho}{r^2}\hat{\mathbf{e}}_r = -\frac{GM\rho}{(x^2+y^2+z^2)^{3/2}}(x\hat{\mathbf{e}}_x + y\hat{\mathbf{e}}_y + z\hat{\mathbf{e}}_z). \qquad (4.7)$$

Gravity is a *conservative* force, which means that the gravitational acceleration can therefore be written in terms of a scalar potential Ψ:

$$\mathbf{g} = -\nabla\Psi, \qquad \Psi = -\frac{GM}{|\mathbf{r}-\mathbf{r}'|}, \qquad \mathbf{f} = -\rho\nabla\Psi. \qquad (4.8)$$

It also means that the work done by gravity in displacing a unit of mass will be independent of the path followed by the displacement. Furthermore, the gravitational force field will be free of curl:

$$\nabla \times \mathbf{g} = -\nabla \times \nabla\Psi = 0. \qquad (4.9)$$

In terms of the language from fluid dynamics (as we will discuss in more depth in Section 6.6), there are no 'eddies' in this field. It would indeed be very disorienting to experience gravity, had this not been the case!

Gravity obeys *Gauss' law for gravity*, which states that the area-integrated gravitational acceleration through a closed surface is determined by the mass contained within the volume bounded by the surface:

$$\oint_A \mathbf{g} \cdot d\mathbf{S} = -4\pi GM. \tag{4.10}$$

This expression is completely analogous to Gauss' law from electrostatics, which expresses a relationship between the electric field and total charge. Gauss' law can be demonstrated quickly for the special case of M contained in the centre of a spherical volume,

$$-\oint \frac{GM}{r^2}\hat{\mathbf{e}}_r \cdot d\mathbf{S} = -\int_\theta \int_\phi \frac{GM}{r^2} r^2 \sin\theta \, d\theta \, d\phi = -4\pi GM, \tag{4.11}$$

but applies regardless of the shape of the volume and where within the volume M resides.

It takes only a few steps to go from Gauss' law for gravity to Poisson's equation. Let M be the volume-integrated mass of a continuum mass distribution, i.e.

$$M = \int \rho \, dV. \tag{4.12}$$

We can now apply Gauss' divergence theorem:

$$\oint_S \mathbf{g} \cdot d\mathbf{S} = -4\pi G \int_V \rho \, dV \Rightarrow \int_V \nabla \cdot \mathbf{g} \, dV = -4\pi G \int_V \rho \, dV. \tag{4.13}$$

Writing the acceleration in terms of the gradient of scalar potential Ψ and acknowledging that the result should hold for arbitrary volume V, leads to Poisson's equation

$$\nabla^2 \Psi = 4\pi G\rho. \tag{4.14}$$

This expression tells us what the local impact is of a self-gravitating system on its gravitational potential. It is valid in general, even for the isothermal atmosphere from Section 4.1. However, for that case Poisson's equation collapses into a trivial $0 = 0$. After all, g was approximated as constant, so $\nabla^2 \Psi = 0$. At the same time, ρ on the RHS of Poisson's equation was taken to be zero: the particles of the atmosphere were assumed to only *experience* gravity, and not to be counted as mass charges in their own right.

PROBLEM 4.2
Gauss' law for gravity

Prove that Gauss' law for gravity holds for any distribution of masses, for example by using the divergence theorem.

4.2.2 GENERAL ISOTHERMAL SLAB DENSITY PROFILES

Instead of taking g to be fixed, we can write Equation 4.2 in terms of gravitational potential. If we continue to assume that T is constant, such that $p = A\rho$ where A the constant of proportionality, we have

$$A\frac{\nabla\rho}{\rho} = -\nabla\Psi \Rightarrow \Psi - \Psi_0 = -A\ln\left(\frac{\rho}{\rho_0}\right). \tag{4.15}$$

As with the constant g case, once we have an expression for the gravitational potential Ψ, we have an expression for the density (and therefore pressure) profile:

$$\rho = \rho_0 e^{-\frac{\Psi-\Psi_0}{A}}. \tag{4.16}$$

The profile for Ψ for plane-parallel layers can be determined now that we have a complementary expression for density in terms of gravitational potential that we can apply to Poisson's equation:

$$\frac{d^2\Psi}{dz^2} = 4\pi G\rho_0 e^{-\frac{\Psi-\Psi_0}{A}}. \tag{4.17}$$

With two changes of variables this expression can be cast in a standardized form. Taking $\chi \equiv -\left(\Psi - \Psi_0\right)/A$ and $Z \equiv \sqrt{2\pi G\rho_0/A}z$, we have

$$\frac{d^2\chi}{dZ^2} = -2e^\chi. \tag{4.18}$$

We will not derive this here, but the general solution to Equation 4.18 is given by

$$\chi = -2\ln\left[C_1\cosh\left(\frac{Z}{C_1}+C_2\right)\right], \tag{4.19}$$

where C_1 and C_2 are two constants that are fixed by applying boundary conditions reflecting the physical set-up under consideration. The corresponding density profile is then given by

$$\rho = \frac{\rho_0}{C_1^2\cosh^2\left(\sqrt{\frac{2\pi G\rho_0}{A}}\frac{z}{C_1}+C_2\right)}. \tag{4.20}$$

For example, consider a fully self-gravitating plane that is not resting on top of a massive foundation (e.g. a galactic disc rather than an atmosphere on top of a planet). Picking $z = 0$ to be the plane of symmetry we set the gravitational potential at $z = z_0 = 0$ equal $\Psi = \Psi_0$, such that $\chi(Z = 0) = 0$. On account of the symmetry, $z = 0$ represents an extremum of the gravitational potential, so $d\chi/dZ = 0$ at $z = 0$. When implementing these boundary conditions, it turns out[1] that $C_1 = 1$ and $C_2 = 0$ in Equation 4.19. In the case of a planetary atmosphere, $\chi_0 = 0$ can still be used, while for a planet of mass M and radius R we have

$$\left.\frac{d\chi}{dZ}\right|_0 = -\left(2\pi G\rho_0 A\right)^{-\frac{1}{2}}g = -\left(2\pi G\rho_0 A\right)^{-\frac{1}{2}}\frac{GM}{R^2}. \tag{4.21}$$

---

[1]Under these boundary conditions, [9] show how Equation 4.18 is solved analytically with some judicious substitutions of variables and the interested reader is referred to that textbook for details (so far we have chosen our variable definitions to match theirs).

The general expressions for constants C_1 and C_2 can be computed to be

$$C_1 = \left(x^2 + 1\right)^{-\frac{1}{2}}, \qquad C_2 = \ln\left[x + \left(x^2 + 1\right)^{\frac{1}{2}}\right], \qquad x \equiv \frac{1}{\sqrt{2\pi G \rho_0 A}} \frac{MG}{2R^2}. \quad (4.22)$$

It is not worth delving into the derivation of this result, because this general solution is of little practical interest: it would require a scenario with so much cumulative mass in the atmosphere that it already begins to make a difference to the gravitational profile while the plane-parallel assumption of zero curvature still applies. More interesting perhaps is to explore its limiting behaviour. Our ad-hoc variable x scales linearly with the planet mass M, so for the galactic disc case where this mass is absent, we can take the limit $M \to 0$ and find:

$$C_1 = (0 + 1)^{-\frac{1}{2}} = 1, \qquad C_2 = \ln\left[0 + (0 + 1)^{\frac{1}{2}}\right] = 0, \qquad x = 0. \quad (4.23)$$

The corresponding density profile obeys

$$\rho = \frac{\rho_0}{\cosh^2\left(\sqrt{\frac{2\pi G \rho_0}{A}} z\right)}, \quad (4.24)$$

all of which matching our earlier statements for this scenario. On the other hand, for a large mass M, we have

$$C_1 \approx x^{-1}, \qquad C_2 \approx \ln(2x), \qquad x \gg 1. \quad (4.25)$$

We can plug this into Equation 4.20 for the density profile. We can use the identity $\cosh(y) = \left(e^y + e^{-y}\right)/2 \sim e^y/2$ for large y:

$$\begin{aligned}
\rho &\approx \rho_0 x^2 4 \left[\exp\left(\frac{zMG}{A2R^2} + \ln(2x)\right) + \exp\left(-\frac{zMG}{A2R^2} - \ln(2x)\right)\right]^{-2}, \\
&\approx \rho_0 x^2 4 \left[2x \exp\left(\frac{zMG}{A2R^2}\right)\right]^{-2}, \\
&\approx \rho_0 \exp\left(-\frac{zMG}{AR^2}\right). \quad (4.26)
\end{aligned}$$

In other words, taking the large M limit takes us back to the hydrostatic atmosphere expression of Equation 4.5 where the self-gravity of the atmosphere could be ignored.

Our general expressions for constants C_1 and C_2 suggest that this limit applies if we can safely assume $x \gg 1$. Other than on G and some mathematical constants, x depends on M, R and surface pressure $p_0 = \rho_0 A$. It should therefore be simple to verify this assumption, for example in the case of the earth's atmosphere (assuming for the sake of argument that we are not already completely convinced by the fact that the total mass of the Earth's atmosphere is six orders of magnitude smaller than the total mass of the Earth). In this case, $M = M_\oplus = 5.974 \times 10^{27}$ g, $R = R_\oplus = 6.378 \times 10^8$ cm and $p_0 = 1$ atm $= 1.013 \times 10^6$ Ba (the latter being the cgs unit 'Barye' for pressure). This indeed results in $x \approx 752 \gg 1$. Even for Venus, with its extremely high atmospheric pressure, this limit safely applies. In this case, $M \approx 0.815 M_\oplus$, $R \approx 0.95 R_\oplus$ and $p_0 \approx 93 \times 10^6$ Ba, leading to $x \approx 70 \gg 1$ (see Figure 4.1).

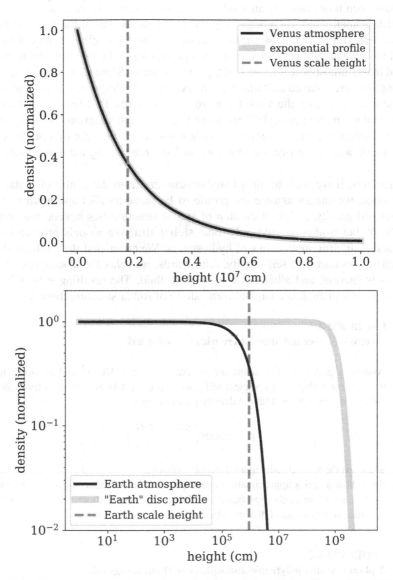

Figure 4.1 Top figure shows the isothermal slab model applied to Venus. The difference between the full expression and an exponential profile is negligible, never exceeding 10^{-5}. The bottom figure shows the Earth's atmosphere, now on a logarithmic scale. The self-gravitating disc model is shown for comparison, obtained by setting $C_1 = 1$ and $C_2 = 0$ in the Earth atmosphere model.

4.3   IDEALIZED STELLAR STRUCTURE MODELS

We have seen how planetary atmospheres can be modelled when assuming a plane-parallel geometry and when neglecting the gravitational pull of the mass within the atmosphere. Both limiting assumptions are quite reasonable, given, respectively, how small the scale height is relative to the planet's radius and how small the mass contained in the atmosphere relative to the planetary mass. Some permutations in which assumptions are retained and which are relaxed lead to simple solutions, and we defer these variations on the same theme to two exercises. Problem 4.3 covers what happens if we no longer neglect curvature but retain the assumptions of an isothermal atmosphere and no self-gravity. Problem 4.4 covers the case of a plane-parallel atmosphere without self-gravity but now no longer assuming the atmosphere to be isothermal.

However, if we wish to model stellar structure from the centre of a star to its outer edge, we cannot assume the profile to be plane-parallel and neither can we neglect self-gravity. A full discussion of stellar structure lies beyond the scope of this book, but isothermal and polytropic stellar structure models provide us with nice and insightful applications of hydrostatics. We cover first the isothermal case, including curvature and self-gravity. Afterwards, we relax the assumption that the fluid is isothermal and allow for a polytropic fluid. The resulting expression, the *Lane-Emden* equation, is a famous cornerstone of stellar structure theory.

PROBLEM 4.3
A curved isothermal atmosphere without self-gravity

i) Assume a gravitational acceleration of magnitude $g = MG/r^2$ and an isothermal atmosphere with $p = A\rho$. Neglect self-gravity (i.e. treat M as a constant throughout the atmosphere). Show that the density profile obeys

$$\rho = \rho_0 \exp\left(\frac{MG}{Ar} - \frac{MG}{Ar_0}\right),$$

where ρ_0 the mass density at a reference radius r_0.

ii) Show how a series approximation of the previous result reduces (to lowest order) to the case of an isothermal plane-parallel atmosphere without self-gravity when considered only at small distances beyond the reference radius r_0.

PROBLEM 4.4
A plane-parallel polytropic atmosphere without self-gravity

i) Assume plane-parallel symmetry, gravitational acceleration g independent of height z and a polytropic gas obeying $p = K\rho^{\hat{\gamma}}$, where K a constant of proportionality. Assume further that the adiabatic exponent $\hat{\gamma} \neq 1$ (i.e. the atmosphere is not isothermal). Show that the normalized density profile of the atmosphere obeys

$$\frac{\rho}{\rho_0} = \left(1 - \frac{z}{z_{max}}\right)^{\frac{1}{\hat{\gamma}-1}},$$

terminating at a maximum height z_{max} given by

$$z_{max} = \frac{\rho_0^{\hat{\gamma}-1}\hat{\gamma}K}{g(\hat{\gamma}-1)}.$$

ii) Compare the above density profile to that of an isothermal atmosphere of equal total mass. Show that the scale height H of the latter obeys

$$H = \frac{\hat{\gamma}-1}{\hat{\gamma}}z_{max}.$$

iii) Plot profiles of both atmospheric models on a single graph to show their similarity (e.g. by using **Python**). Take $\hat{\gamma} = 7/5$, representative of the Earth's atmosphere.

4.3.1 ISOTHERMAL SPHERES

In the case of an isothermal sphere, Equations 4.15 and 4.16 still apply, so finding the density profile again amounts to solving for the gravitational potential profile from Poisson's equation. Poisson's equation, however, takes on different form in spherical symmetry:

$$\frac{1}{r^2}\frac{d}{dr}\left(r^2\frac{d\Psi}{dr}\right) = 4\pi G\rho. \tag{4.27}$$

Setting the gravitational potential to zero ($\Psi_0 = 0$) at the centre of the star then leads to

$$\frac{2}{r}\frac{d\Psi}{dr} + \frac{d^2\Psi}{dr^2} = 4\pi G\rho_0 e^{-\frac{\Psi}{A}}. \tag{4.28}$$

This expression becomes less cluttered if we introduce some ad-hoc variables[2], $w \equiv \Psi/A$ and $z^2 \equiv r^2(4\pi G\rho_0/A)$:

$$\frac{2}{z}\frac{dw}{dz} + \frac{d^2w}{dz^2} = e^{-w}. \tag{4.29}$$

To find the density profile, we need to integrate this equation assuming a suitable set of boundary conditions. At the centre of the star, we have $w = 0$ (ensuring that $\rho = \rho_0$ at this point) and can assume $dw/dz = 0$ (to help avoid a singularity at the LHS of Equation 4.29 that does not occur at the RHS when $w = 0$).

Unfortunately, an exact solution for w is not available. This means that we will have to change tactics from attempting to find a full analytical expression to solving the equation numerically and/or exploring its limiting behaviour. Let us say we want to solve 4.29 numerically. We can rewrite the expression as two coupled first-order differential equations:

$$\frac{dv}{dz} + \frac{2}{z}v = e^{-w}, \qquad v = \frac{dw}{dz}, \tag{4.30}$$

---

[2]With these ad-hoc variables we follow the notation from [20].

with initial conditions $w = 0$, $v = 0$ at $z = 0$. In the case of `Python`, these equations can be integrated with `odeint` (available in the `Python` module `SciPy`):

However, when set to start from $z = 0$, numerical integration will fail on account of the singularity in Equation 4.29 at this point. This singularity can be avoided by starting from a non-zero z position, and if we stick to a very small z value we can approximate the deviations of w and v from their values at $z = 0$ by a series expansion, e.g.

$$w = \alpha_0 + \alpha_1 z + \alpha_2 z^2 + \dots, \qquad v = \alpha_1 + 2\alpha_2 z + \dots. \qquad (4.31)$$

The boundary conditions imply that α_0 and α_1 are both zero, leaving to lowest order that $w \approx \alpha_2 z^2$ and $v \approx \alpha_2 z$. If this is substituted in Equation 4.29 and the RHS is approximated too via a series expansion, the equation can be solved for α_2:

$$\frac{d}{dz}\left(\alpha_2 z^2\right) + \frac{2}{z}\left(2\alpha_2 z\right) \approx 1 - \alpha_2 z^2 \Rightarrow \alpha_2 \approx \frac{1}{6}. \qquad (4.32)$$

So for small z, $w \approx z^2/6$ while $v \approx z/3$.

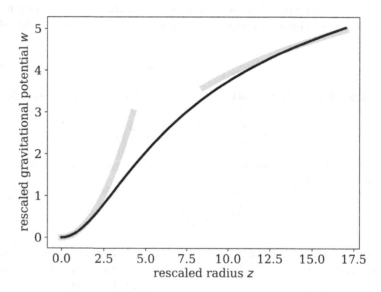

Figure 4.2 Density profile for an isothermal sphere, along with approximate solutions for large and small z values.

The numerical solution for $w(z)$ is shown in Figure 4.2, along with approximated solutions at large and small z values. The small z approximation is just the one we used to avoid the singularity at $z = 0$. For the large z approximation we instead expand ρ/ρ_0 in terms of $\bar{z} \equiv 1/z$. If we apply the boundary conditions $\rho = 0$ and $d\rho/d\bar{z} = 0$ at $\bar{z} = 0$ (anticipating that the isothermal sphere solution will have infinite radius, something we will return to below), we arrive at $\rho/\rho_0 \approx \bar{\alpha}_2 \bar{z}^2$ at the lowest order for the series expansion, implying $w \approx 2\ln z - \ln \bar{\alpha}_2$. As with the small z approximation, the coefficient can be determined from substituting this solution into Equation 4.29, revealing $\bar{\alpha}_2 = 2$. The figure indeed shows how

$$w \approx 2\ln z - \ln 2, \qquad \frac{\rho}{\rho_0} \approx \frac{2}{z^2}, \tag{4.33}$$

is approached asymptotically for large z.

This large z limit z^{-2} dependency on radius for the density profile of an isothermal sphere means that their total mass content will continue to grow with increasing radius, given that $dM = 4\pi r^2 \rho dr \rightarrow constant$. Isothermal sphere solutions will therefore need to have a truncation point imposed, which in turn will set the scale for ρ_0. As stand-alone solutions to stellar structure, their use might be limited, but they remain applicable to describe cores of stars where these can be assumed to be non-degenerate and isothermal.

4.3.2 POLYTROPIC STELLAR STRUCTURE MODELS

Finally, let us drop the remaining constraint that T is constant, and instead consider models for which

$$p = K\rho^{\hat{\gamma}} = K\rho^{1+\frac{1}{n}}. \tag{4.34}$$

Here $\hat{\gamma}$ is the polytropic exponent, and we will consistently label n the polytropic *index* in order to distinguish it from $\hat{\gamma}$ even if the words *exponent* and *index* are interchangeable mathematical terminology. The isothermal sphere model is obtained when setting $n \equiv \infty$.

We start again from Equation 4.1, but because T is not constant in general, we cannot proceed along the lines of Equations 4.15 and 4.16. Instead, we use the polytropic relation Equation 4.34 and spherical symmetry to obtain:

$$\nabla p = \rho \mathbf{g} \Rightarrow \frac{d\Psi}{dr} = -K\hat{\gamma}\rho^{\hat{\gamma}-2}\frac{d\rho}{dr}, \tag{4.35}$$

with a straightforward solution

$$\Psi - \Psi_0 = -\frac{K\hat{\gamma}}{\hat{\gamma}-1}\left(\rho^{\hat{\gamma}-1} - \rho_0^{\hat{\gamma}-1}\right). \tag{4.36}$$

As boundary conditions for Ψ_0 and ρ_0, we can set the gravitational potential to zero at the edge of the star determined by $\rho_0 = 0$ (the gravitational potential will be negative throughout the star as a consequence). We now have an expression for the density,

$$\rho = \left(\frac{-\Psi}{K(n+1)}\right)^n, \tag{4.37}$$

that we can use in Poisson's equation 4.27 in order to obtain:

$$\frac{d^2\Psi}{dr^2} + \frac{2}{r}\frac{d\Psi}{dr} = 4\pi G \left(\frac{-\Psi}{K(n+1)}\right)^n. \tag{4.38}$$

This equation is similar to Equation 4.28 for the isothermal sphere case. It can be decluttered in a similar fashion. If we introduce a scale factor Ψ_c for Ψ to both sides of the equation and shift the constant terms around, we get

$$\left(\frac{(n+1)^n K^n}{4\pi G}\right)(-\Psi_c)^{-n}\left(\frac{d^2\Psi}{dr^2}+\frac{2}{r}\frac{d\Psi}{dr}\right)=\left(\frac{\Psi}{\Psi_c}\right)^n. \tag{4.39}$$

This in turn suggests how we can use a dimensionless fractional potential Ψ/Ψ_c and absorb the constants in the distance measure:

$$w \equiv \frac{\Psi}{\Psi_c}, \qquad z^2 \equiv \frac{4\pi G}{(n+1)^n K^n}(-\Psi_c)^{n-1} r^2. \tag{4.40}$$

Once applied, we get

$$\frac{d^2w}{dz^2}+\frac{2}{z}\frac{dw}{dz}+w^n=0, \tag{4.41}$$

which is the polytrope analog to Equation 4.29. In a slightly rephrased version, this becomes the well-known *Lane-Emden* equation for the stellar structure of a polytrope:

$$\frac{1}{z^2}\frac{d}{dz}\left(z^2\frac{dw}{dz}\right)+w^n=0. \tag{4.42}$$

At the centre of the star where $z=0$, we have $w=1$ (do not confuse this w with the auxiliary variable w for the isothermal sphere). A singularity at $z=0$ is avoided if $v=dw/dz=0$ at this point. Like the isothermal sphere case, no general solution exists for the Lane-Emden equation, at least not for arbitrary value of the polytropic index n. Exceptions are $n=0$, $n=1$ and $n=5$.

PROBLEM 4.5
The dimensionless Lane-Emden equation

i) Show that the terms in the Lane-Emden equation (Equation 4.42) are dimensionless.
ii) What does this equation being dimensionless imply when the Lane-Emden equation is deployed in stellar structure modelling?

Figure 4.3 shows the solution to the Lane-Emden equation for a range of n values. Note that at small radius, all profiles look similar. But the higher n, the larger the z value for which the solution crosses $w=0$ and the star will be more extended as a result. Once $n=5$, the stellar radius becomes infinite. For the purpose of solving the Lane-Emden equation or estimating its behaviour at small radius, we can apply the same series expansion approach that we used in the isothermal case. Substituting $w \approx \alpha_0 + \alpha_1 z + \alpha_2 z^2 + \alpha_3 z^3 + \alpha_4 z^4$ in Equation 4.41, will lead to

$$w(z) \approx 1 - \frac{1}{6}z^2 + \frac{n}{120}z^4. \tag{4.43}$$

Note how the z^4 term is the first term in this expansion to depend on n, explaining why the profiles look similar at small z.

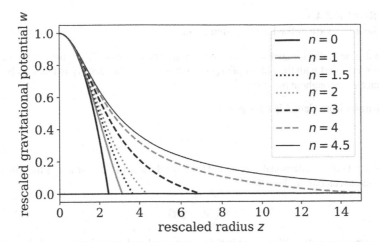

Figure 4.3 Solutions to the Lane-Emden equation for various values of n. The higher n, the larger the z value for which the solution crosses $w = 0$.

For the purpose of obtaining the exact solutions for $n = 0$ and $n = 1$, we can introduce $\chi \equiv wz$ in the Lane-Emden equation to obtain an alternative expression of the form

$$\frac{d^2\chi}{dz^2} = -\frac{\chi^n}{z^{n-1}}, \tag{4.44}$$

which for $n = 1$ reduces to

$$\frac{d^2\chi}{dz^2} = -\chi. \tag{4.45}$$

The boundary condition $\chi = 0$ at $z = 0$ implies the solution

$$\chi(z) = \sin z \rightarrow w(z) = \frac{\sin z}{z}, \qquad \text{if } n = 1. \tag{4.46}$$

Solving for $n = 0$ and imposing boundary conditions $\chi = 0$ and $w = 1$ at $z = 0$, leads to

$$w(z) = 1 - \frac{1}{6}z^2, \qquad \text{if } n = 0, \tag{4.47}$$

with the RHS, in this case, providing the full solution rather than the first terms in a series expansion. We discuss the $n = 0$ case further in exercise 4.6. For $n = 5$ we provide the solution without derivation[3]:

$$w(z) = \frac{1}{\sqrt{1 + z^2/3}}, \qquad \text{if } n = 5. \tag{4.48}$$

Note how this solution approaches zero asymptotically for large z.

---

[3] A derivation can be found in [9].

PROBLEM 4.6
The $n = 0$ case of the Lane-Emden equation

In a spherically symmetric polytropic stellar structure model, setting $n = 0$ can be reconciled with the polytropic relation $\rho \propto p^{\frac{n}{n+1}}$ if the density is taken to be constant throughout the star.

i) Starting from Gauss' law for gravity

$$\oint \mathbf{g} \cdot d\mathbf{S} = -4\pi GM,$$

where G the gravitational constant, M the total mass of the star and $\mathbf{g}$ the gravitational acceleration, show that

$$\Psi = \Psi_c + \rho G \frac{4\pi}{6} r^2,$$

where r the radius, Ψ the gravitational potential and the subscript c refers to the centre of the star.

ii) Apply the same change of variables as in the main text,

$$w \equiv \frac{\Psi}{\Psi_c}, \qquad z^2 \equiv \frac{4\pi G}{(n+1)^n K^n} (-\Psi_c)^{n-1} r^2,$$

to show that at the edge of the star where $w = 0$, we have $z = \sqrt{6}$.

iii) Show that the pressure p_c at the centre of a star with radius R is given by

$$p_c = \frac{1}{2} \frac{\rho GM}{R}.$$

You might find it more straightforward to work directly from $\nabla p = \rho \mathbf{g}$ rather than the Lane-Emden equation.

4.3.3   THE PHYSICS OF POLYTROPES

Once we have a solution to the Lane-Emden equation we have a general result for a given n that can be applied to any actual realization of that polytrope. The dimensionless variables of the Lane-Emden equation are a function of the central values of the fluid state for the star, T_c, p_c and ρ_c. From these, global properties of the star can be derived, such as total mass M, stellar radius R and total energy of the star E. Alternatively, we work with the global properties (that are more easily accessible as observables) and from these deduce the internal physics of the star. Key variables that occur in the definitions of w and z are K and Ψ_c (see Equation 4.40). Combining the ideal gas law with our starting assumption Equation 4.34 for polytropes, we have

$$K = \frac{k_B T_c}{\mu m} \rho_c^{1-\hat{\gamma}} = \frac{k_B T_c}{\mu m} \rho_c^{-\frac{1}{n}}. \tag{4.49}$$

For the potential we have from Equation 4.37 and the definition of w:

$$-\Psi = \rho^{\frac{1}{n}} K(n+1) = -\Psi_c w. \tag{4.50}$$

Using these expressions along with Equation 4.40 allows us to recast expressions for the mass M and radius R in terms of K and ρ_c. If we leave K explicit, we obtain for the mass:

$$M = \int_0^R 4\pi \rho r^2 dr = \left(\frac{(n+1)}{G(4\pi)^{\frac{1}{3}}}\right)^{\frac{3}{2}} K^{\frac{3}{2}} \rho_c^{\frac{3-n}{2n}} \int_0^{z_{max}} w^n z^2 dz. \quad (4.51)$$

The remaining integral is dimensionless. If instead we use the ideal gas as applied in Equation 4.49, we find

$$M = \left(\frac{(n+1)k_B}{G\mu m (4\pi)^{\frac{1}{3}}}\right)^{\frac{3}{2}} \rho_c^{-\frac{1}{2}} T_c^{\frac{3}{2}} \int_0^{z_{max}} w^n z^2 dz. \quad (4.52)$$

Also from Equation 4.40 we can obtain expressions for the stellar radius. These are

$$R = \left(\frac{n+1}{4\pi G}\right)^{\frac{1}{2}} K^{\frac{1}{2}} \rho_c^{\frac{1-n}{2n}} z_{max}, \quad (4.53)$$

if K is left intact and

$$R = \left(\frac{(1+n)k_B}{4\pi G\mu m}\right)^{\frac{1}{2}} T_c^{\frac{1}{2}} \rho_c^{-\frac{1}{2}} z_{max}. \quad (4.54)$$

if the central temperature T_c is made explicit.

PROBLEM 4.7
Mass and radius of polytropes

Derive Equations 4.52 and 4.54, starting from $M = 4\pi \int_0^R \rho r^2 dr$ and the right expression of Equations 4.40 respectively.

Equations 4.51–4.54 allow for scaling relations between solutions even if their central densities and/or other properties differ. We can solve Equations 4.52 and 4.54 for density and equate the result in order to find out how mass varies with radius across stars with a similar central temperature T_c:

$$M \propto R^1, \text{ similar central temperature.} \quad (4.55)$$

It is not a stretch to consider similar temperatures at the centres of stars rather than similar densities, e.g. when the central temperatures are dictated by the local conditions for releasing energy and heating the fluid (i.e. similar nuclear processes), rather than the global mass distribution providing the local gravitational compression of the star. A roughly linear relation between stellar mass and radius is indeed observed in nature, so this type of modelling is at least semi-realistic, if still strongly oversimplified.

On the other hand, we can have a change in the global mass budget of a star after it has established its initial equilibrium following some polytropic equation of state (i.e. fixing K, rather than T_c). This subsequent change could be caused for example by mass loss from a stellar wind or mass gain from accretion flow (where external gas is pulled onto the star by gravity). In this case, we can compare Equations 4.51 and 4.53 and find:

$$M \propto R^{\frac{3-n}{1-n}}, \text{ similar central } K. \tag{4.56}$$

In the case of $\hat{\gamma} = 5/3$, $n = 3/2$ we have $M \propto R^{-3}$. If $\hat{\gamma} = 4/3$, $n = 3$ we have $M \propto R^0$. This suggests a star would first shrink strongly while it accumulates mass and that this process stabilizes as the equation of state becomes more extreme (i.e. stronger internal radiation pressure, or massive particles starting to move relativistically and beginning to resemble photons dynamically). The limiting case of the latter is represented by the *Chandrasekhar mass*. Assuming K to be fixed at a value representing the physics of a completely degenerate relativistic gas (where the Fermi-exclusion principle provides the counter-pressure against gravity), this represents the case where a further increase in central density will not further impact the total mass M. For example, no observed white dwarfs stars so far have masses beyond this limit, which in their case stands at $M = 1.46\ M_\odot$. The analog for neutron stars (the Tolman–Oppenheimer–Volkoff limit) lies around $2.1 - 2.3\ M_\odot$. Note that this value is sensitive to the equation of state of the exotic matter at the core of a neutron star (generalizing the role of K for basic polytropes).

In addition to mass M and radius R, we can compute the total energy of a polytropic stellar model. For a hydrostatic star, this total energy E will be the sum of internal energy E_{int} and gravitational energy E_g. The gravitational energy over the full volume can be defined in term of cumulative mass $m(r)$ up to radius r (recall Section 3.3) by integrating the gravitational potential over the full volume of the star:

$$E_g = -\int_V \frac{Gm\rho}{r}dV = -\int_0^R \frac{Gm}{r}\rho 4\pi r^2 dr = -\int_0^M \frac{Gm}{r}dm. \tag{4.57}$$

The internal energy results from integrating over the internal energy density, which is linked to pressure via the EOS for a polytropic fluid:

$$E_{int} = \int_V e dV = \int \frac{4\pi r^2 p}{\hat{\gamma}_{ad} - 1}dr. \tag{4.58}$$

By manipulating the expression for internal energy we can show how it connects to the gravitational energy E_g. We start with

$$E_{int} = \int \frac{4\pi r^2 p}{\hat{\gamma}_{ad} - 1}\frac{dr}{dm}dm = \int \frac{4\pi p}{\hat{\gamma}_{ad} - 1}\frac{1}{3}\frac{d\left(r^3\right)}{dm}dm. \tag{4.59}$$

Then we integrate by parts and use that r^3/m vanishes at the centre while p is zero at the outer edge of the integration domain. We obtain

$$E_{int} = -\frac{4\pi}{3\left(\hat{\gamma}_{ad} - 1\right)}\int_0^M \frac{dp}{dm}r^3 dm = \frac{1}{3\left(\hat{\gamma}_{ad} - 1\right)}\int_0^M \frac{Gm}{r}dm. \tag{4.60}$$

Here we made use of Equation 3.30 for the hydrostatic case in our substitution. In the last term we can recognize E_g multiplied by a pre-factor, and we can conclude that

$$E_{int} = -\frac{1}{3(\hat{\gamma}_{ad} - 1)} E_g \Rightarrow 3(\hat{\gamma}_{ad} - 1) E_{int} + E_g = 0. \tag{4.61}$$

This result is actually a specific case of the well-known *virial theorem* that plays a role in various branches of theoretical astrophysics. In general, this theorem describes a relationship between kinetic and gravitational energy of a system. In our hydrostatic case, all kinetic energy resides in internal motion while the bulk component is zero.

We have now reduced the problem of computing the total energy $E = E_{int} + E_g = \frac{3-n}{3} E_g$ to computing E_g. We begin the latter with another round of partial integration:

$$E_g = -\int_0^M \frac{Gm}{r} dm \tag{4.62}$$

$$= \int_0^{M/R} Gm d\left(\frac{m}{r}\right) - G\frac{M^2}{R} \tag{4.63}$$

$$= \int_0^M G\frac{m}{r} dm - \int_0^R G\frac{m^2}{r^2} dr - G\frac{M^2}{R} \tag{4.64}$$

$$= -E_g - \int_0^R G\frac{m^2}{r^2} dr - G\frac{M^2}{R}, \tag{4.65}$$

leading to

$$E_g = \frac{1}{2}\int_0^R \frac{Gm^2}{r^2} dr - \frac{GM^2}{2R}. \tag{4.66}$$

To sort out the remaining integral, we take a detour via the gravitational potential $\Psi = -Gm/r$, leading to

$$E_g = -\frac{1}{2}\int_0^R \frac{d\Psi}{dr} m dr - \frac{GM^2}{2R}. \tag{4.67}$$

A rewrite and yet another partial integration nets us

$$E_g = -\frac{1}{2}\int_{\Psi_c}^0 m d\Psi - \frac{GM^2}{2R} = \frac{1}{2}\int_0^M \Psi dm - \frac{GM^2}{2R}. \tag{4.68}$$

The partial integration is made simple by the boundary values $m = 0$ at the centre and $\Psi = 0$ at the outer radius. The potential can be cast in terms of fluid quantities through application of Equation 4.37, from which we obtain

$$\Psi = -\rho^{\frac{1}{n}} K(n+1) = -(n+1)\frac{p}{\rho} = -(n+1) p \frac{dr}{dm} 4\pi r^2. \tag{4.69}$$

This last expression begins to resemble the integrand of the E_{int} integral (see Equation 4.59), and if we make use of that, we obtain

$$E_g = -\frac{1}{2}(n+1)(\hat{\gamma}_{ad} - 1) E_{int} - \frac{GM^2}{2R} = -\frac{1}{6}(n+1)E_g - \frac{GM^2}{2R}. \tag{4.70}$$

We therefore eventually conclude

$$E_g = -\frac{3}{5-n}\frac{GM^2}{2R} \Rightarrow E = -\frac{1}{5-n}\left(\frac{3\hat{\gamma}_{ad}-4}{\hat{\gamma}_{ad}-1}\right)\frac{GM^2}{2R}. \tag{4.71}$$

Note how a polytrope with $\hat{\gamma}_{ad} = 4/3$ will have a total energy of zero, regardless of its values for M and R.

4.4   STELLAR STRUCTURE MODELLING USING MASS COORDINATES

Treating the subject in much further depth here would go beyond the scope of this book, but realistic modelling of stellar structure and evolution can be taken far beyond the confines of polytropic models. We will limit ourselves to a brief discussion to wrap up the topic and to illustrate how various concepts from other branches of physics can interact with a fluid dynamics system.

Recall the introduction of *mass* coordinates from Section 3.3. These provide us with a truly Lagrangian approach to spherically symmetric systems (for which $dX/dt \rightarrow \partial X/\partial t$, given our default assumption for the path along which full derivatives are evaluated), and form the basis of many implementations of stellar structure models. In Section 3.3, we established two conservation laws covering the mechanics of the fluid (leaving its thermodynamical evolution to be covered later):

$$\frac{\partial r(m,t)}{\partial m} = \frac{1}{4\pi r^2 \rho}, \tag{4.72}$$

$$\frac{\partial p(m,t)}{\partial m} = -\frac{1}{4\pi r^2}\frac{\partial^2 r(m,t)}{\partial t^2} - \frac{Gm}{4\pi r^4}. \tag{4.73}$$

Pressure p and density ρ connect through an equation of state. If this EOS is of the form $p = p(\rho)$ then expressions 4.72 and 4.73 are sufficient to determine the system, as shown for the polytropes above. In mechanical equilibrium, the time derivative of r in Equation 4.73 can be set to zero.

Realistically, pressure depends on (specific) internal energy as well as density, while the internal energy in turn is influenced by various processes that occur in stars such as nuclear reactions, heat conduction and production of radiation and neutrinos. We therefore need to add an expression that captures energy conservation in stars. Taking the same starting point as in Section 2.1.4, where we first derived an expression for energy conservation in a fluid, we take our cue from the first law of thermodynamics:

$$\frac{d\varepsilon}{dt} = \dot{\varepsilon}_n - \dot{\varepsilon}_\nu - \frac{\partial L_r}{\partial m} - p\frac{d\mathcal{V}}{dt}, \tag{4.74}$$

where we introduce various means to generate heat and extract internal energy from the fluid. The rate of change of the specific internal energy due to nuclear processes is denoted by $\dot{\varepsilon}_n$. The term $\dot{\varepsilon}_\nu$ denotes a loss of energy from the production of neutrinos. Under normal conditions, these neutrinos pass through the stellar fluid unaffected, generating a total neutrino flux

$$L_\nu = 4\pi \int_0^R \rho \dot{\varepsilon}_\nu r^2 dr = \int_0^M \dot{\varepsilon}_\nu dm. \tag{4.75}$$

Note how this implies $\dot{\varepsilon}_v = \partial L_v / \partial m$. The similarly shaped term $\partial L_r / \partial m$ captures the net total of *all* forms of energy transport through a sphere of radius r that are not already accounted for explicitly by the other RHS terms in Equation 4.74. These include radiation, conduction and an approximation to convection. The latter mechanism is not automatically accounted for in a spherically symmetric stellar evolution model, whereas a full multidimensional approach to solving Euler's equations would be extremely numerically challenging for relatively little gain in accuracy over (sophisticated) spherically symmetric models. Hence its inclusion in the total luminosity term. At the boundary of the star, where the fluid density becomes zero, all non-radiation contributions drop away and L_r can be identified with the photon luminosity L.

Stellar structure models typically solve for $\partial L_r / \partial m$ rather than ε. In the stationary case and in the absence of neutrino emission, Equation 4.74 reduces to

$$\frac{\partial L_r}{\partial m} = \dot{\varepsilon}_n, \tag{4.76}$$

which indicates that the energy flux is set completely by the nuclear energy released per second. More generally, we have

$$\frac{\partial L_r}{\partial m} = \dot{\varepsilon}_n - \dot{\varepsilon}_v - \frac{d\varepsilon}{dt} - p\frac{d\mathcal{V}}{dt}. \tag{4.77}$$

This expression can be rewritten into the following form:

$$\frac{\partial L_r}{\partial m} = \dot{\varepsilon}_n - \dot{\varepsilon}_v - c_p\frac{dT}{dt} + \frac{\delta}{\rho}\frac{dP}{dt}, \tag{4.78}$$

where the auxiliary quantity δ is defined according to

$$\delta \equiv -\left(\frac{\partial \ln \rho}{\partial \ln T}\right)_p. \tag{4.79}$$

Because getting from Equation 4.77 to Equation 4.78 is an exercise in applying thermodynamic relations that adds little insight into the matter at hand, we do not cover it here (but include it in appendix A.5 for the sake of completeness). The auxiliary quantity δ is useful to store information about the EOS of the fluid. Note that in the simple case of an ideal gas, $\delta = 1$.

Equation 4.78 introduces the temperature T as a variable into the system of equations to be solved. We therefore need either a closed-form EOS expression for T or a differential equation that couples to the others. In terms of mass coordinates we can start from $\partial T / \partial m$ and apply the chain rule to get

$$\frac{\partial T}{\partial m} = \frac{\partial p}{\partial m}\frac{dT}{dp}, \tag{4.80}$$

which we can combine with Equation 4.73 under the assumption of mechanical equilibrium:

$$\frac{\partial T}{\partial m} = -\frac{Gm}{4\pi r^4}\frac{dT}{dp} = -\frac{GmT}{4\pi r^4 p}\frac{d\ln T}{d\ln p}. \tag{4.81}$$

The physics that sets the temperature profile is encapsulated in the final factor of the expression above, which is usually given a separate definition referred to by the symbol '∇' (*not* to be confused with the nabla operator!):

$$\nabla \equiv \frac{d \ln T}{d \ln p}. \tag{4.82}$$

The system of stellar structure equations to solve (Equations 4.72, 4.73 and 4.78) is thus expanded by an expression

$$\frac{\partial T}{\partial m} = -\frac{GmT}{4\pi r^4 p} \nabla. \tag{4.83}$$

In most of the interior of a star the polytrope approach of the preceding sections is very reasonable (and represents a limiting case of very efficient convection settling the star into equilibrium). Using the definition of the polytropic exponent $\hat{\gamma}$, we have in this case

$$\nabla \sim \nabla_{ad} = \left(\frac{\partial \ln T}{\partial \ln p} \right)_{ad} = \frac{\hat{\gamma} - 1}{\hat{\gamma}}. \tag{4.84}$$

The adiabatic gradient ∇_{ad} can be used in a criterion to test for convective stability, and we will briefly return to this in Section 10.1. We will not provide them here, but expressions for ∇ under conditions where either conduction or radiation dominates can be shown to be very similar in structure to each other and straightforward to combine[4].

Finally, nuclear reactions will affect the relative abundancies of elements in the stellar interior and introduce a need for bookkeeping of the different species. We have for the relative mass fractions X_i of species i (e.g. ^{1}H, ^{4}He, ^{12}C, ^{14}N, ^{16}O, ... up to a total of I species) a series of expressions

$$\frac{\partial X_i}{\partial t} = \frac{m_i}{\rho} \sum_j (r_{ji} - r_{ij}), \qquad i = 1 \ldots I. \tag{4.85}$$

Here r_{ij} are the reaction rates (per unit volume) going from species i to j and m_i the mass of the nuclei of species i. The relative mass fractions must add up to unity:

$$\sum_i X_i = 1. \tag{4.86}$$

This completes the set of equations that one needs to solve when modelling stellar structure. What remains is determining the boundary conditions of the system and applying an appropriate numerical method. Unfortunately, the boundary conditions in practice are not all specified at the same point. Some of the variables take on a particular known value at the centre of the star (i.e. $m = 0$, $r = 0$, $L_r = 0$). To avoid singularities, these can be implemented at a small offset from the centre (like we did in Section 4.3.1). But other boundary conditions (like pressure and density) are

---

[4]Again, see [20] for details.

instead set at the outer boundary of the star. Also, stars typically have different layers, corresponding to different physics dominating ∇ and stellar models are often solved by linking these at their boundaries.

Whatever numerical approach is used therefore needs to take this into account. A venerable approach to solving boundary values problems is the *shooting method*[5], where the missing conditions at each boundary position are estimated and systematically improved upon via an iterative procedure. A more advanced approach along similar lines is the *Henyey method*[6], but we will limit our in-depth discussions of numerical methods in this book to methods for solving problems in hydro*dynamics* rather than hydrostatics (see Chapter 13). Unlike numerical stellar structure models, hydrodynamics simulations tend to start from a single completely specified boundary (i.e. the fluid at time $t = 0$) and then evolve from there.

[5]For a discussion of the shooting method, see e.g. [42].
[6]Discussions of the Henyey method can for example be found in [20, 3].

5 Sound Waves and Sub-/supersonic Flow

Causal contact across a fluid is not instantaneous. Instead, it relies on the propagation speed with which local changes are communicated throughout a fluid. In the absence of additional carriers of information such as photons or magnetic fields, this must be done through pressure differences hopping across fluid parcels from the source of a local change or perturbation to the receiver. For minor perturbations, what we are describing here are *sound waves*. And while sound waves may steepen to form shock waves that represent the propagation of steeper and more abrupt changes within the fluid (to be discussed later in Chapter 7), sound waves remain the principal mechanism by which causal contact is sustained through a fluid. As a result, an understanding of sound waves is useful to estimate things as diverse as the dynamical response timescale of a star to changing conditions or the optimal time step size to take a numerical solver for Euler's equations forward in time. The study of sound waves is therefore an important aspect of terrestrial and astrophysical fluid dynamics.

5.1 THE WAVE NATURE OF SOUND

For the sake of simplicity, we start our analysis of sound waves by considering a fluid otherwise at rest, i.e. a *hydrostatic* fluid with $\rho = \rho_0$, $p = p_0$ and $\mathbf{v} = 0$. Here ρ_0 and p_0 are constants. No external forces are assumed to act on the fluid ($\mathbf{f} = 0$). We now *perturb* the state of this fluid by adding time-dependent perturbation terms $\rho_1 \ll \rho_0$, $p_1 \ll p_0$, giving rise to a velocity component $|\mathbf{v}_1| \ll c_s$, where c_s is the speed of sound[1]. We will later check from the first two imposed conditions whether the third is met, once we have defined this sound speed c_s. It can be shown that these perturbations propagate through the fluid in the form of waves, that is, the fluid dynamical equations when applied to these perturbations can be recast in the form of wave equations. To show this, let us first plug $\rho = \rho_0 + \rho_1$, $p = p_0 + p_1$ and $\mathbf{v} = \mathbf{v}_1$ into the continuity equation and velocity equation,

$$\frac{\partial \rho}{\partial t} + \nabla \cdot (\rho \mathbf{v}) = 0, \tag{5.1}$$

$$\frac{\partial \mathbf{v}}{\partial t} + \mathbf{v} \cdot \nabla \mathbf{v} = -\frac{1}{\rho} \nabla p. \tag{5.2}$$

We can approximate the impact of the perturbations on the fluid by using a *first-order* approximation. We do so by dropping all terms that are multiples of more than

---

[1]Note that $\mathbf{v}_1$ is a vector. The first component of a generic vector $\mathbf{v}$ is written as v^1 (i.e. not in **bold**), or v_1 in terms of covariant rather than contravariant components. The first component of this vector $\mathbf{v}_1$ is $(v_1)^1$, or $(v_1)_1$.

DOI: 10.1201/9781003095088-5

82

one small perturbation term, under the previously stated assumption that $p_1/p_0 \ll 1$, $|\mathbf{v_1}| \ll c_s$ and $\rho_1 \ll \rho_0$. This leaves us with:

$$\frac{\partial \rho_1}{\partial t} + \rho_0 \nabla \cdot \mathbf{v_1} = 0, \tag{5.3}$$

$$\frac{\partial \mathbf{v_1}}{\partial t} + \frac{1}{\rho_0} \nabla p_1 = 0, \tag{5.4}$$

where we additionally assume that the fluid at rest varies spatially over length scales far larger than the ones introduced by the perturbation (i.e. $\nabla \rho_0 \ll \rho_1 \ll \rho$ and $\nabla p_0 \ll p_1 \ll p$). A single equation can be obtained by combining these two expressions and eliminating the velocity term. This can be achieved by taking the partial time derivative of the first, multiplying the second equation by ρ_0 and then taking the divergence of the second equation. Subtracting the two equations at that stage leads to

$$\frac{\partial^2 \rho_1}{\partial t^2} - \nabla^2 p_1 = 0. \tag{5.5}$$

Now we make an additional assumption. Let us say that the perturbation triggering a sound wave leaves the entropy of the system intact, so $s_1 = 0$ (later this section, we will contrast this assumption with the assumption that the sound wave is isothermal instead). This means that if p is approximated as a Taylor series around the unperturbed state $p_0(\rho, s)$ truncated at first order, the derivative in the series expansion can be be taken at constant s, leaving a series in terms of ρ_1 rather than ρ_1, s_1 both:

$$p_1 \approx \left(\frac{\partial p}{\partial \rho} \right)_0 \rho_1. \tag{5.6}$$

We use this to define the sound speed c_s according to

$$c_s^2 \equiv \left(\frac{\partial p}{\partial \rho} \right)_0. \tag{5.7}$$

Our physical reason for this definition will make sense in a minute (you can already check that both sides of this definition at least match in dimension). Note that the sound speed defined this way from the fluid at rest is a constant throughout the fluid and can be taken outside of any derivative. Our resulting expression for the evolution of ρ_1 now reads

$$\frac{\partial^2 \rho_1}{\partial t^2} - c_s^2 \nabla^2 \rho_1 = 0. \tag{5.8}$$

This is indeed a wave equation and a justification for our definition of c_s, which can be seen from this equation to be the propagation velocity linked to the perturbation ρ_1. A wave equation for pressure can readily be shown to be

$$\frac{\partial^2 p_1}{\partial t^2} - c_s^2 \nabla^2 p_1 = 0. \tag{5.9}$$

Based on the sound speed, the behaviour of a fluid can be divided in to two regimes: *supersonic* flow, for which $v > c_s$ and *subsonic* flow for which $v < c_s$. As with the Knudsen number from Section 2.2.1, we can define a dimensionless similarity parameter, in this case, the *Mach* number $\mathcal{M}$:

$$\mathcal{M} \equiv \frac{v}{c_s}. \tag{5.10}$$

What breaks the scale-invariance of Euler's equations here and allows for the construction of a dimensionless self-similarity parameter can be traced back to the introduction of an equation of state that will dictate $dp/d\rho$. It is important to note that labelling a flow as sub- or supersonic is a relative statement. If *everything* moves with a velocity v, then stating whether the flow is supersonic or not reduces to merely a statement about a reference frame (which should not impact the physics of the system). So a flow being supersonic is only meaningful if the flow occurs in the presence of flow that is not and vice versa.

For an ideal gas and adiabatic (or isentropic) sound waves, we have the following expression for the speed of sound:

$$c_s^{\text{ad}} = \left(\frac{5p}{3\rho}\right)^{1/2} = \left(\frac{5k_B T}{3\mu m_H}\right)^{1/2} \propto \rho^{1/3} \qquad \text{(ideal adiabatic).} \tag{5.11}$$

On the other hand, a sound wave can also be isothermal. Considering p to be a function $p(\rho, T)$, we can deploy a series expansion for p_1 in terms of ρ rather than ρ, T and again end up with the same definition for sound speed as in Equation 5.7. The sound speed then takes the form

$$c_s^{\text{iso}} = \left(\frac{p}{\rho}\right)^{1/2} = \left(\frac{k_B T}{\mu m}\right)^{1/2} \qquad \text{(ideal isothermal).} \tag{5.12}$$

Sound waves can also behave differently from the unperturbed medium. For example, sound propagates approximately adiabatically in air even while the earth's atmosphere can be reasonably described using an isothermal model. This is a consequence of a mismatch between the timescale for thermal equilibrium to (re-)establish itself and the timescale associated with the frequency of the wave. It depends on the main mechanism for establishing thermal equilibrium (e.g. radiation, conduction) at which wave frequency and in which direction a turnover occurs from sound waves of one type to the other. The case of thermal conduction is discussed further in Problem 5.4

As a final note, we remark that truly *incompressible fluids* represent are a limiting case where no sound waves can occur. After all, if $d\rho \equiv 0$, then the sound speed is formally infinite and any sound-like perturbation would spread through the system instantaneously.

PROBLEM 5.1
The sound of thunder

On a warm summer day ($T = 27°$ Celsius), you see a flash of lightning at the horizon and hear thunder about 3 seconds later. The earth atmosphere consists predominantly of molecular nitrogen (N_2). It is therefore a diatomic gas with polytropic index $\hat{\gamma} = 1.4$, and mean molecular weight $\mu = 28.97$ (close to 28.01 for pure nitrogen), if we take the mass unit $m = 1.6605 \times 10^{-24}$ g (i.e. one atomic mass unit). What is the distance to the thunder storm? How cold does it need to be for the storm to be twice as close for the same three second gap?

5.2 ACOUSTIC WAVES

Let us first concentrate on *plane acoustic waves*[2] along x, that is wave solutions to Equations 5.8 and 5.9 that propagate in a single direction with a speed-of-sound c_s. These are independent of y, z, and we thus have $\rho_1 = \rho_1(x,t)$, $p_1 = p_1(x,t)$, $\mathbf{v}_1 = v_1(x,t)\hat{\mathbf{e}}_x$ (with $\hat{\mathbf{e}}_x$ the *unit* vector in the x-direction). Let us also introduce a velocity potential Φ that obeys $\mathbf{v}_1 = \nabla\Phi$. It then follows that

$$\frac{\partial^2\rho_1}{\partial t^2} - c_s^2\frac{\partial^2\rho_1}{\partial x^2} = 0, \quad \frac{\partial^2 p_1}{\partial t^2} - c_s^2\frac{\partial^2 p_1}{\partial x^2} = 0, \quad \frac{\partial^2\Phi}{\partial t^2} - c_s^2\frac{\partial^2\Phi}{\partial x^2} = 0. \quad (5.13)$$

PROBLEM 5.2
The velocity potential

Derive the third of Equations 5.13. If the velocity can be expressed as the gradient of a scalar field, what does this mean for the *vorticity* of $\mathbf{v}$? Demonstrate that this also follows from the first order approximation of the perturbations of the fluid state applied in Section 5.1.

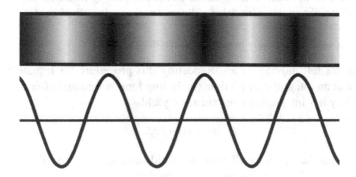

Figure 5.1 Sound waves are *longitudinal* waves. The bottom figure shows the density as a function of position relative to its baseline value for a sound wave, while the top figure uses a colour-coding to convey sounds waves are longitudinal waves.

The planar waves equations admit solutions of the type $\Phi = f_R(x - c_st) + f_L(x + c_st)$ (using Φ as an example, p and ρ are similar), where f_R and f_L arbitrary functions of their arguments. For plane waves in an inviscid fluid obeying Euler's equations,

---

[2]The name *acoustic* has its etymological root in the Greek term for *hearing*.

the shape of the perturbation will remain unchanged while it propagates (in other words, the *group velocity* and *phase velocity* are equal). Sound waves travelling in the positive x-direction can be described with f_R, sound waves moving in the negative x-direction can be described with f_L. Note that

$$\frac{\partial f_R}{\partial x} = \frac{\partial (x - c_s t)}{\partial x} \frac{d f_R}{d (x - c_s t)}, \quad \frac{\partial f_R}{\partial t} = \frac{\partial (x - c_s t)}{\partial t} \frac{d f_R}{d (x - c_s t)}. \tag{5.14}$$

There is no distinction between a partial and a full derivative of $f_{L,R}$ with respect to $(x - c_s t)$, since there are no other variables left to keep fixed. It follows that

$$\partial f_R / \partial t = -c_s \partial f_R / \partial x, \tag{5.15}$$

and in a similar manner for f_L that $\partial f_L / \partial t = +c_s \partial f_L / \partial x$. These equations tell us that if we want to know the fluid state change over time at a given fixed location, it will be the superposition of a right-moving profile coming in from the left from a distance $c_s \Delta t$ and a left-moving profile coming in from the right from the same distance.

If we now take another look at Equation 5.4, we can see that in terms of Φ it says

$$\nabla \left(\frac{\partial \Phi}{\partial t} + \frac{p_1}{\rho_0} \right) = 0, \tag{5.16}$$

which has to hold everywhere, including far from the perturbations. As a consequence of the latter, the constant of integration is zero when spatially integrating both sides to get rid of the ∇-term on the LHS. Combining that with Equation 5.15, it follows that right-moving (positive x-direction, those described by f_R) waves obey

$$p_1 = \rho_0 v_1 c_s, \tag{5.17}$$

and similar for left-moving waves. Repeating this procedure for Equation 5.3, and requiring that an integral over all time (including long before and after the perturbation) similarly has integration constant zero, yields

$$p_1 = \rho_0 v_1 / c_s. \tag{5.18}$$

This tells us that the waves for all perturbed fluid variables ρ_1, p_1 and $\mathbf{v}_1$ are *in phase*. It furthermore confirms our initial assumption that $v_1 = c_s \rho_1 / \rho_0 \ll c_s$. Finally, note also that *plane waves* with nonzero components to $\mathbf{v}_1$ only in the direction of motion (i.e. that of c_s) describe sound waves as *longitudinal* waves (see Figure 5.1).

As a concrete example of a solution to the wave equation, consider a monochromatic plane wave propagating in a direction $\hat{\mathbf{n}}$, described by

$$\rho_1 = \mathrm{Re}\, C \exp\left[i\left(\omega t - \mathbf{k} \cdot \mathbf{x}\right)\right], \tag{5.19}$$

$$p_1 = \mathrm{Re}\, c_s^2 C \exp\left[i\left(\omega t - \mathbf{k} \cdot \mathbf{x}\right)\right], \tag{5.20}$$

$$v_1 = \mathrm{Re}\, \frac{c_s}{\rho_0} C \exp\left[i\left(\omega t - \mathbf{k} \cdot \mathbf{x}\right)\right]. \tag{5.21}$$

where C a constant setting the amplitude. In this solution $\mathbf{k} = k\hat{\mathbf{n}}$, for a wave with *wavenumber* k ([rad] [cm^{-1}] in cgs units). The *wavelength* λ of the wave is then given by $\lambda = 2\pi/k$ ([cm]), the angular frequency $\omega = c_s k$ ([rad] [s^{-1}]), and the period $\tau = 2\pi/\omega$ ([s]). For a plane wave, we can easily get rid of the vector nature of $\mathbf{k}$ and $\mathbf{x}$ again by orienting our coordinate frame such that $\hat{\mathbf{n}}$ e.g. lies along the positive x-axis.

The phase velocity, tracing where $d(\omega t - kx) = 0$ for a fixed set of ω and k such that $\omega t - kx$ remains constant, leads to $dx/dt = \omega/k$ and is indeed equal to the speed of sound. The group velocity is defined by $d\omega/dk$, and therefore too is equal to the sound speed (strictly speaking the concept of a group velocity does not really apply to a monochromatic wave, where the derivative of ω with respect to k is ill-defined, i.e. there is no 'group' of frequency components).

5.2.1 WAVE PACKETS AND FOURIER ANALYSIS

In this section we dwell a bit more on the wave equation. The previous section showed that any function f_R, f_L is a valid solution to the wave equation as long as the argument of the function has the form $x \pm c_s t$. Monochromatic waves can be used as building blocks to form any arbitrarily shaped *wave packet*, a localized disturbance of the fluid state. A wave packet can be analyzed using *Fourier transforms* to reveal its constituent harmonics. For a given wave packet $\rho_1(x,t)$, the coefficients $A(k)$ of these harmonics are given by (we will omit the Re , it being understood that for physical solutions we drop the imaginary part):

$$\rho_1(x,t) = \frac{1}{\sqrt{2\pi}} \int_{-\infty}^{\infty} A(k)e^{i(kx-\omega(k)t)}dk. \tag{5.22}$$

Let us first demonstrate this for a monochromatic wave of frequency ω, where the Fourier transform presumably reveals no other harmonics than that single frequency ω:

$$\rho_1(x,t) = \frac{1}{\sqrt{2\pi}}e^{i(kx-\omega t)}, \tag{5.23}$$

sets up our monochromatic wave of frequency ω (this is not actually a *localized* wave packet, since ρ_1 does not go to zero for large $|x|$). The inverse Fourier transform can be used at $t = 0$ to determine the coefficients of the harmonics:

$$A(k') \equiv \frac{1}{\sqrt{2\pi}} \int_{-\infty}^{\infty} \rho_1(x,t=0)e^{-ik'x}dx = \frac{1}{2\pi} \int_{-\infty}^{\infty} e^{i(k-k')x}dx = \delta(k-k'). \tag{5.24}$$

That the integral can be seen as defining a delta function, is a key part of Fourier analysis[3]. This confirms our assumption that there is indeed a single harmonic, i.e. the one for which $k' = k$, with all other harmonics having zero amplitude.

---

[3]A more in-depth treatment of this topic lies beyond the scope of the current discussion. Delta functions in this form pop up in all sorts of branches of physics whenever wave phenomena are discussed, such as quantum mechanics and classical dynamics of harmonic oscillators, if not explicitly then implicitly.

A different example would be a Gaussian-shaped wave-packet, i.e.

$$\rho_1(x,t) = C_1 e^{-(x-c_s t)^2/C_2^2}, \tag{5.25}$$

with C_1 and C_2 two constants that respectively set the amplitude and width of the wave packet. Obtaining the amplitudes of the harmonics via the usual manner, we obtain

$$A(k) = \frac{1}{\sqrt{2\pi}} \int_{-\infty}^{\infty} C_1 e^{-x^2 C_2^{-2} - ikx} dx, \tag{5.26}$$

$$= \frac{C_1}{\sqrt{2\pi}} \int_{-\infty}^{\infty} e^{-\left(xC_2^{-1} + ikC_2/2\right)^2 - k^2 C_2^2/4} dx, \tag{5.27}$$

$$= \frac{C_1}{\sqrt{2\pi}} e^{-k^2 C_2^2/4} C_2 \int_{-\infty}^{\infty} e^{-\left(xC_2^{-1} + ikC_2/2\right)^2} d\left(xC_2^{-1} + ikC_2/2\right), \tag{5.28}$$

$$= \frac{C_1}{\sqrt{2}} e^{-k^2 C_2^2/4} C_2 \left[\frac{1}{\sqrt{\pi}} \int_{-\infty}^{\infty} e^{-y^2} dy \right], \tag{5.29}$$

$$= \frac{C_1}{\sqrt{2}} e^{-k^2 C_2^2/4} C_2. \tag{5.30}$$

In the second step, we merely rewrote the integral. In the third step, we changed the integration variable and corrected for that with the extra factor C_2 in front. The slight shift into the complex plane can be shown to make no difference at $\mathrm{Re}\, xC_2^{-1} + ikC_2/2 = \pm\infty$, justifying the change of variables to an integration variable that we labelled y (this is an application of Cauchy's theorem for integrals in the complex plane). The integral in square brackets is a standard one and evaluates to one.

What this integration tells us is that a Gaussian wave packet also consists of a Gaussian distribution of harmonics. The larger C_2, the more narrow the wave packet, but the larger the distribution of harmonics that makes up the wave packet. As far as sound waves go, a Gaussian wave packet as defined above is of limited relevance, since sound waves typically merely *perturb* the existing fluid rather than add extra fluid. More interesting solutions are therefore those for which integrating ρ_1 over a sufficiently large domain V equals zero, i.e.

$$\int \rho_1 dV = 0. \tag{5.31}$$

5.2.2 THE ENERGY OF AN ACOUSTIC SOUND WAVE

As stated above, the perturbations describing sound waves do not add net mass to the system. Sounds waves do, however, increase the amount of energy in the system in a manner that does not average to zero over a larger volume covering the perturbation. For the kinetic energy we have $\frac{1}{2}\rho v_1^2$, with v_1 set by the perturbation. We can drop the third order term $\frac{1}{2}\rho_1 v_1^2$ and keep $\frac{1}{2}\rho_0 v_1^2$. This remaining term might be second order (and thus very small), but there is no remaining first order term in kinetic energy to compare against, so this is the lowest order that matters.

We will see that this kinetic energy is also not negligible relative to the internal energy for a total perturbation energy $\mathscr{E}_1 = e_1 + \rho_0 v_1^2/2$ (assuming a pressure perturbation of p_1 to lead to an internal energy perturbation e_1), even if the e_1 term artificially suggests otherwise. For starters, that part of the internal energy $e \equiv \rho\varepsilon$ driven by the mass fluctuation will average out to zero in the same manner the density does:

$$\int \rho_1 \varepsilon_0 dV = 0, \tag{5.32}$$

for an unperturbed specific internal energy ε_0. Whether the remaining internal energy ends up exactly compensating for the kinetic energy and therefore a total energy perturbation averaging to zero or whether this remainder leads to a positive contribution of larger, similar or smaller order than the kinetic term, remains to be demonstrated. For this purpose, let us expand the internal energy in a Taylor series around the unperturbed density:

$$\rho\varepsilon \approx \rho_0\varepsilon_0 + \left.\frac{\partial\rho\varepsilon}{\partial\rho}\right|_0 p_1 + \frac{1}{2}\left.\frac{\partial^2(\rho\varepsilon)}{\partial\rho^2}\right|_0 \rho_1^2. \tag{5.33}$$

For the case of adiabatic perturbations, we can indeed use density as the single perturbation variable, as per the first law of thermodynamics $d\varepsilon = dq - pd\rho^{-1}$ with $dq = 0$. We can also use this to obtain the partial derivatives:

$$\frac{\partial\rho\varepsilon}{\partial\rho} = \varepsilon + \rho\frac{\partial\varepsilon}{\partial\rho} = \varepsilon + \frac{p}{\rho} = h, \tag{5.34}$$

$$\frac{\partial^2(\rho\varepsilon)}{\partial\rho^2} = \frac{p}{\rho^2} + \frac{\partial p}{\partial\rho}\frac{1}{\rho} - \frac{p}{\rho^2} = \frac{1}{\rho}\frac{\partial p}{\partial\rho}. \tag{5.35}$$

Evaluating these derivates at the unperturbed state, this leads us to

$$\rho\varepsilon \approx \rho_0\varepsilon_0 + h_0\rho_1 + \frac{1}{2}\frac{c_s^2}{\rho_0}\rho_1^2. \tag{5.36}$$

The first term on the RHS is just the unperturbed internal energy density, the second averages out to zero like Equation 5.32. The third term describes a quantity (the 'compressional energy') that is positive everywhere and therefore cannot compensate for the equally positive kinetic energy density of the perturbation. We can therefore adopt the following measure of total energy density for a sound wave:

$$\mathscr{E}_1 = \frac{1}{2}\frac{c_s^2}{\rho_0}\rho_1^2 + \frac{1}{2}\rho_0 v_1^2. \tag{5.37}$$

According to Equation 5.18, this expression can be simplified for a plane wave to just $\mathscr{E}_1 = \rho_0 v_1^2$. Furthermore, since the wave propagates with the speed of sound, we can expect an associated energy flux of $\mathscr{E}_1 c_s \hat{\mathbf{n}}$ for a plane wave propagating in a direction $\hat{\mathbf{n}}$ (this can also be derived in a few steps, see exercise 5.3). The corresponding conservation equation reads

$$\frac{\partial\mathscr{E}_1}{\partial t} + \nabla \cdot (\mathscr{E}_1 c_s \hat{\mathbf{n}}) = 0. \tag{5.38}$$

PROBLEM 5.3
The energy flux of a sound wave

Starting from the flux term $(\mathscr{E} + p)\mathbf{v} = h\rho\mathbf{v}$ from Euler's equation for conservation of energy, demonstrate that the energy flux of an acoustic plane wave reduces to $\mathscr{E}_1 c_s \hat{\mathbf{n}}$. Justify why certain terms can be dropped during your derivation. Note that the thermodynamic identities can be found in appendix A, if you wish to express the adiabatic perturbation in specific enthalpy h in terms of a series expansion in p_1.

The quantity $\mathscr{E}_1$ captures the instantaneous energy density of a sound wave. Of greater practical interest is arguably the *average* energy density computed over a full period τ of the sound wave. Assuming a plane acoustic wave that obeys $v_1 = |v_1|\sin(\omega t - c_s x)$, we have

$$\langle \mathscr{E}_1 \rangle \equiv \frac{\rho_0}{\tau}\int_0^\tau (v_1)^2\,dt = \frac{\rho\,|v_1|^2}{\omega\tau}\int_0^\tau \sin^2(y)\,dy, \tag{5.39}$$

where we evaluate the perturbation at the origin without loss of generality. Instead of a sine wave we could have used a cosine wave or a combination of both as our starting point, but this too has no impact on the outcome. Performing the integral and using that $\tau\omega = 2\pi$ results in

$$\langle \mathscr{E}_1 \rangle = \frac{1}{2}\rho_0\,|v_1|^2. \tag{5.40}$$

Similarly, the average flux obeys

$$\langle \mathscr{E}_1 c_s \rangle = \frac{1}{2}\rho_0\,|v_1|^2\,c_s. \tag{5.41}$$

5.2.3   SPHERICAL ACOUSTIC WAVES

Plane acoustic waves are triggered by perturbations occurring along an entire plane. But if the perturbation in an isotropic medium is localized, one would instead expect this to lead to a spherically expanding sound wave. In this case, the ∇-terms in the wave equation need to be considered in spherical symmetry:

$$\frac{\partial^2 p_1}{\partial t^2} - \frac{c_s^2}{r^2}\frac{\partial}{\partial r}\left(r^2\frac{\partial p_1}{\partial r}\right) = 0, \tag{5.42}$$

and likewise for density and velocity. It is straightforward to check by substitution in the wave equation that the general solution is given by

$$p_1 = \frac{f_{out}(r - c_s t)}{r} + \frac{f_{in}(r + c_s t)}{r}. \tag{5.43}$$

Here, the labels *out* and *in* label sound waves respectively expanding from the origin and converging on the origin. Interestingly enough, unlike the case for plane acoustic waves, the different perturbations ρ_1, v_1 and p_1 are not everywhere in phase. This can be seen from Equation 5.16, along with the definition for Φ as the potential of the velocity perturbation. If we take a solution for Φ in the form of a monochromatic spherically expanding wave, $\Phi = C\exp\left[i\left(\omega t - kr\right)\right]/r$, we have

$$p_1 = -\rho_0 \frac{\partial \Phi}{\partial t} = -\rho_0 i\omega\Phi, \tag{5.44}$$

and

$$v_1 = \frac{\partial \Phi}{\partial r} = -ik\Phi - \frac{\Phi}{r}. \tag{5.45}$$

When these are combined, this reveals that

$$v_1 = \frac{p_1}{\rho_0 \omega}\left(k - \frac{i}{r}\right). \tag{5.46}$$

The appearance of the r^{-1} term is new relative to plane waves. It tells us that at small radii, where this term dominates, the velocity perturbation and the pressure perturbation are out of phase by $90°$. At large radii, this extra term becomes negligible and the wave front starts to resemble a plane wave (as it should once the curvature becomes negligible).

PROBLEM 5.4
Between adiabatic and isothermal sound waves

For the case of thermal conduction, it is possible to demonstrate that assuming adiabatic sound waves is appropriate in the vast majority of terrestrial conditions. This exercise will guide you through the argument[4]. We require some thermodynamics and manipulations of thermodynamic relations (see appendix A).

i) Confirm that the following expression holds:

$$ds = \left(\frac{\partial s}{\partial \rho}\right)_{p,0}\left(d\rho - \frac{dp}{c_s^2}\right),$$

where the differentials are small perturbations away from the non-moving baseline fluid state (i.e., $ds \equiv s_1$ etc., as in our derivation of the wave equation).

ii) Starting from the heat equation (see also Equation 2.31)

$$\rho T \frac{ds}{dt} = \mathscr{K}\nabla^2 T,$$

where *mathcalK* is assumed constant, derive an expression for the heat conduction affecting the perturbed state:

$$\rho_0 T_0 \left(\frac{\partial s_1}{\partial t}\right) = \mathscr{K}\nabla^2 T_1.$$

---

[4]This argument can also be found e.g. in [40].

iii) Further rewrite the result from (ii) to obtain the following expression:

$$\rho_0 c_p \frac{\partial}{\partial t}\left(\rho_1 - \frac{p_1}{c_s^2}\right) = \mathscr{K}\nabla^2\left(\rho_1 - \frac{p_1}{c_T^2}\right).$$

Here we introduced an analog to the sound speed for the case of *isothermal* perturbations: $c_T^2 \equiv (\partial p/\partial \rho)_{T,0}$. Note that c_p still refers to the specific heat capacity at constant pressure, not to a measure of sound speed. To obtain the answer, use the result from (i), along with it analog expression for T_1:

$$T_1 = \left(\frac{\partial T}{\partial \rho}\right)_{p,0}\left(\rho_1 - \frac{p_1}{c_T^2}\right) \tag{5.47}$$

iv) We can eliminate references to the density perturbation from the result from (iii) in order to obtain an expression for the pressure perturbation only. This can be done by making use of the expression $\partial^2\rho_1/\partial t^2 - \nabla^2 p_1 = 0$ (this general expression follows early on when deriving the wave equations, before we committed to either an isotropic or isothermal perturbation; do double-check if this is not obvious). Continuing from (iii), derive the expression

$$\frac{\partial}{\partial t}\left(\nabla^2 - \frac{1}{c_s^2}\frac{\partial^2}{\partial t^2}\right)p_1 = \frac{\mathscr{K}}{\rho_0 c_p}\nabla^2\left(\nabla^2 - \frac{1}{c_T^2}\frac{\partial^2}{\partial t^2}\right)p_1.$$

v) Consider plane waves (such that we can simplify $\nabla^2 = \partial^2/\partial x^2$ etc.) and a trial plane wave solution $p_1 = \text{Re}\, C'\exp(-i\omega t + ikx)$. Show that this requires the following *dispersion relation* to hold:

$$\frac{k^2 - \omega^2/c_s^2}{k^2 - \omega^2/c_T^2} = \frac{\mathscr{K}}{\rho_0 c_p}\frac{k^2}{i\omega}.$$

A solution to this expression for fixed ω will, in general, require an imaginary term for k. When applied to the trial solution for p_1, a damping term will result (leading to *attenuation* of the sound wave).

vi) Assuming an adiabatic plane wave solution to be approximately correct for a given angular frequency ω, what does this imply for its value relative to $\omega_{TC} \equiv \rho_0 c_p c_s^2/\mathscr{K}$?

vii) Conversely, assuming an isothermal plane wave solution to be approximately correct for a given frequency ω, what does this imply for its value relative to ω_{TC}? (in this case, viscosity can no longer be neglected and would lead to a large amount of attenuation).

viii) Compute the turnover frequency $\nu_{TC} \equiv \omega_{TC}/2\pi$ for air under normal conditions. Assume a pressure of 1 atm (in cgs units: 1.013×10^6 dyne cm^{-2}, where 'dyne' the cgs unit of force). Other relevant quantities are the same as in exercise 5.1, that is, the earth atmosphere is a diatomic gas with polytropic index $\hat{\gamma} = 1.4$ and mean molecular weight $\mu = 28.97$ (reflecting predominantly nitrogen), if we take the mass unit $m = 1.6605 \times 10^{-24}$ g. Boltzmann's constant $k_B = 1.38 \times 10^{-16}$ erg / K. Take the coefficient of thermal conductivity to be equal to $\kappa = 2.6 \times 10^3$ erg / (K cm s). You should find a value vastly above the upper limit of human hearing (which is approximately 2×10^4 Hz).

5.3 SOUND WAVES IN A GRAVITATIONAL FIELD

So far we have assumed a homogeneous base state and have ignored the potential impact of gravity on the propagation of sound waves. It his however not difficult to envisage circumstances where a gravitational field is both responsible for an initial non-homogeneous fluid profile and has potential impact on the evolution of a fluid perturbation, for example planetary and stellar atmospheres. Let us therefore consider in some depth the case where an external gravitational field is present that exerts a constant gravitational force $\mathbf{g}$. We assume a steady base state of non-moving fluid to be dictated by the interplay between gravitational force and pressure p_0 according to

$$\frac{\partial p_0}{\partial z} = -\rho_0 g, \tag{5.48}$$

where g the magnitude of the gravitational acceleration, which we assume to be pointing in the negative z-direction. If we follow along the same steps as in Section 5.1, but allow for the presence of gravity and an inhomogeneous fluid profile in the z-direction, we obtain

$$\frac{\partial \rho_1}{\partial t} = -v_1^z \frac{\partial \rho_0}{\partial z} - \rho_0 \frac{\partial v_1^z}{\partial z}, \tag{5.49}$$

$$\frac{\partial v_1^z}{\partial t} = -\frac{1}{\rho_0} \frac{\partial p_1}{\partial z} + \frac{\rho_1}{\rho_0} g, \tag{5.50}$$

after some approximating and subtracting the base state from the second equation. The same assumptions as usual apply between perturbed state and base state, $\rho_1/\rho_0 \ll 1$ and $p_1/p_0 \ll 1$. The equations can be combined in the same manner as done in Section 5.1. Allowing for an inhomogeneous base state produces a slew of extra terms at first, but luckily most of these cancel and we are left with

$$\frac{\partial^2 \rho_1}{\partial t^2} - \frac{\partial^2 p_1}{\partial z^2} + g \frac{\partial \rho_1}{\partial z} = 0. \tag{5.51}$$

PROBLEM 5.5
Deriving the wave equation in a gravitational field

Derive Equation 5.51 from Equations 5.49 and 5.50 to confirm that indeed various terms cancel each other.

One way of proceeding at this point is by specializing to the case of an isothermal atmosphere and assuming the ideal gas law to apply. The sound speed (derived from the base state) obeys

$$c_s^2 \equiv \left. \frac{\partial p}{\partial \rho} \right|_0 = \frac{k_B T}{\mu m}. \tag{5.52}$$

Whether we describe an atmosphere or not, under isothermal conditions this sound speed is independent of height z. For an isothermal atmosphere with scale height H

(see Section 4.1), we also have $g = c_s^2/H$. The wave equation can be further modified to

$$\frac{\partial^2 \rho_1}{\partial t^2} - c_s^2 \frac{\partial^2 \rho_1}{\partial z^2} + g \frac{\partial \rho_1}{\partial z} = 0. \tag{5.53}$$

If this were an ordinary wave equation (i.e. $g = 0$) and the initial perturbation were independent of x and y, this expression would allow for independent planar wave solutions of the form

$$\rho_1(z,t) = \bar{\rho}_1 \exp\left[i\omega t - ikz\right], \tag{5.54}$$

where $\bar{\rho}_1$ sets the amplitude, which propagate unaltered through the fluid as before. The presence of the additional term complicates the *dispersion relation* (that is, the relation between angular frequency ω and wavenumber k). But since any general solution can be built up out of Fourier components, we can still meaningfully plug Equation 5.54 into Equation 5.53 and see what happens. We get

$$-\omega^2 \rho_1 + k^2 c_s^2 \rho_1 - ik\rho_1 g = 0. \tag{5.55}$$

This is a quadratic equation in wavenumber k that can be rewritten and solved to obtain

$$k^2 - \frac{ik}{H} - \frac{\omega^2}{c_s^2} = 0 \Rightarrow k = \frac{iH^{-1} \pm \sqrt{-H^{-2} + 4\omega^2/c_s^2}}{2}. \tag{5.56}$$

In the homogeneous limit where $H \rightarrow \infty$, this reduces to the expected $k = \omega/c_s$. But for finite H values, a few options appear. One possibility is that the term within the square root is negative. In this case, we might as well write

$$k = \frac{i}{2H} \pm i\sqrt{\left(\frac{1}{2H}\right)^2 - \frac{\omega^2}{c_s^2}}. \tag{5.57}$$

The term within the square root is now positive and the expression emphasizes that, in this case, k is altogether imaginary. But a solution with a purely imaginary wavenumber is not going to propagate through space. Instead, what results is a perturbation oscillating in place with an angular frequency still set by ω and with a z-dependent amplitude set in part by k:

$$\rho_1 = \bar{\rho}_1 \exp\left[\frac{z}{2H} \pm z\sqrt{\left(\frac{1}{2H}\right)^2 - \frac{\omega^2}{c_s^2}}\right] \exp\left[i\omega t\right] \tag{5.58}$$

The other option is for the term in the square root in Equation 5.56 to be positive. In this case, we have

$$\rho_1 = \bar{\rho}_1 \exp\left[\frac{z}{2H}\right] \exp\left[i\left(\omega t \pm \sqrt{\frac{\omega^2}{c_s^2} - \left(\frac{1}{2H}\right)^2} z\right)\right]. \tag{5.59}$$

Now the wavenumber is not purely imaginary, and the wave does propagate. It also grows exponentially in the upwards direction, in response to the exponentially decreasing density of the atmosphere. As a consequence, a perturbation ρ_1 will not remain small indefinitely and at some point $\rho_1 \sim \rho_0$. Dropping terms ρ_1/ρ_0 beyond the linear term is no longer a valid approximation. When a perturbation becomes non-linear, a shock develops. We will discuss shock waves separately in Chapter 7. The transition frequency between a stationary perturbation and a wave propagating through an atmosphere can be found by setting the term in the square root of Equation 5.56 equal to zero. The resulting frequency,

$$\omega = \frac{c_s}{2H},\tag{5.60}$$

is known as the Brunt-Väisälä frequency.

There are a number of velocities associated with the waves. Similar to the density perturbation, we can take the velocity perturbation v_1^z to be of the form

$$v_1^z = \bar{v}_1^z \exp\left[i\omega t - ikz\right].\tag{5.61}$$

Using this in Equation 5.49, we can obtain

$$i\omega\rho_1 = -v_1^z \frac{\partial \rho_0}{\partial z} + i\rho_0 v_1^z k \Rightarrow i\omega\rho_1 = \frac{\rho_0}{H} v_1^z + i\rho_0 v_1^z k.\tag{5.62}$$

Using the terms that the density and velocity perturbations have in common, this can be rewritten as

$$i\omega \frac{\bar{\rho}_1}{\rho_0} = \bar{v}_1^z \left(\frac{1}{H} + ik\right),\tag{5.63}$$

which in turn can be combined with the dispersion relation from Equation 5.56 to show that

$$\bar{v}_1^z = \frac{\bar{\rho}_1}{\rho_0} \frac{c_s^2 k}{\omega}.\tag{5.64}$$

For acoustic waves in a homogeneous environment, this reduces to $\bar{v}_1^z = c_s\bar{\rho}_1/\rho_0$, as expected. In the presence of gravity, phase velocity ω/k is no longer equal to the sound speed in general. Also, the wavenumber k can be partially or fully complex, which means that $\bar{v}_1^z$ both might not capture the full amplitude of the velocity perturbation and include a phase shift.

The phase velocity and the group velocity are defined in terms of the real part of k. If we define $k_{real} \equiv \operatorname{Re} k$ for a minute, we can define phase velocity and group velocity respectively according to $v_{phase} \equiv \omega/k_{real}$ and $v_{group} \equiv \partial\omega/\partial k_{real}$. Without gravity and in a homogeneous environment, both velocities are equal (as they are in any scenario where ω and k_{real} are linearly proportional to each other) and equal to the sound speed. Starting from Equation 5.56, we have

$$k_{real} = \sqrt{\frac{\omega^2}{c_s^2} - \left(\frac{1}{2H}\right)^2},\tag{5.65}$$

which thanks to the definition of the phase velocity can relatively easily be recast in the form

$$v_{phase} = c_s\sqrt{1 + \left(\frac{1}{2k_{real}H}\right)^2}.\tag{5.66}$$

The phase velocity tells us how fast a wave of given frequency moves. But in a wave packet that changes shape while in motion because phase and group velocity are not equal (as is the case here), we need the group velocity to tell us how fast the wave packet moves. Using a slightly recast form of the same starting point, we have

$$k_{real}^2 = \left(\frac{\omega}{c_s}\right)^2 + \left(\frac{1}{2H}\right)^2 \Rightarrow 2k_{real}dk = c_s^{-2}2\omega d\omega. \qquad (5.67)$$

This tells us that $v_{phase}v_{group} = c_s^2$ and v_{group} is given by

$$v_{group} = \frac{c_s}{\sqrt{1 + \left(\frac{1}{2k_{real}H}\right)^2}}. \qquad (5.68)$$

The group velocity of a wave packet centred around frequency ω in the presence of gravity is therefore smaller than that of an acoustic wave of the same frequency, with the Brunt-Väisälä frequency leading to the limiting case of zero velocity.

6 Properties and Kinematics of Fluid Flow

6.1 AN OVERVIEW OF TERMINOLOGY

In this chapter we delve more deeply into the kinematics and properties of fluid flows. By now, we have encountered various terms to label fluid behaviour under different conditions and constraints. Before embarking on a discussion of various scenarios that play out under particular constraints, it might be useful to take stock of terminology that so far has been introduced in a scattered manner and add a few terms that have not yet come up in the text.

First of all, it is important to distinguish between conditions that are assumed to hold throughout the fluid under consideration or merely along fluid parcel trajectories. This is the difference between **polytropic** flow obeying $p \propto \rho^{\hat{\gamma}}$ along a parcel trajectory and a **polytrope** obeying $p = K\rho^{\hat{\gamma}}$ everywhere as introduced in Section 3.4. In other words, if the polytropic property is merely asserted to hold along the flow, then K is not necessarily the same constant throughout the fluid.

Fluid flow can be restricted in various other ways. Slightly more general than polytropic flow, we can have **barotropic** flow for which $p = p(\rho)$, that is, the pressure is a function only of density (just not necessarily proportional to $\rho^{\hat{\gamma}}$). When this property applies in a fluid rather than along a flow, the fluid is a **barotrope**. For barotropes, constant pressure (**isobaric**, with $dp = 0$) regions coincide with constant density (**isopycnic**, $d\rho = 0$) regions. If this is not the case, the fluid is instead termed **baroclinic**.

Flows and fluid regions can be subjected to other constraints as well. In an **incompressible** fluid the density of fluid parcels remains unchanged. As a consequence, these fluids have divergence-free velocities. **Isothermal** flow has constant temperature with $dT = 0$. **Isentropic** flow has $ds = 0$. **Adiabatic flow** obeys $dq = 0$. If the flow is **steady**, $\partial \mathbf{v}/\partial t = 0$. Euler's equations describe **inviscid** fluids that completely lack viscosity (in Chapter 9, we discuss viscosity in detail).

The above covers some of the relevant terminology one is likely to encounter in generic (astrophysical) fluid dynamics. As we will see later on, there are some further additions to be made to this list, related to concepts we have not yet introduced (e.g. the vorticity of fluid flow).

6.2 STREAMLINES, STREAKLINES AND PATH LINES

The motion of fluid parcels will follow the velocity field $\mathbf{v}(\mathbf{x}, t)$, and we can employ various tools to describe and visualize fluid parcel *kinematics*. First, consider *path lines*, which trace the paths over time of individual fluid parcels (for an illustration of the concept, see Figure 6.1). If $\mathbf{r}$ is the position vector associated with a given parcel,

DOI: 10.1201/9781003095088-6

Figure 6.1 *Path lines*: The condensation trail from an airplane provides a visualization of the path line describing the path of the plane, with the plane playing the role of a fluid parcel.

its path obeys

$$\frac{d\mathbf{r}}{dt} = \mathbf{v}(\mathbf{r},t). \tag{6.1}$$

Instead of following the motion of a single fluid parcel over time, we can also map the local displacements of all parcels throughout the velocity field at the same instant in time. The curve that one obtains (from a given starting point) by moving tangentially along each local velocity vector that it encounters, is called a *stream-line*. If the velocity field does not change over time, these curves will coincide with path lines for individual parcels. As mentioned in the previous section, such flows are called *steady flows*, and for steady flows the partial time derivative $\partial/\partial t$ of all conserved variables will be zero. But, in general, this is not the case, and path lines and streamlines will follow different trajectories. For an illustration of a streamline, see Figure 6.2.

A streamline can be generated by parametrizing a curve $\mathbf{r}$ using an arbitrary parameter s, and applying the condition that it be locally tangential to the velocity field everywhere:

$$\frac{d\mathbf{r}}{ds} = \mathbf{v}(\mathbf{r},t), \tag{6.2}$$

where t is kept fixed rather than used as the parameter to describe the curve (as is the case for parcel paths). Isolating ds in the three expressions covered by the single

Figure 6.2 *Streamlines*: A river carves a path through erosion if it follows a velocity vector field that does not change much over time (at least in direction), and the canyon path therefore also serves as a visualization of a streamline at a given instant in time.

vector expression and comparing tells us that

$$\frac{dx}{v^x} = \frac{dy}{v^y} = \frac{dz}{v^z},$$
(6.3)

in Cartesian coordinates. In the case of curvilinear coordinates such as spherical coordinates it is important to keep track of the distinction between normalized and natural basis vectors, i.e.

$$\begin{aligned}
\mathbf{v}\,ds &= \left(\dot{r}\mathbf{e}_r + \dot{\theta}\mathbf{e}_\theta + \dot{\phi}\mathbf{e}_\phi\right)ds \\
&= \left(\dot{r}\hat{\mathbf{e}}_r + \dot{\theta}r\hat{\mathbf{e}}_\theta + \dot{\phi}r\sin\theta\hat{\mathbf{e}}_\phi\right)ds \\
&= \left(v^r\hat{\mathbf{e}}_r + v^\theta\hat{\mathbf{e}}_\theta + v^\phi\hat{\mathbf{e}}_\phi\right)ds,
\end{aligned} \tag{6.4}$$

where the velocities v^r, v^θ and v^ϕ all have dimension of length over time (unlike $\dot{\theta}$ and $\dot{\phi}$, where some of the dimensionality is absorbed in the natural basis vectors). With a displacement in spherical coordinates

$$d\mathbf{r} = \hat{\mathbf{e}}_r dr + r\hat{\mathbf{e}}_\theta d\theta + r\sin\theta\hat{\mathbf{e}}_\phi d\phi, \tag{6.5}$$

the analog of Equation 6.3 becomes

$$\frac{dr}{v^r} = \frac{r d\theta}{v^\theta} = \frac{r\sin\theta\, d\phi}{v^\phi}. \tag{6.6}$$

Figure 6.3 *Streaklines*: A sprinkler provides a visualization of streaklines where the droplets share a common origin but do not necessarily share path lines.

A third option is to consider again a single instant in time t and to draw a curve that connects the positions of all parcels that have passed a given point or will do so in the future. If this wording sounds abstract, these curves are nevertheless easy to visualize and/or produce experimentally: Figure 6.3 shows a demonstration in the form of droplets from a sprinkler. While actual the droplet path lines will be radially outwards from the sprinkler, the spiral pattern of droplets forms a *streakline* made up of droplets sharing a point of origin in the nozzle of the hose. Streaklines can also be demonstrated for example by injecting ink into water for some time. Again, for steady flow the difference between streaklines, parcel paths and streamlines dissappears.

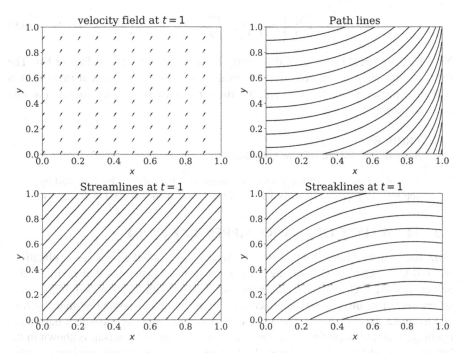

Figure 6.4 Fluid parcel kinematics resulting from a velocity vector field rotating counter-clockwise, obeying $[\mathbf{v}(\mathbf{x},t)]^T = [\cos(t) \quad \sin(t)]$. Path lines are for parcels passing $x = 0$ at $t = 0$.

Figure 6.4 shows another illustration of path lines, streamlines and streaklines. Here, we assume a vector field that rotates counter-clockwise according to $[\mathbf{v}(\mathbf{x},t)]^T = [\cos(t) \quad \sin(t)]$. Parcel paths will curve counter-clockwise over time as they follow the local vector field (top right figure). A particle starting at coordinates r_0^x, r_0^y at time t_0 will obey

$$\begin{bmatrix} r^x(t) \\ r^y(t) \end{bmatrix} = \begin{bmatrix} r_0^x + \sin(t) - \sin(t_0) \\ r_0^y - \cos(t) + \cos(t_0) \end{bmatrix}. \tag{6.7}$$

In the case of streamlines, t is not the variable that parametrizes the distance along the curve but is kept constant instead. For our rotating vector field, streamline curves are instead described by a parameter s starting at s_0:

$$\begin{bmatrix} r^x(s) \\ r^y(s) \end{bmatrix} = \begin{bmatrix} r_0^x + \cos(t)(s - s_0) \\ r_0^y + \sin(t)(s - s_0) \end{bmatrix}. \tag{6.8}$$

As shown in the bottom left image of Figure 6.4, the streamlines are just straight lines with a slope determined by t.

For streaklines, the relevant expression is identical to that for path lines but with the understanding that t is kept fixed and t_0 is varied in order to account for all past and future particles that will have passed $\mathbf{r}_0$ at different points in time:

$$\begin{bmatrix} r^x(t_0) \\ r^y(t_0) \end{bmatrix} = \begin{bmatrix} r_0^x + \sin(t) - \sin(t_0) \\ r_0^y - \cos(t) + \cos(t_0) \end{bmatrix}. \tag{6.9}$$

Note how the streaklines curve differently from the streamlines in Figure 6.4. The further to the right in the bottom right image, the earlier the parcel at that position will have passed through $x = 0$ and the further clockwise will its direction at that point of passing have been.

PROBLEM 6.1
Streamlines and curvilinear coordinates

Show that Equation 6.6 applies to streamlines in spherical coordinates. Start by applying the product rule to the RHS of $d\mathbf{r} = d\,(r\hat{\mathbf{e}}_r)$, in the expression $d\mathbf{r}/ds = \mathbf{v}$.

6.3   FLOW LINES OF AN INCOMPRESSIBLE FLUID

Flow lines can be a powerful tool to gain direct insight into the physical implications of fluid flow patterns. Various theorems exist in fluid dynamics that facilitate this and we will highlight, in particular, *Bernoulli's theorem* and *Kelvin's circulation theorem*. But before discussing these theorems, let us first pluck some low-hanging fruit and examine the flow lines of an incompressible fluid moving through a narrowing opening, such as a water flow through a narrowing river. This set-up is shown in the left image of Figure 6.5. Fluid parcels starting out further distant from one another on the left, will exit the snapshot on the right closer to each other. In steady state, the

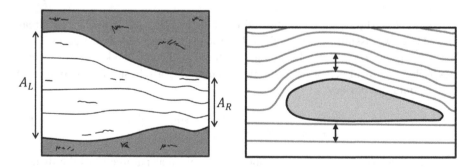

Figure 6.5 Two illustrations of stream lines. Left image shows streamlines in a river converging as the river narrows. Right image shows a side view of an airplane wing. The shape of the upside of the wing creates a flow pushing stream lines closer.

continuity equation reads $\nabla \cdot (\rho \mathbf{v}) = 0$, which implies

$$\int \nabla \cdot (\rho \mathbf{v}) \, dV = \int \rho \mathbf{v} \cdot d\mathbf{A} \Rightarrow \rho_L v_L \Delta A_L = \rho_R v_R \Delta A_R, \tag{6.10}$$

for the setup from the figure, with a volume V and a surface A (we are basically following the reasoning of our original derivation of continuity in reverse, applying the divergence theorem). Here v_L and v_R are the velocity components perpendicular to the areas A_L and A_R from the figure. In basic terms, what goes in on the left, must come out on the right. It immediately follows that

$$\frac{v_R}{v_L} = \frac{\rho_L}{\rho_R} \frac{\Delta A_L}{\Delta A_R}. \tag{6.11}$$

Many flows (like water) can indeed be approximated reasonably well as being *incompressible flows*, where the sizes of fluid parcels remain unchanged as they follow flow lines (i.e. $\rho_L \approx \rho_R$). In this case, it follows that $v_R > v_L$ in order to maintain the equality when $A_R < A_L$. In everyday life, this implies that a river will flow faster as it narrows or becomes more shallow. In terms of streamlines and steady-state incompressible flow, this means that we can directly interpret streamlines coming together as indicating higher flow velocity. As it turns out, incompressible flow is closely related to subsonic flow, something we will return to below.

6.4 BERNOULLI'S EQUATION

Bernoulli's equation is an insightful conservation law that holds along a streamline under certain conditions. To derive it, we start from Euler's law for fluid velocity, assuming both a steady flow and an external force that can be expressed in terms of a potential:

$$\frac{\partial \mathbf{v}}{\partial t} + \mathbf{v} \cdot \nabla \mathbf{v} = -\frac{\nabla p}{\rho} + \frac{\mathbf{f}}{\rho} \Rightarrow \mathbf{v} \cdot \nabla \mathbf{v} = -\frac{\nabla p}{\rho} - \nabla \Psi. \tag{6.12}$$

As usual, you are free to mentally substitute gravity around a massive source for the generic conservative force $\mathbf{f}$, in which case the potential obeys $\Psi = -GM/r$. Applying a vector identity (Equation B.2 in this case) allows us to rewrite the LHS to obtain

$$\nabla \left(\frac{1}{2} v^2 \right) - \mathbf{v} \times (\nabla \times \mathbf{v}) = -\frac{\nabla p}{\rho} - \nabla \Psi, \tag{6.13}$$

which can be trivially reorganized into

$$\nabla \left(\frac{1}{2} v^2 + \Psi \right) + \frac{\nabla p}{\rho} = \mathbf{v} \times (\nabla \times \mathbf{v}). \tag{6.14}$$

It would be nice if we could place the pressure term neatly within the parentheses along with the potential and velocity-squared terms, but the ρ^{-1} prohibits this in the general case. The issue disappears if we limit ourselves to incompressible flow, where $d\rho = 0$ and ρ a constant along path lines:

$$\nabla \left(\frac{1}{2} v^2 + \frac{p}{\rho} + \Psi \right) = \mathbf{v} \times (\nabla \times \mathbf{v}) \text{ (along incompressible streamline).} \tag{6.15}$$

Alternatively, we express the problem in terms of specific enthalpy h (see appendix A). Typically, fluid flows are isentropic and adiabatic along flow lines ($T ds = dq = 0$), unless something 'extreme' happens such as shocks impacting the flow or additional physics is included such as particle production through radioactive decay or a connection to an external environment (e.g. a radiation field). The specific enthalpy differential then reduces according to

$$dh = T ds + \frac{dp}{\rho} \rightarrow \frac{dp}{\rho}, \tag{6.16}$$

and we have

$$\nabla \left(\frac{1}{2} v^2 + h + \Psi \right) = \mathbf{v} \times (\nabla \times \mathbf{v}) \text{ (along adiabatic streamline).} \tag{6.17}$$

Given that we are considering flow along a streamline, we might as well limit ourselves to just the component of this vector equation along said streamline. This we can achieve by taking an inner product with $\mathbf{v}$ to get a quantity projected along $\mathbf{v}$ and thus the direction of the streamline (i.e. we take the 'directional derivative'):

$$\mathbf{v} \cdot \nabla \left(\frac{1}{2} v^2 + h + \Psi \right) = \mathbf{v} \cdot (\mathbf{v} \times (\nabla \times \mathbf{v})). \tag{6.18}$$

The combination of inner and cross product of $\mathbf{v}$ on the RHS means that the RHS will always be zero. This leads us to Bernoulli's equation:

$$\frac{1}{2} v^2 + h + \Psi = \text{constant (along adiabatic streamline).} \tag{6.19}$$

This equation expresses a balance between different forms of specific energy. In the case of an incompressible fluid, it can also be cast directly in terms of energy densities:

$$\frac{1}{2} \rho v^2 + p + \rho \Psi = \text{constant (along a streamline of incompressible fluid).} \tag{6.20}$$

A physical implication of Bernoulli's equation is shown in the right image of Figure 6.5. This illustrates the streamlines along an airplane flying through the air at a steady velocity (allowing us to assume at least temporarily steady flow in the reference frame of the airplane). Underneath, the wing is flat and the streamlines are unaffected. The upside of the wing is convex, pushing the airflow along streamlines that pass each other more closely as a result. Assuming the airplane is flying at subsonic velocity, we can reasonably approximate the airflow as incompressible flow. For the narrowing riverbed we had concluded that streamlines moving closer to each other implied an increase in velocity. This conclusion, based on the continuity equation, still holds for the airplane wing case. However, increasing v affects the kinetic energy term of Bernoulli's equation and one of the other terms therefore needs to change as well to compensate and ensure the total remains constant. Since presumably the potential term in this set-up is externally defined (e.g. gravity, if not assumed to be negligible altogether), this will have to be the pressure term. The resulting pressure difference between the two sides of the airplane wing will help lift the airplane upward.

6.5 THE DE LAVAL NOZZLE

The *De Laval nozzle* provides us with another insightful example of the behaviour of fluid flow in a scenario of converging and diverging streamlines. In the case of the De Laval nozzle we consider a slightly more extended setup than in the earlier river example: this time we do not only force the fluid through a narrowing volume, but include a widening part at the other end as well (see Figure 6.6). This hourglass-like shape is well known in the design of rocket engines for reasons that will become apparent in this section.

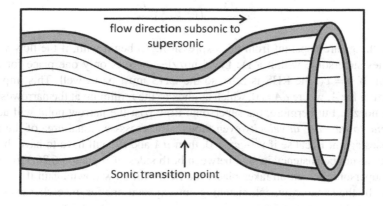

Figure 6.6 Sketch of a De Laval nozzle, subsonic flow enters on the left, passes the sonic transition point in the middle and exists supersonically on the right.

Repeating our first few steps from the discussion of Bernoulli's equation above but assuming the force to be negligible, we have

$$\mathbf{v} \cdot \nabla \mathbf{v} = -\frac{\nabla p}{\rho}, \tag{6.21}$$

from the momentum equation. Once again, we assume a steady flow. We further limit ourselves to *barotropic* flow where the density is a function of the pressure alone and vice versa ($p = p(\rho)$, which is less restrictive than assuming the flow to be incompressible). Under this assumption, we can write

$$\mathbf{v} \cdot \nabla \mathbf{v} = -\frac{1}{\rho}\frac{dp}{d\rho}\nabla\rho = -c_s^2\frac{\nabla\rho}{\rho}, \tag{6.22}$$

where we used the definition of the sound speed c_s. If we then deploy the vector identity Equation B.2 again, but this time straight-up assume the flow to be irrotational ($\nabla \times \mathbf{v} = 0$), we conclude

$$\nabla\left(\frac{1}{2}v^2\right) = -c_s^2\frac{\nabla\rho}{\rho}. \tag{6.23}$$

For any given choice of direction, this implies

$$v\,dv = -c_s^2\frac{d\rho}{\rho}. \tag{6.24}$$

At this point we can refer back to the river example, where we recognized that the continuity equation requires that $d(\rho v A) = 0$, with v the fluid flow velocity through an area A. Any deviation from zero would lead to a local net gain or loss of fluid in contradiction to our steady-state requirement. Applying the product rule and dividing the result by $\rho v A$, we have

$$\frac{d\rho}{\rho} + \frac{dv}{v} + \frac{dA}{A} = 0, \qquad (6.25)$$

which we can use to eliminate $d\rho/\rho$ from Equation 6.24 and arrive at

$$c_s^2 \frac{dA}{A} = \left(v^2 - c_s^2\right) \frac{dv}{v}. \qquad (6.26)$$

A few things are apparent from this expression. To begin with, if the flow velocity becomes supersonic within a De Laval nozzle, there is only one place for this to happen. If $v = c_s$, the RHS is zero, so the LHS must be as well. This implies an extremum for A, where $dA = 0$, which applies to the minimum at the narrowest point of the nozzle. Furthermore, the supersonic flow past the narrow point will actually *continue to increase in velocity* even though the cross-sectional area of the nozzle increases again. Because if $v^2 - c_s^2 > 0$, then dA and dv will have to have the same sign to ensure consistency in sign between both sides of Equation 6.26. Conversely, if no supersonic transition takes place, the flow will slow down again if the nozzle widens by the same argument, matching our expectations for the earlier example of incompressible flow in a river.

The river example represented a special case of Equation 6.25, where $d\rho = 0$ had been imposed. For supersonic flow through a nozzle having the flow remain incompressible would not be possible, given that (according to an argument rooted in the momentum equation) dv and dA then share the same sign and cannot cancel each other out even though Equation 6.25 (derived from continuity) demands that the LHS terms add up to zero. The conclusion is therefore that the supersonic flow will necessarily be *compressible*. It is the greater compressibility of fluid parcels in supersonic flow that is responsible for the perhaps counter-intuitive result of increasing flow velocity while flow lines diverge. This compressibility behaviour can be traced back to Equation 6.24, which can be rewritten as

$$\frac{d\rho}{\rho} = -\frac{v^2}{c_s^2} \frac{dv}{v} = -\mathcal{M}^2 \frac{dv}{v}, \qquad (6.27)$$

showing the strength of the response in compression to a change in velocity to be strongly proportional to the Mach number of the flow.

Given that the bulk velocity of the flow is a frame-dependent quantity whereas the compressibility is not, this proportionality cannot be claimed to hold in general. We had indeed already made particular choice of reference frame, one in which the nozzle itself is at rest. We also assumed *steady* flow, and this too is not something that automatically survives a change in reference frame (see Problem 2.8). The De Laval nozzle and Bernoulli's equation can of course be considered from the perspective of a different frame. The reasoning would merely become slightly more cumbersome once you have to starting accounting for things such as piston-like work done on the fluid by the moving walls in the nozzle[1]. But ultimately this will only cosmetically alter the outcome of the analysis and the physics of the system will remain the same.

The De Laval nozzle is probably more familiar to engineers than to astrophysicists, given its ongoing application in rocket engines that work using the combustion of hot gas[2]. However, jetted outflows that reach highly supersonic velocities are quite common in astrophysics. Examples include the jets associated with active galactic nuclei (AGN) and jets associated with gamma-ray bursts (GRB), the latter type of jet being triggered when a massive stars collapses or two neutron stars merge. While the assumptions used in modelling the De Laval nozzle are not an exact fit to these scenarios and do not account for features in astrophysical jets such as large-scale magnetic fields, a lack of long-term steady state flow or flow velocities becoming relativistic, the model can still be instructive. GRB jets, for example, also need to pass through a dense environment (the envelope of a massive star, the debris from the merger of neutron stars) and experience counter-pressure bending their streamlines[3].

PROBLEM 6.2
Analytical expressions for De Laval nozzle flow

The boundary conditions of De Laval nozzle flow are provided by cross-sectional area A, incoming mass flux $\dot{M} = \rho v A$ and the known shape of the nozzle. Here, we explore analytical expression for isothermal and polytropic flows through the nozzle.

i) Assume isothermal flow, an ideal gas of known composition, and a known value for T. In this case, the sound speed is constant at $c_s^2 = k_B T/(m/\mu)$. Using Bernoulli's equation and continuity, show that the velocity through a De Laval nozzle can be written in the form

$$\frac{v(x)^2}{c_s^2} = 1 + 2\ln\left(\frac{v(x)}{c_s}\frac{A(x)}{A_m}\right),$$

where A_m the cross-section at the sonic transition point where the nozzle is at its narrowest.

ii) For the isothermal case, make a plot of $\mathcal{M}$ versus A/A_m for $\mathcal{M}$ between $1/2$ and 2, using e.g. Python. You will have to find a workaround for the fact that the expression for v above is an *implicit* expression for the velocity rather than an analytical solution.

iii) Now relax the isothermal assumption to merely assuming the gas to be barotropic, that is, obeying $p = K\rho^{\hat{\gamma}}$. Show that, in this case, the velocity profile is given by

$$\frac{1}{2}\left[\left(\frac{v}{c_{s,m}}\right)^2 - 1\right] + \frac{1}{\hat{\gamma}-1}\left[\left(\frac{A}{A_m}\right)^{1-\hat{\gamma}}\left(\frac{v}{c_{s,m}}\right)^{1-\hat{\gamma}} - 1\right] = 0.$$

Here we assumed that $\hat{\gamma} \neq 1$. In the general polytropic case the sound speed is no longer fixed and $c_{s,m}$ refers specifically to the sound speed at $A = A_m$.

---

[1] This particular set-up is discussed in [37].

[2] On the other hand, it is a staple of text books covering fluid dynamics involving an astrophysics perspective, e.g. [48, 9, 47, 43].

[3] See e.g. [56] for an explicit parallel being drawn between a De Laval nozzle and a GRB jet.

iv) For the polytropic case with $\hat{\gamma} = 5/3$, make a plot of $v/c_{s,m}$ versus A/A_m for $v/c_{s,m}$ between $1/2$ and $5/3$, using e.g. `Python`. Similar to (ii), you are working from an implicit expression. The profile asymptotes to infinite A for $v/c_{s,m} \to 2$.

6.6 VORTICITY

We continue our exploration of the behaviour of fluid parcels along streamlines by considering the *vorticity* of a fluid flow. We define the vorticity ω as

$$\omega \equiv \nabla \times \mathbf{v}. \tag{6.28}$$

Roughly speaking, the more eddies (loosely defined as semi-circular currents) in a flow, the higher its vorticity. Thanks to Stokes' theorem,

$$\int_S (\nabla \times \mathbf{v}) \cdot d\mathbf{S} = \oint_l \mathbf{v} \cdot d\mathbf{r}, \tag{6.29}$$

we know that vorticity can be connected across scales, as illustrated in Figure 6.7 (see also Section 1.1.5). Mathematically, larger eddies can be seen as contained within smaller eddies, down to the scale of fluid parcels. The term on the RHS of Equation 6.29 is also known as the **circulation** in the flow. For an inviscid fluid, circulation is a conserved quantity and eddies are conserved within fluid parcels as illustrated in Figure 6.8. We demonstrate this below.

Figure 6.7 An illustration of Stokes theorem connecting large and small eddies.

The circulation is given by the integrated velocity along a closed loop or string. To check whether it is conserved, we need to look at its change over time using the co-moving derivative:

$$\frac{d}{dt} \oint_{l_{string}} \mathbf{v} \cdot d\mathbf{r}_{string} = \frac{d}{dt} \oint_{l_{string}} \left(v_x dx_{string} + v_y dy_{string} + v_z dz_{string} \right). \tag{6.30}$$

We are at this point placed for the usual Lagrangian dilemma when it comes to analyzing moving fluid parcels, which is that our parcel coordinates and size change

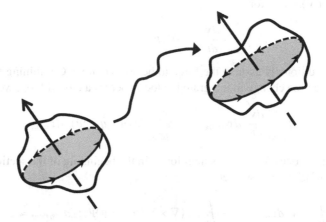

Figure 6.8 An area element contained within a fluid parcel and co-moving with the parcel. Arrows indicate the vorticity and orientation of the element.

while we take the integral. And, as usual, the answer is to explicitly work in terms of a time change relative to a fixed starting point:

$$
\begin{aligned}
\frac{d}{dt} \oint_{l_{string}} \mathbf{v} \cdot d\mathbf{r}_{string} &= \frac{d}{dt} \oint_{l_0} \left(v_x \left| \frac{dx_{string}}{dx_0} \right| dx_0 + v_y \left| \frac{dy_{string}}{dy_0} \right| dy_0 + v_z \left| \frac{dz_{string}}{dz_0} \right| dz_0 \right) \\
&= \oint_{l_0} \left(\frac{dv_x}{dt} \left| \frac{dx_{string}}{dx_0} \right| dx_0 + v_x \left| \frac{dv_x}{dx_0} \right| dx_0 + \ldots \right) \\
&= \oint_{l_{string}} \frac{d\mathbf{v}}{dt} \cdot d\mathbf{r}_{string} + \oint_{l_{string}} \mathbf{v} \cdot d\mathbf{v} \\
&= \oint_{l_{string}} \frac{d\mathbf{v}}{dt} \cdot d\mathbf{r}_{string} + \oint_{l_{string}} d\left(\frac{1}{2} v^2 \right) \\
&= \oint_{l_{string}} \frac{d\mathbf{v}}{dt} \cdot d\mathbf{r}_{string} + 0.
\end{aligned}
\tag{6.31}
$$

Here we used that a closed loop integral of a scalar quantity always evaluates to zero. Euler's velocity equation in Lagrangian form is given by

$$
\frac{d\mathbf{v}}{dt} = -\frac{\nabla p}{\rho} + \frac{\mathbf{f}}{\rho}.
\tag{6.32}
$$

According to the definition of specific enthalpy, we have

$$
\nabla h = \nabla q + \frac{\nabla p}{\rho},
\tag{6.33}
$$

in any given direction (note that the closed string does not generally align with a path line, so we do not want to limit ourselves to path lines of the flow when considering

the gradient ∇). Therefore,

$$\frac{d\mathbf{v}}{dt} = -\nabla\left(h - q + \Psi\right), \qquad (6.34)$$

where we additionally assume the force to be conservative. Combining this with our result for the time change in integrated velocity across a closed loop, we obtain

$$\oint_{l_{string}} \frac{d\mathbf{v}}{dt} \cdot d\mathbf{r}_{string} = -\oint_{l_{string}} \nabla\left(h - q + \Psi\right) \cdot d\mathbf{r}_{string}. \qquad (6.35)$$

Finally using Stokes' theorem as mentioned in the beginning of the section to remind us that the RHS is zero, we get

$$\frac{d}{dt} \oint_{l_{string}} \mathbf{v} \cdot d\mathbf{r}_{string} = -\int_{S_{string}} \left(\nabla \times \nabla\left(h - q + \Psi\right)\right) \cdot d\mathbf{S}_{string} = 0. \qquad (6.36)$$

(note that the curl of ∇ is always zero). This result is known as *Kelvin's circulation theorem*. It tells us that under certain conditions (inviscid gas flow experiencing only conservative forces), the local circulation of the flow is preserved. A famous example is Jupiter's great red spot (see Figure 6.9).

Figure 6.9 An image of Jupiter that prominently shows the great red spot and other vortices. Image taken by the Wide Field Camera on the Hubble Space Telescope, 2019, image credit NASA / STScI.

We can also manipulate the expressions above to arrive at an expression directly for the time evolution of the vorticity ω. Applying Stokes' theorem once more and realizing that the surface element size is arbitrary, we obtain:

$$\oint_{l_{string}} \frac{d\mathbf{v}}{dt} \cdot d\mathbf{r}_{string} = 0 \Rightarrow \int_{S_{string}} \left(\nabla \times \frac{d\mathbf{v}}{dt} \right) \cdot d\mathbf{S} = 0 \Rightarrow \nabla \times \frac{d\mathbf{v}}{dt} = 0. \quad (6.37)$$

The result gets us close to the time evolution of the vorticity, but due to the placement of the full time derivative in this expression we are not quite there yet. Taking another few steps:

$$\nabla \times \frac{d\mathbf{v}}{dt} = \nabla \times \left(\frac{\partial \mathbf{v}}{\partial t} + \mathbf{v} \cdot \nabla \mathbf{v} \right) = \frac{\partial \omega}{\partial t} + \nabla \times (\mathbf{v} \cdot \nabla \mathbf{v}) = 0. \quad (6.38)$$

That term involving a curl still looks somewhat opaque, and we will again make use of the same vector identity as earlier (appendix Equation B.2) to improve matters:

$$\mathbf{v} \cdot \nabla \mathbf{v} = \frac{1}{2} \nabla v^2 - \mathbf{v} \times (\nabla \times \mathbf{v}) = \frac{1}{2} \nabla v^2 - \mathbf{v} \times \omega. \quad (6.39)$$

Of these resulting two terms, the first is zero when its curl is taken, so after application of the vector identity our end result reads

$$\frac{\partial \omega}{\partial t} = \nabla \times (\mathbf{v} \times \omega). \quad (6.40)$$

This expression is known as *Helmholtz's equation*. It once more confirms that if we had a fluid under certain conditions (again, inviscid polytropic gas flow) with no eddies to begin with ($\omega = 0$), no eddies will emerge within the fluid if not externally introduced at some point. Suchs flows are called *irrotational*. Once complications such as viscosity are introduced, this result no longer holds and eddies can emerge and dissipate in the flow.

6.7 POTENTIAL FLOW, IRROTATIONAL FLOW AND INCOMPRESSIBLE FLOW

If a flow indeed obeys $\omega = \nabla \times \mathbf{v} = 0$ and is therefore irrotational, its flow velocity can be written as the gradient of a potential function Φ_v:

$$\mathbf{v} = -\nabla \Phi_v \Rightarrow \nabla \times \mathbf{v} = -\nabla \times \nabla \Phi_v = 0. \quad (6.41)$$

Irrotational flow is therefore also known as *potential flow*. Streamlines in potential flow will be open-ended by definition (a closed streamline is basically an eddy, after all, and we just argued against the existence of eddies in potential flow).

Potential flow that is also *barotropic* obeys a strong form of Bernoulli's equation. A barotropic equation of state implies

$$\frac{\nabla p}{\rho} = \nabla \int \frac{1}{\rho} dp. \quad (6.42)$$

(Just try it e.g. for a polytrope obeying $p = K\rho^{\hat{\gamma}}$.) So without specializing to directions along path lines we have that

$$\nabla\left(\frac{1}{2}v^2 + \int \frac{dp}{\rho} + \Psi\right) = 0, \tag{6.43}$$

with the RHS zero by definition for potential flow. Having this condition be valid throughout the steady flow fluid, even if only approximately, significantly simplifies the numerical modelling of the fluid.

Finally, let us take another look at *incompressible* flow, which we defined earlier as maintaining a constant density ρ along path lines. Recall that this implies that the velocity field is also divergence free:

$$\frac{d\rho}{dt} + \rho\nabla \cdot \mathbf{v} = 0 \Rightarrow \nabla \cdot \mathbf{v} = 0. \tag{6.44}$$

In the case of a fluid that is *both* irrotational and incompressible, we have

$$\nabla \cdot \mathbf{v} = 0 \Rightarrow -\nabla \cdot \nabla\Phi_v = 0 \Rightarrow \nabla^2\Phi_v = 0. \tag{6.45}$$

This is *Laplace's equation*. It has the same mathematical structure as the gravitational potential in a vacuum, which means that we can easily adapt (numerical) methods for computing gravitational fields to incompressible fluid problems and vice versa. This is more relevant to terrestrial fluid mechanics than to astrophysical applications (i.e. water instead of dilute gases in space), on account of the requirement that the flow be approximately incompressible.

7 Shock Waves

7.1 THE SHOCK-JUMP CONDITIONS

In Chapter 5 we discussed *sound waves*, the mechanism by which small disturbances propagate through a fluid. Mathematically, we assumed these perturbations of the fluid state to be sufficiently small that maintaining only linear perturbation terms was a valid approximation for the purpose of deriving the properties of sound waves. It is however certainly feasible for fluid flow to exceed the sound speed. In astrophysical settings this is even quite common, with gravity pulling surrounding plasma towards massive objects such as black holes with a velocity that reaches the free-fall velocity as the plasma motion becomes increasingly *supersonic* (i.e. 'faster than sound'). Other examples of supersonic flow occur in astrophysical explosions, such as novae or supernovae, driven by pressure. A third astrophysical scenario is the supersonic flow of jets launched by electro-magnetic forces, occurring in e.g. X-ray binaries, active galactic nuclei and gamma-ray bursts. When a supersonic flow encounters subsonic flow, shock fronts are formed. In this case, first-order perturbation theory is no longer appropriate and the solution to the conservation laws of fluid dynamics becomes non-linear. Similarly, when the non-linear terms in a propagating sound wave grow without dissipating away, the fluid profile steepens locally and a shock front can be formed.

From a mathematical perspective it is perfectly feasible to have solutions to Euler's equations (and the viscous Navier-Stokes equations) that are *discontinuous* across space. In terms of physical interpretation, these amount to the aforementioned shock fronts. While exhibiting rich behaviour on the micro-physical level, they can be considered as infinitesimally thin on the macro-physical level. Here we discuss how to treat shocks in fluid dynamics. Throughout this discussion it is important to keep in mind that a shock is like a sound wave in that it *outruns the motions of the fluid parcels*. While shocks convey information, this information jumps from parcel to parcel—it is not *advecting* (i.e. co-moving) with the flow.

A key concept is that of *shock-jump conditions*, connecting the fluid properties immediately behind a shock front to those in front of the shock. To derive the shock-jump conditions, we first move to a reference frame aligned with the local direction of the shock at the point of interest. Because we are considering strictly local properties of the fluid, we can use Cartesian coordinates without loss of generality even if the shock profile is curved at larger scales (e.g. in the form of a spherical blast wave from an explosion). Let us say we rotate our lab frame such that the local shock velocity U points in the positive x-direction. We will quickly see that it is practical to also have the reference frame itself move with velocity U, but for now we limit ourselves to merely a convenient rotation of the reference frame.

The relationships between different frames are illustrated in Figure 7.1 for the case of a non-moving upstream region, along with standard terminology for the 'upstream' (unshocked) and 'downstream' (shocked) regions. We will indicate the un-

DOI: 10.1201/9781003095088-7

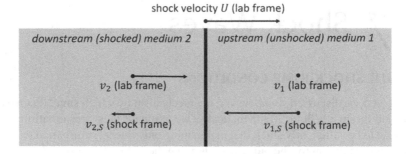

Figure 7.1 A shock front and relative velocity directions. The lengths of the vectors correspond to a strong shock running into a quiescent medium with $\hat{\gamma} = 5/3$.

shocked region with subscript 1 and the shocked region with subscript 2, not to be confused with vector indices. There will be no shock discontinuity in the y- or z-direction and the flow velocity in these directions will be unaffected by the shock front. We therefore have our first, trivial, jump conditions: $v_2^y = v_1^y$ and $v_2^z = v_1^z$.

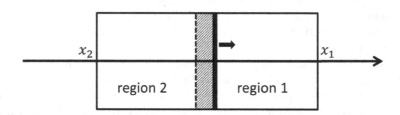

Figure 7.2 An integration interval covering the width of the shock. The shaded interval denotes the region $U dt$ covered by a shock front travelling to the right with shock velocity U during a time interval dt.

To obtain a jump condition in the x-direction, we are going to consider a small distance along x that is wide enough to span the width of the shock. As said, our continuum approach assumes a shock to be a singular discontinuity in space; in reality the width of a shock happens to be governed by similar length scales as those that set a limit on the validity of the fluid approximation, such as the mean free path in a gas. For now we assume the interval along x to be frozen in place but wide enough to fully contain a displacement $U dt$ of the shock front (see Figure 7.2) during time interval dt.

We will start from mass conservation, spatially integrated over the interval. We have

$$\int \frac{\partial \rho}{\partial t} dx + \int \frac{\partial (\rho v^x)}{\partial x} dx + \int \frac{\partial (\rho v^y)}{\partial y} dx + \int \frac{\partial (\rho v^z)}{\partial z} dx = 0. \qquad (7.1)$$

The first term is the partial differentiation with respect to time of the total mass-per-area M_x contained within the interval. For the second term, the integration and partial differentiation with respect to x cancel each other out, leaving their values at the boundaries (like Gauss' theorem in one dimension):

$$\frac{\partial M_x}{\partial t} - \rho_2 v_2^x + \rho_1 v_1^x + \int \frac{\partial (\rho v^y)}{\partial y} dx + \int \frac{\partial (\rho v^z)}{\partial z} dx = 0. \qquad (7.2)$$

We are going to take the integration domain to be as small as possible, shrinking it to a size dx, which will allow us to approximate all integrands as being constant throughout the domain—barring a discontinuous jump across the shock. For the term involving a mass flux in the y-direction, we get

$$\int_{x_2}^{x_1} \frac{\partial (\rho v^y)}{\partial y} dx \approx \frac{\partial (\rho_2 v_2^y)}{\partial y} dx_2 + \frac{\partial (\rho_1 v_1^y)}{\partial y} dx_1, \qquad (7.3)$$

where we split the integration domain in two separate intervals dx_1 and dx_2, one for each side of the shock, obeying $dx = dx_1 + dx_2$.

For the content of M_x, taking a very small domain means that we have on the left side of the shock front an approximately constant density ρ_2 and on the right side an approximately constant density ρ_1. However, because the shock front is in motion within this interval, the respective sizes of these sides change over time as indicated in Figure 7.2. Region 2 grows, while region 1 shrinks and during a time interval dt the mass content of both sides changes according to

$$dM_x = U\rho_2 dt - U\rho_1 dt. \qquad (7.4)$$

Unlike the spatial partial derivatives, the partial derivative with respect to time is not taken at fixed value for a single instant in time (i.e. $dt \neq 0$), so this change needs to be taken into account. Taking the time derivative of M_x then cancels the dt term in the above. Altogether, this means that we have

$$U\rho_2 - U\rho_1 - \rho_2 v_2^x + \rho_1 v_1^x + \frac{\partial (\rho_2 v_2^y)}{\partial y} dx_2 + \frac{\partial (\rho_1 v_1^y)}{\partial y} dx_1 +$$
$$\frac{\partial (\rho_2 v_2^z)}{\partial z} dx_2 + \frac{\partial (\rho_1 v_1^z)}{\partial z} dx_1 = 0, \qquad (7.5)$$

If we take the limit of a *vanishingly* small dx (and therefore of dx_1 and dx_2 as well), the flux divergence terms in the y- and z-directions drop out. We are left with

$$\rho_2 (v_2^x - U) = \rho_1 (v_1^x - U). \qquad (7.6)$$

This is the shock-jump condition from mass conservation across the shock. In a reference frame co-moving with the shock velocity, we get

$$\rho_2 v_{2,S}^x = \rho_1 v_{1,S}^x, \qquad (7.7)$$

denoting quantities in the shock frame with a subscript S. If the fluid flow had been steady in this frame to begin with, this outcome would have followed directly: in this frame, time derivative $\partial M_x/\partial t = 0$ and the shock advects with the flow in the y- and z-directions such that $v_S^y = v_S^z = 0$. This demonstrates that for the purpose of deriving the shock-jump conditions, it is permissible to regard the flow as steady in the frame of the shock, even if only instantaneously so. Physically, it means that the time scales associated with passage of the shock are far smaller than those otherwise associated with the fluid (i.e. changes in the flow take far longer to become noticeable than the small time interval we need to establish a jump condition). In terms of a shock-crossing time of characteristic length scales, this amounts to a assuming $dx/U \ll H/U$, where dx the width of the shock and H a characteristic macroscopic scale of the fluid (the factor H in the hydrostatic isothermal atmosphere from Equation 4.5 being an example of a characteristic fluid scale). For the purpose of deriving the remaining shock-jump conditions, we will therefore move to the shock frame first and consider the flow to be steady in that frame.

Applying this analysis to the equation for momentum density conservation, we then get

$$\lim_{x_1 \to x_2} \int_{x_1}^{x_2} \frac{\partial}{\partial x}\left(\rho v_S^x v_S^x + p\right) dx = \lim_{x_1 \to x_2} \int_{x_1}^{x_2} f_S dx = \lim_{dx \downarrow 0} f_S dx = 0, \qquad (7.8)$$

leading to

$$\left(\rho_2 v_{2,S}^x v_{2,S}^x + p_2\right) - \left(\rho_1 v_{1,S}^x v_{1,S}^x + p_1\right) = 0. \qquad (7.9)$$

Finally, for energy,

$$\left(\mathscr{E}_2 + p_2\right) v_{2,S}^x - \left(\mathscr{E}_1 + p_1\right) v_{1,S}^x = 0. \qquad (7.10)$$

Together with $v_{1,S}^y = v_{2,S}^y$ and $v_{1,S}^z = v_{2,S}^z$, equations 7.7, 7.9 and 7.10 provide a complete set of five shock-jump conditions (also known as *Rankine-Hugoniot* relations).

If the shock velocity $\mathbf{U}$ is known, the shock-jump conditions can be used to obtain a complete fluid description of the shocked (or *downstream*) medium (i.e. ρ_2, p_2 and $\mathbf{v}_2 \equiv \mathbf{U} + \mathbf{v}_{2,S}$) in terms of the unshocked (or *upstream*) medium (i.e. ρ_1, p_1 and $\mathbf{v}_1 \equiv \mathbf{U} + \mathbf{v}_{1,S}$). For this purpose, the shock-jump conditions as derived can be rewritten in all sorts of physically insightful forms.

In terms of the specific volume $\mathscr{V} \equiv 1/\rho$, we can rewrite the momentum jump condition

$$p_1 + \left(v_{1,S}^x\right)^2 / \mathscr{V}_1 = p_2 + \left(v_{2,S}^x\right)^2 / \mathscr{V}_2, \qquad (7.11)$$

which we can combine with continuity $v_{1,S}^x/\mathscr{V}_1 = v_{2,S}^x/\mathscr{V}_2$ to find the following flow velocities:

$$\left(v_{1,S}^x\right)^2 = \mathscr{V}_1^2 \frac{p_2 - p_1}{\mathscr{V}_1 - \mathscr{V}_2}, \qquad (7.12)$$

$$\left(v_{2,S}^x\right)^2 = \mathscr{V}_2^2 \frac{p_2 - p_1}{\mathscr{V}_1 - \mathscr{V}_2}. \qquad (7.13)$$

With that in hand, one can derive the jumps in magnitude of velocity in the x-direction,

$$\left|v_{1,S}^x\right| - \left|v_{2,S}^x\right| = \sqrt{(p_2 - p_1)(\mathscr{V}_1 - \mathscr{V}_2)}, \qquad (7.14)$$

velocity squared,

$$\frac{1}{2}\left(v_{1,s}^x\right)^2 - \frac{1}{2}\left(v_{2,s}^x\right)^2 = \frac{1}{2}(p_2 - p_1)(\mathscr{V}_1 + \mathscr{V}_2), \tag{7.15}$$

and specific internal energy

$$\varepsilon_2 - \varepsilon_1 = \frac{1}{2}(p_2 + p_1)(\mathscr{V}_1 - \mathscr{V}_2). \tag{7.16}$$

PROBLEM 7.1
More shock-jump conditions

Derive Equations 7.14, 7.15 and 7.16. The specific internal energy jump follows from combining the velocity squared equation and the full energy jump equation (Equation 7.10).

If we assume the gas to be a perfect gas, we can expand our set of useful shock jump conditions. Recall the adiabatic equation of state for a perfect gas (Equation 2.45):

$$p = (\hat{\gamma}_{ad} - 1)\rho\varepsilon \Rightarrow \varepsilon = \frac{p\mathscr{V}}{\hat{\gamma}_{ad} - 1}. \tag{7.17}$$

We can apply this relation to Equation 7.16, to obtain an expression for the pressure jump across a shock front:

$$\frac{p_2}{p_1} = \frac{(\hat{\gamma}_{ad} + 1)\mathscr{V}_1 - (\hat{\gamma}_{ad} - 1)\mathscr{V}_2}{(\hat{\gamma}_{ad} + 1)\mathscr{V}_2 - (\hat{\gamma}_{ad} - 1)\mathscr{V}_1}. \tag{7.18}$$

This same equation can be rewritten as

$$\frac{\mathscr{V}_2}{\mathscr{V}_1} = \frac{(\hat{\gamma}_{ad} + 1)p_1 + (\hat{\gamma}_{ad} - 1)p_2}{(\hat{\gamma}_{ad} - 1)p_1 + (\hat{\gamma}_{ad} + 1)p_2} = \frac{\rho_1}{\rho_2} = \frac{v_{2,s}^x}{v_{1,s}^x}. \tag{7.19}$$

At this point we can draw some physically interesting conclusions about the behaviour of shocks. Note for example that Equation 7.18 has a difference between two terms in its denominator, while Equation 7.19 has a sum. This means that while it is possible for p_2/p_1 to approach infinity (the limit of a strong shock), this is not possible for either the density or velocity ratio. In fact, in the limiting case of a strong shock, the density ratio would approach a finite value corresponding to an infinite pressure jump:

$$\frac{\mathscr{V}_1}{\mathscr{V}_2} = \frac{\rho_2}{\rho_1} = \frac{\hat{\gamma}_{ad} + 1}{\hat{\gamma}_{ad} - 1}. \tag{7.20}$$

For an ideal gas with $\hat{\gamma}_{ad} = 5/3$, this corresponds to a density jump of only a factor 4. Later, we will see that relativistic shocks manage to circumvent this restriction, allowing for even more extreme astrophysical fireworks.

PROBLEM 7.2
Polytropic jump conditions

Derive Equations 7.18 and 7.19 from Equation 7.16.

7.2   COMPRESSION SHOCKS, RAREFACTION WAVES AND CONTACT DISCONTINUITIES

In the context of shocks, we distinguish between three types of flow discontinuities: *compression shocks*, *rarefaction waves* and *contact discontinuities*, depending on the values that the ratio p_2/p_1 takes. Compression shocks and contact discontinuities are both first-order spatial discontinuities, involving a sudden jump directly in the fluid state. Rarefaction waves are second-order discontinuities, where the spatial derivatives of the fluid state jump suddenly while the fluid remains continuously connected across the rarefaction wave profile.

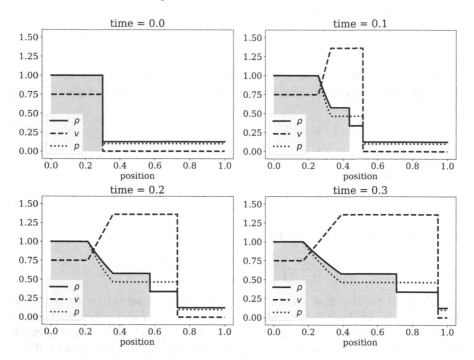

Figure 7.3   Fluid profile for a one-dimensional *shock tube* problem, shown at a range of times. A rarefaction wave runs to the left, stretching across an increasing interval in the shock tube. To the right of the rarefaction, a contact discontinuity in density is pushed to the right, but shows no discontinuity in pressure and velocity. A compression shock moves to the right the fastest, increasing fluid density, pressure and velocity in its wake. The grey area indicates fluid to the left of the contact discontinuity.

Figure 7.3 provides examples of all three discontinuities by showing a series of snapshots from a *shock tube* problem[1], where two constant fluid states are set up divided by a barrier that gets removed at $t = 0$. The left fluid state at $x < 0.3$ has $(\rho, p, v) = (1, 1, 0)$, while the right fluid state initially has $(\rho, p, v) = (0.125, 0.1, 0)$.

---

[1]This is an example of a 'Sod' shock tube problem [49] commonly used to test numerical fluid dynamics algorithms and named after the academic who introduced them as such.

The fluid as a whole is a perfect gas with $\hat{\gamma}_{ad} = 7/5$. Because of the pressure and velocity differences between the left and right states, a right-moving compression shock wave is triggered immediately upon removal of the barrier. This shock front compresses the fluid it shocks by increasing its density. It also raises the pressure of the fluid it overtakes and accelerates the flow in its wake. Any fluid to the right of the compression shock has not yet felt the impact of the removal of the barrier.

The leftmost feature of the shock tube is provided by a left-moving rarefaction wave, which rarefies the left state fluid and communicates the removal of the barrier to the left at the speed of sound. Because the rarefaction wave does not come with a first-order discontinuity, it extends over a widening distance. In the figure, the rarefaction pattern covers the part where the fluid state curves are not horizontal but curved to connect the onset and end of rarefaction. Any fluid to the left of the rarefaction wave is not yet affected by the removal of the barrier.

Because this is a one-dimensional example, fluid parcels have no means of overtaking each other. The contact discontinuity will keep marking the contact point between the 'rightmost' fluid parcel of the left state and the 'leftmost' fluid parcel of the right state, even while it gets pushed rightwards. In the figure, all fluid associated with the left state is indicated in grey. Because the flow velocity of the left initial state is not zero, 'grey' fluid is continuously added through the leftmost boundary of the shock tube (if not for this, the total area in grey would have remained constant, signifying the total mass in the left state). Even though no mass can move across a contact discontinuity, energy and momentum can still be exchanged across this boundary. These will not be discontinuous across a contact discontinuity.

Shock tubes allow for a range in shock behaviour, depending on how the initial states are set up. As a mathematical problem, finding an analytical solution to the one-dimensional set of Euler's equations in the case of two constant fluid states brought into contact is known as solving a Riemann problem. Its solution is known exactly, which makes shock tubes an extremely useful tool in numerical fluid dynamics. Shock tubes can be used as a benchmark for code performance against a known solution that nicely encapsulates a range of possibilities of extreme flow patterns (in particular, shocks). The Riemann problem also turns out to be a useful component of numerical algorithms for solving Euler's equations more generally, and we will therefore revisit shock tubes in Chapter 13 when discussing this topic.

Shock tube-like flow patterns occur in astrophysics as well. At a basic level, stellar winds are spherical versions of shock tubes, where the gas released by the star according to a loss term $\dot{M}$ make up the 'left' state and the interstellar medium the 'right' state. Also, whenever a gas is suddenly released into a highly different environment, a forward shock will propagate outwards while a rarefaction wave or compression shock propagates inwards. Examples of this are provided by the emerging blast waves from supernovae and long gamma-ray bursts. Both are explosive events that can be triggered by the collapse of certain massive stars, with gamma-ray burst blast waves able to reach relativistic flow velocities.

In the case of gamma-ray bursts, as the ejecta emerge from the progenitor stellar envelope, a forward shock will run into the interstellar medium. Or, rather, first into the stellar wind environment previously produced by the star before it collapsed. The latter is what one would expect, although the observational data has so far remained inconclusive about the radial distribution of gas around the sources of gamma-ray bursts. At the same time, a reverse shock will run back into the ejecta. This is not necessarily a rarefaction wave, but might be a second compression shock instead, depending on the fluid conditions in the ejecta. Since both shocks have the potential to shock-accelerate the local electron population at the microscopical level, both regions are expected to emit radiation as these electrons interact with shock-generated small-scale magnetic fields. Of these two emitting regions, the reverse shock is arguably the more interesting one, as its signal is produced by the collapsing object rather than the excited external environment and might therefore signal the properties of the burster to the observer. But since these reverse shocks only last briefly until they cross the ejecta completely, the challenge is to get on target as soon as possible with telescopes capable of capturing the emission. As of writing, there are various completely autonomous robotic telescopes and satellites active that can do just that, including the 2.0 meter Liverpool optical Telescope on the Canary Islands in Spain, and the Swift optical / UV / X-ray satellite launched in 2004. In Sections 7.4 and 7.6 we will discuss respectively blast waves and shock-acceleration of charged particles in more depth.

7.3 THE ENTROPY CHANGE ACROSS A SHOCK

If $p_2 > p_1$, we have a compression shock that increases the internal energy of the fluid. Compression shocks also increase the *entropy* of the fluid, as can be shown purely from a macroscopic analysis. Generally, one might expect an increase in entropy to be traceable to some microscopical mechanism at work, changing the nature of the fluid. In principle, this is no different for shocks, but because in our macroscopic view shock fronts are infinitesimally thin, this process is hidden from our sight. Nevertheless, the shock-jump conditions will allow us to quantify the jump in entropy as well.

To show this change in entropy does require another foray into thermodynamics. We start by manipulating the first law of thermodynamics for a reversible process,

$$T ds = d\varepsilon + p d\rho^{-1} \Rightarrow T ds = \left(\frac{\partial \varepsilon}{\partial T} \right)_V dT + p d\rho^{-1}. \tag{7.21}$$

Note that the left term on the RHS is indeed taken at constant volume (or, equivalently, density), given that changes in volume are grouped into the right term on the RHS. We consider a perfect gas evolving adiabatically outside of the shock discontinuity. We take Joule's second law as given by Equation A.28 from our appendix on thermodynamics, which states $\varepsilon = c_V T$, connecting specific internal energy and temperature through the specific heat capacity at constant volume. It then follows that

$$ds = \frac{c_V}{T} dT + \frac{p}{T} d\rho^{-1}. \tag{7.22}$$

If we combine Joule's second law with the equation of state for a perfect adiabatic gas Equation 2.45, stating $p = (\hat{\gamma}_{ad} - 1)\rho\varepsilon$, we have

$$p = \rho(\hat{\gamma}_{ad} - 1)c_{\mathcal{V}}T, \tag{7.23}$$

which we can use in our expression for entropy change:

$$ds = c_{\mathcal{V}}d\ln T + c_{\mathcal{V}}(\hat{\gamma}_{ad} - 1)d\ln\rho^{-1}. \tag{7.24}$$

Integrating both sides across a shock front yields an expression for the entropy jump:

$$s_2 - s_1 = c_{\mathcal{V}}\ln\left[\left(\frac{T_2}{T_1}\right)\left(\frac{\rho_1}{\rho_2}\right)^{\hat{\gamma}_{ad}-1}\right]. \tag{7.25}$$

We can re-apply the EOS to eliminate T_2/T_1, followed by using Equation 7.19 to eliminate ρ_2/ρ_1, in order to obtain a right hand side in terms of the pressure jump p_2/p_1 only. This can be shown to lead to

$$s_2 - s_1 = c_{\mathcal{V}}\ln\left[\left(\frac{p_2}{p_1}\right)\left(\frac{(\hat{\gamma}_{ad} - 1)p_2/p_1 + (\hat{\gamma}_{ad} + 1)}{(\hat{\gamma}_{ad} + 1)p_2/p_1 + (\hat{\gamma}_{ad} - 1)}\right)^{\hat{\gamma}_{ad}}\right]. \tag{7.26}$$

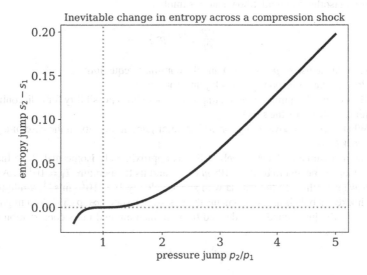

Figure 7.4 The jump in entropy across a shock front for various values of p_2/p_1. Here, we have used $c_{\mathcal{V}} = 1$ and $\hat{\gamma}_{ad} = 5/3$.

This entropy difference across the shock is illustrated in Figure 7.4. It illustrates how our entropy jump expression indeed encodes that $s_2 > s_1$ as soon as $p_2 > p_1$. The entropy equation 7.26 and the figure also tell us that first-order rarefaction shock discontinuities with $p_2/p_1 < 1$ do not exist. They would lead to a decrease in entropy and thus be in violation of the second law of thermodynamics. Instead, a rarefaction inevitably spreads through the fluid in the form of a more gentle *wave*, where the conditions in the fluid change without a first-order discontinuity in pressure.

Finally, there is the possibility that $p_2 = p_1$. On the one hand, this can correspond to no shock at all (and $\rho_2 = \rho_1$ as well). On the other hand, we can still posit the existence of a discontinuity, one that advects with the flow velocity. In this case, $v_{1,s}^x = v_{2,s}^x = 0$, and the only shock-jump condition that remains of the original five is that $p_2 = p_1$. This still allows for an arbitrary jump in density, and these jumps are the *contact discontinuities* that we identified earlier in the shock tube set-up.

PROBLEM 7.3
An isothermal shockwave

For this exercise, you will need atomic mass $m = 1.6605402 \times 10^{-24}$ g. Sometimes the shocked regions of a fluid are isothermal, for example when steep gradients produced during the passage of a shock induce a connection to an external *thermalizing* mechanism, such as a radiation field. For this exercise, we consider shock waves in a fully *isothermal* fluid.

i) Along flow lines, we have

$$\frac{d}{dt}\left(\frac{p}{\rho}\right) = 0,$$

for an isothermal fluid. Show that this implies

$$\frac{\partial p}{\partial t} + \nabla \cdot (p\mathbf{v}) = 0.$$

Hint: make use of $p = (p/\rho)\rho$ and the continuity equation.

ii) Provide the corresponding shock-jump condition.

iii) How does your jump-condition imply that there is *no* possibility for a discontinuous jump in T across the shock?

iv) What does the above imply for the potential jump in density in the case of a *strong* shock?

v) Let us assume a planetary nebula that is approximately isothermal. Take the density of the nebula to be $\rho_1 = 10^3$ m cm^{-3} and its temperature $T_1 = 10^4$ K. A shock wave runs through the nebula with shock velocity $U = 10^7$ cm s^{-1}, while the unshocked nebula is at rest. Compute the post-shock state (p_2, ρ_2, v_2). You might want to use the jump condition (derived from momentum and mass conservation across the shock)

$$\left(v_{1,s}^x\right)^2 = \mathcal{V}_1^2 \frac{p_2 - p_1}{\mathcal{V}_1 - \mathcal{V}_2},$$

to obtain ρ_2.

PROBLEM 7.4
A rarefaction shock?

In this Problem, we provide another demonstration of the impossibility of rarefaction shock waves. Consider the interstellar medium at rest, i.e. $v_1 = 0$ (we will consider only the *x*-direction, so we will drop the *x* index). Further, assume a temperature of 10 K, and a density of $\rho = 1m_p$ cm^{-3}, i.e. about one proton per cubic centimeter.

First, show that $v_1 = 0$ implies the following for the shock velocity U:

$$U^2 = (p_2 - p_1)(\mathscr{V}_1 + \mathscr{V}_2)\left(1 - \frac{v_{2S}^2}{v_{1S}^2}\right)^{-1}. \tag{7.27}$$

Now compute U^2, $c_{s,1}^2$ and $c_{s,2}^2$ for two cases: $p_2/p_1 = 100$ (a compression shock) and $p_2/p_1 = 1/100$ (the impossible rarefaction shock). Note that $c_{s,1}$ and $c_{s,2}$ as always denote sound speeds in the frame co-moving with fluid—do not confuse subscript s ('sound') with subscript S ('Shock frame').

Can you reason why the resulting three values for the two cases demonstrate that compression shocks can happen in nature, but rarefaction shocks decay quickly?

7.4 BLAST WAVES

Blast waves are a key application of shock wave theory and are produced both by terrestrial explosions and astrophysical explosions such as supernovae (see Figure 7.5 for a visually striking example). In many cases the explosion presents a *self-similar* problem leading to a fluid profile that can be solved for analytically (as we will see in Section 7.5). For now, we will postpone a full explanation and introduction

Figure 7.5 Supernova remnant 0509-67.5, shown in a composite image combining X-ray data from the Chandra X-ray observatory (responsible for most of the hazy structure within the bubble) and optical data from the Hubble Space Telescope (sharp blast wave profile and surrounding stars. The blast wave was produced by a supernova of type Ia (an exploding white dwarf star). The bubble has a radius $R = 10^{19}$ cm $= 11.5$ light years and velocity $U = 5 \times 10^8$ cm/s $= 5 \times 10^3$ km/s. Image credit NASA/CXC/SAO/J.Hughes et al (X-rays) and NASA/ESA/Hubble Heritage Team (STScI/AURA) (optical).

of the concept of self-similarity, focusing instead on a simplified spherical blast wave model where the blast wave is approximated by a narrow homogeneous shock front rather than a complete radial fluid profile. This simplified approach is already quite useful in a practical sense[2]. Furthermore, a lot of concepts involved in formulating more detailed solutions to the blast wave problem can be introduced here as well.

Consider the case of a sudden spherical release of a large amount energy in a small volume. We will assume the fluid where the explosion occurs (be it the earth's atmosphere, or the interstellar medium surrounding an exploding star) to be homogeneous and at rest ($v_1 = 0$). We will assume the blast wave to be adiabatic and obeying a polytropic gas law, so we can apply a $\hat{\gamma} = \hat{\gamma}_{ad}$ polytropic exponent. We focus on the blast wave while it is still fronted by a *strong* shock, such that $p_2 \gg p_1$. If the shock has a velocity U, the radial velocities will obey $v_{2,s} = v_2 - U$ and $v_{1,s} = -U$. The adiabatic gas flow jump conditions for a strong shock apply, and for the density jump we have

$$\rho_2 = \frac{\hat{\gamma}+1}{\hat{\gamma}-1}\rho_1 \rightarrow \rho_2 = 4\rho_1, \text{ if } \hat{\gamma} = \frac{5}{3}, \tag{7.28}$$

from Equation 7.20. This can be used in the shock-jump expression for the jump in momentum from the continuity jump Equation 7.7 to show that the velocity jump across a strong shock is given by

$$v_2 = \frac{2}{\hat{\gamma}+1}U \rightarrow v_2 = \frac{3}{4}U, \text{ if } \hat{\gamma} = \frac{5}{3}. \tag{7.29}$$

This velocity is as expected. The fluid in the wake of the shock gains a velocity comparable to that of the shock itself, while still being outpaced by the shock:

$$v_{2,s} = -\frac{\hat{\gamma}-1}{\hat{\gamma}+1}U \rightarrow v_{2,s} = -\frac{1}{4}U \text{ if } \hat{\gamma} = \frac{5}{3}. \tag{7.30}$$

To obtain an expression for p_2, we can neglect the contribution of p_1 in Equation 7.9, the jump condition derived from momentum conservation. Plugging in our previous jump condition results, we then obtain

$$p_2 = \frac{2}{\hat{\gamma}+1}\rho_1 U^2 \rightarrow p_2 = \frac{3}{4}\rho_1 U^2, \text{ if } \hat{\gamma} = \frac{5}{3}. \tag{7.31}$$

This jump condition does show that all terms in Equation 7.9 are of order p_2, except for p_1. Together with the strong shock assumption ($p_2 \gg p_1$), this justifies after the fact our neglecting of p_1 in this expression. Since for a polytropic gas the sound speed is given by $c_s^2 = \hat{\gamma}p/\rho$, we can quickly construct a sound-speed jump condition from the previous results:

$$c_{s,2}^2 = \frac{2\hat{\gamma}(\hat{\gamma}-1)}{(\hat{\gamma}+1)^2}U^2 \rightarrow c_{s,2}^2 = \frac{5}{16}U^2, \text{ if } \hat{\gamma} = \frac{5}{3}. \tag{7.32}$$

---

[2]When coupled to an approximation for their non-thermal emission (typically synchrotron emission), blast wave models like this are for example used routinely to predict the time evolution of the fading emission from supernovae remnants and gamma-ray burst afterglows.

The post-shock kinetic energy density follows directly from combining the jump conditions for density and velocity:

$$\frac{1}{2}\rho_2 v_2^2 = \frac{2}{(\hat{\gamma}+1)(\hat{\gamma}-1)}\rho_1 U^2 \rightarrow \frac{1}{2}\rho_2 v_2^2 = \frac{9}{8}\rho_1 U^2, \text{ if } \hat{\gamma} = \frac{5}{3}. \tag{7.33}$$

Assuming the blast wave to be adiabatic[3], we can use the equation of state $p = (\hat{\gamma}-1)e$ to determine the post-shock internal energy density e_2:

$$e_2 = \frac{2}{(\hat{\gamma}+1)(\hat{\gamma}-1)}\rho_1 U^2 \rightarrow e_2 = \frac{9}{8}\rho_1 U^2, \text{ if } \hat{\gamma} = \frac{5}{3}. \tag{7.34}$$

That the post-shock internal energy density is the same as the kinetic energy density tells us that the blast wave generates a large amount of shock-heating of the unshocked medium.

The jump conditions derived above tell us about the local state of the fluid behind the shock. All jump conditions link to shock velocity U, which is a global measure of the state of the blast wave. In order to determine U, shock radius R and effective blast wave width ΔR, we can turn to some global statements of conservation. To begin with, while we deposited an energy E into the fluid, we did not (by construction) add any mass to the system. Therefore, the total mass M contained within the blast wave at radius R must be identical to all the swept-up mass provided by the surroundings. If we approximate the blast wave profile by a narrow shell ($\Delta R \ll R$) of homogeneous density ρ_2, this allows us to easily estimate its value from mass conservation. That is,

$$M = \rho_1 \frac{4\pi}{3}R^3 = 4\pi\rho_2 R^2 \Delta R, \tag{7.35}$$

can be solved to reveal

$$\Delta R = \frac{\hat{\gamma}-1}{\hat{\gamma}+1}\frac{1}{3}R \rightarrow \Delta R = \frac{R}{12}, \text{ if } \hat{\gamma} = \frac{5}{3}. \tag{7.36}$$

Unless energy is removed from the blast wave (by radiation, for example), the total energy within the blast wave should remain a constant E. This provides us with a second global conservation law that we can use to determine R:

$$E = 4\pi R^2 \Delta R\left(e_2 + \frac{1}{2}\rho_2 v_2^2\right) = 4\pi R^2 \Delta R \frac{4}{(\hat{\gamma}-1)(\hat{\gamma}+1)}\rho_1 U^2. \tag{7.37}$$

This can be rewritten to show

$$R^3 U^2 = \frac{E}{\rho_1}\frac{3(\hat{\gamma}+1)^2}{16\pi}. \tag{7.38}$$

---

[3]We are here discussing the fluid *behind* the shock front, not the infinitesimally small shock front itself, where an increase in entropy is unavoidable (as discussed in Section 7.2). So even if the flow is adiabatic within the shock itself, it will not be isentropic in this region.

Since $U \equiv dR/dt$, we have a differential equation for R,

$$R^{\frac{3}{2}} dR = \left(\frac{E}{\rho_1} \frac{3\,(\hat{\gamma}+1)^2}{16\pi} \right)^{\frac{1}{2}} dt. \tag{7.39}$$

This can be solved (with boundary condition $R = 0$ at $t = 0$):

$$R = \left(\frac{25}{4} \frac{3\,(\hat{\gamma}+1)^2}{16\pi} \right)^{\frac{1}{5}} \left(\frac{Et^2}{\rho_1} \right)^{\frac{1}{5}} \rightarrow R \approx 1.2 \left(\frac{Et^2}{\rho_1} \right)^{\frac{1}{5}}, \text{ if } \hat{\gamma} = \frac{5}{3}. \tag{7.40}$$

While the numerical pre-factor in this expression had to be computed, the fact that R relies on the other variables E, ρ_1 and t the way it does, should not have come as a surprise. Shock-wave radius R is a measure of distance, and the *only* way in which to combine the other variables in a manner that produces a measure of distance (i.e. eliminates mass and time dimensions, while scaling to a linear dependence on mass), is the one given in this equation. This application of dimensional analysis provides a quick means to establish the self-similarity of the explosion problem, and we will return to it in Section 7.5. The power-law dependence of R on time means that the shock velocity obeys

$$U = \frac{2}{5} \frac{R}{t}. \tag{7.41}$$

The blast wave model presented above is quite basic, although effective, telling us how the hot regions of the blast wave evolve over time (which can be plugged into subsequent emission estimates for explosive astrophysical sources, as mentioned in footnote 2). A full solution of the fluid profile of a strong explosion in a medium with $\hat{\gamma} = 5/3$ would lead to $R \approx 1.15 \left(Et^2/\rho_1 \right)^{1/5}$. This pre-factor is very close to the value 1.2 computed above, showing that the shell model is indeed quite accurate already.

The shell model is still unphysical, however. To name one thing, realistically, the inner regions of the blast wave are not expected to be a complete vacuum. Instead, the gas behind the shell will be highly dilute, but still contain a certain amount of pressure to support the shell in front. Within the context of a simple shell model expanding in a homogeneous environment, it can be shown that this pressure value will be about half the pressure of the shell itself[4]. We will see a similar asymptotic value at small radius confirmed by the self-similar solution in a homogeneous environment discussed below.

The model can be extended in various ways. In the case of a supernova blast wave, it might be of interest to replace the assumption that the environment is homogeneous by a density profile that is a function of radius (representing e.g. a stellar wind). When doing so, do keep in mind that the shock-jump conditions above refer to jumps relative to the fluid state immediately ahead of the shock and that, in particular, ρ_1 should be considered a function of radius R, e.g. $\rho_1 = \rho_{ref} \times \left(R/R_{ref} \right)^{-2}$,

---

[4]See e.g. [57] for a description of a blast wave model including pressure behind the shell front.

where we use subscript 'ref' to indicate reference values. Note that changing the density profile impacts the solution to the global conservation laws. Another conceivable addition could be extending the time period over which the energy is injected, modelling an ongoing wind or other process. This can be implemented by setting $E = E_{ref}\left(t/t_{ref}\right)^{-\alpha}$ in Equation 7.37, where α some value representative of how the ongoing release of energy decays over time. Yet a third option would be to add a significant initial mass to the explosion. In supernovae, for example, a substantial amount of stellar mass is blown away with the blast wave[5]. A massive shell model including an ejecta mass M_{ej} can be implemented by adding a kinetic energy term $M_{ej}v_2^2/2$ for cold ejecta to Equation 7.37, which will lead to

$$E = \left(\frac{1}{2}M_{ej} + M\right)\frac{4U^2}{(\hat{\gamma}+1)^2}. \tag{7.42}$$

At this level of approximation, the details of the interaction between the original ejecta and the swept-up mass are not included (in terms of Section 7.2, the crossing of the reverse shock or rarefaction wave is assumed to have been completed, such that the entire ejecta is in thermal contact with the entire swept-up mass). Looking at Equation 7.42, initially the shock will coast along at constant velocity U, given that $M \ll M_{ej}$ while M_{ej} and E are presumed constant. Around $M \sim M_{ej}/2$, the balance of energy will shift and the swept-up material will start to dictate the dynamics of the system. Eventually, $M_{ej} \ll M$, and the model will revert to Equation 7.37. Equation 7.42 gives us blast wave velocity U as a function of swept-up mass M (and thus of R), but does not provide us directly with $R(t)$, i.e. the radius as a function of time. For that we need to integrate $U = dR/dt$ as extracted from Equation 7.42, like we did in Equation 7.39 for the basic shell model of Equation 7.37. This can still be done analytically, but in practice just solving the differential equation using e.g. Python might be more useful. Figure 7.6 shows the evolution of the shock velocity for a supernova blast wave that starts out as a massive cold shell ($M \sim M_\odot$ with kinetic energy $E \sim 10^{51}$ erg), running into an interstellar medium of density $\rho = m_p$ cm^{-3}.

7.5 SELF-SIMILAR EXPLOSIONS

The concept of *self-similarity* is a very powerful one in mathematics and physics. For various scenarios, such as the case of the large concentrated energy release leading to a spherical blast wave, self-similarity makes it possible to determine the fluid profile of the resulting blast wave in full, going far beyond the simplified homogeneous shell approximation. Another example of a self-similar fluid profile is the shock tube set-up from Figure 7.3. The key element of self-similarity is in its name: there exists a constraint on the manner in which the coordinates occur in the fluid solution, which means that if one variable is scaled, the others will follow in a manner that renders

---

[5]See e.g. [28] for a nice visual representation of mass and kinetic energy in supernovae of type Ib/c, showing that sometimes the ejecta obtain even near light-speed velocities. Typically, Ib/c supernovae resulting from massive star collapse confer a kinetic energy of about 10^{51} erg onto ejecta containing mass $M_\odot$, i.e. one solar mass. The resulting velocity v is a few percent of light speed, following $E = M_{ej}v^2/2$.

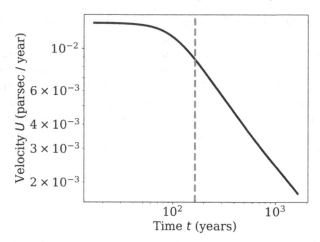

Figure 7.6 Velocity evolution of a massive supernova remnant. The vertical grey line indicates a rough estimate of the balance point where $M_{ej}/2 = M$, under the simplifying assumption that $R = Ut$ and U unchanged from its initial value set by $E = Mv_2^2/2$. The polytropic exponent $\hat{\gamma}$ has been set to 5/3.

the new fluid profile similar to its unscaled self. Dimensional analysis often is a quick route to establishing self-similarity, but it is not a prerequisite to it.

To make the notion of self-similarity more concrete and to demonstrate its effectiveness, we here treat the *Sedov-Taylor* solution to the strong explosion problem[6]. Note however that this section will add little physics to the discussion on blast waves from the preceding section, adding a mostly mathematical approach, and readers not particularly interested in self-similarity can safely skip this section.

When setting up a self-similar model in the case of a massive energy release in a homogeneous medium, we have the model coordinates r, t, and parameters E and ρ_0. All carry a dimension: in cgs-units, we have r in [cm], t in [s], E in [erg] = [g] [cm]^2 [s]$^{-2}$ and ρ_1 in [g] [cm]$^{-3}$. As mentioned in the previous section, the only dimensionally viable combination in an expression for radius R (which is itself not a function of r) is going to be

$$R = C_{ST} \left(\frac{Et^2}{\rho_1} \right)^{\frac{1}{5}},\qquad(7.43)$$

where C_{ST} a dimensionless pre-factor (that happens to be close to unity). Similarly, if the medium is not homogeneous but has a mass profile $\rho(r) \equiv Ar^{-2}$ (a 'stellar

---

[6]The solution to the self-similar point explosion problem was formulated independently in the 1940's by Taylor, Von Neumann and Sedov. L.I. Sedov was the first to provide a full solution, also published in his well-known book on similarity methods [46]. A famous early full treatment can be found in [23]. Here we describe a simple $\rho \propto r^{-2}$ case and general solution due to [55].

wind'), dimensional analysis dictates that

$$R = C_{ST} \left(\frac{Et^2}{A} \right)^{\frac{1}{3}}, \tag{7.44}$$

reflecting that A carries the dimension [g] [cm]$^{-1}$.

PROBLEM 7.5
Self-similar explosions

Consider a self-similar explosion in a homogeneous environment. Suppose that the energy is not injected instantaneously, but grows linearly over time. We would have to replace $E \to C_E t$, with C_E a constant (with dimension!) that sets the growth over time. How would the blast wave radius and velocity depend on time in this case?

Going beyond merely determining the blast wave radius, we can define a self-similar variable ξ:

$$\xi \equiv \frac{r}{R} = C_{ST}^{-1} \left(\frac{r^5 \rho_1}{Et^2} \right)^{\frac{1}{5}}, \text{ (constant medium)}, \tag{7.45}$$

$$\xi \equiv \frac{r}{R} = C_{ST}^{-1} \left(\frac{r^3 A}{Et^2} \right)^{\frac{1}{3}}, \text{ (wind medium)}, \tag{7.46}$$

The parameter ξ is another example of a similarity parameter that appears once the scale invariance of the fluid equations is broken, this time by accounting for the boundary conditions E and ρ_1 (or A). Where the Knudsen number (Section 2.2.1) marked a boundary below which the continuum approximation becomes questionable, ξ marks a boundary below which we cannot assume a 'regime' unaffected by the explosion. Discussing ξ as a similarity parameter in this manner is admittedly somewhat artificial, but hopefully serves the point about how the inclusion of additional physics or explicit scales (energy, density values for the explosion in this case) can be used to break the scale-invariance in Euler's equations and inevitably allows for the formulation of dimensionless parameters than can help guide how to treat a fluid dynamics problem.

As a matter of fact, ξ is the *only* independent dimensionless variable that can be defined according to dimensional analysis, since any alternative will inevitably be writeable as a function solely of ξ. What we have, therefore, is a mapping $r, t \to \xi$ that makes explicit that the fluid state is a function of one variable only. Any differentiation with respect to time or radius occurring in the spherically symmetric conservation laws of fluid dynamics can be cast in terms of ξ. Introducing a variable k for the slope of the medium density profile such that $\rho = Ar^{-k}$, we have

$$\frac{\partial}{\partial t} = -\frac{2}{5-k} \frac{\xi}{t} \frac{\partial}{\partial \xi}, \qquad \frac{\partial}{\partial r} = \frac{\xi}{r} \frac{\partial}{\partial \xi}. \tag{7.47}$$

Before applying this to an actual set of conservation laws, we can already utilize dimensional analysis once again to lay some of the groundwork. Following the standard treatments on this subject, we start by casting expressions for fluid velocity v, density ρ and sound speed c_s in terms of factors carrying dimension and dimensionless functions:

$$v = \frac{2}{5-k}\frac{r}{t}V(\xi), \qquad \rho = \rho_1 G(\xi), \qquad c_s^2 = \frac{4r^2}{(5-k)^2 t^2}Z(\xi). \qquad (7.48)$$

Here by ρ_1 we specifically refer to external density at $r = R$ (for a homogeneous medium, this distinction is irrelevant). The factors carrying dimension are inevitable, following from dimensional analysis. V, G and Z are dimensionless functions that will need to be determined by solving the fluid equations, although we do know their values directly behind the shock front on account of the shock-jump conditions:

$$V_2 = \frac{2}{\hat{\gamma}+1}, \qquad G_2 = \frac{\hat{\gamma}+1}{\hat{\gamma}-1}, \qquad Z_2 = \frac{2\hat{\gamma}(\hat{\gamma}-1)}{(\hat{\gamma}+1)^2}. \qquad (7.49)$$

All this we can now apply to the conservation laws of fluid dynamics. We need three conservation laws in spherical symmetry (along with an equation of state) to fully specify the system, and we use conservation of mass density,

$$\frac{\partial \rho}{\partial t} + \frac{\partial(\rho v)}{\partial t} + \frac{2\rho v}{r} = 0, \qquad (7.50)$$

the velocity equation,

$$\frac{\partial v}{\partial t} + v\frac{\partial v}{\partial r} = -\frac{1}{\rho}\frac{\partial p}{\partial r}, \qquad (7.51)$$

and an equation that encodes the isentropic nature of the fluid flow directly into a third conservation law,

$$\frac{d}{dt}\left(\frac{p}{\rho^{\hat{\gamma}}}\right) = 0. \qquad (7.52)$$

Let us start by applying our Ansätze from Equation 7.48 to mass conservation. We get

$$\frac{\partial}{\partial t}(\rho_1 G) + \frac{\partial}{\partial r}\left(\rho_1 G\frac{2}{5-k}\frac{r}{t}V\right) + 2\rho_1 G\frac{2}{5-k}\frac{1}{t}V = 0. \qquad (7.53)$$

If we indeed want to allow for options with $k \neq 0$, we have to be mindful of ρ_1 being a function of blast wave radius, and thus of time:

$$\frac{\partial \rho_1}{\partial t} = -kA\frac{\partial R}{\partial t}R^{-k-1} = -kA\frac{2}{5-k}\frac{R}{t}R^{-k-1} = -\frac{2k}{5-k}\frac{\rho_1}{t}. \qquad (7.54)$$

Using this, along with the mapping of the partial derivatives onto derivatives with respect to ξ from Equation 7.47, it can be shown that mass conservation can eventually be written in this form (skipping some intermediate steps):

$$\frac{dV}{d\ln\xi} - (1-V)\frac{d\ln G}{d\ln\xi} = -3V + k. \qquad (7.55)$$

What is most important about this equation is that it is a completely dimensionless differential equation in terms of ξ only. If we take the same approach to the velocity equation, we can ultimately derive the following expression:

$$\frac{\hat{\gamma}(V-1)}{Z}\frac{dV}{d\ln\xi}+\frac{d\ln G}{d\ln\xi}+\frac{d\ln Z}{d\ln\xi}=-\frac{\hat{\gamma}}{Z}V^2+\frac{5-k}{2}\frac{\hat{\gamma}}{Z}V-2 \qquad (7.56)$$

From the equation expressing adiabatic flow we obtain:

$$(1-\hat{\gamma})\frac{d\ln G}{d\ln\xi}+\frac{d\ln Z}{d\ln\xi}=-\frac{5-2V-\hat{\gamma}k}{1-V}. \qquad (7.57)$$

In principle, all it takes now for the full solution to be obtained is to solve the self-similar set of equations 7.55, 7.56 and 7.57. This would require some reorganizing of the equations to isolate the three distinct derivatives of V, G and Z and solving the coupled pair of differential inwards from the outer boundary (where the fluid state is known thanks to the shock-jump conditions). In practice, this is still a daunting task if not done via computer (either numerically, e.g. in Python, or using some symbolic manipulation program like Maple or Mathematica). It is however possible to render the task at least somewhat less daunting thanks to some further implications of self-similarity. Consider the energy E_m contained within a sphere of outer radius r_m (or ξ_m, in self-similar coordinates),

$$E_m = \int_0^{r_m} 4\pi r^2\left(e+\frac{1}{2}\rho v^2\right)dr. \qquad (7.58)$$

Using $e = p/(\hat{\gamma}-1) = \hat{\gamma}^{-1}\rho c_s^2/(\hat{\gamma}-1)$, along with $\xi = r/R$ and $\rho_1 = AR^{-k}$, it can be shown that

$$E_m = 4\pi\frac{4}{(5-k)^2}A\frac{R^{5-k}}{t^2}\int_0^{\xi_m}\xi^4 G\left(\frac{Z}{(\hat{\gamma}-1)\hat{\gamma}}+\frac{V^2}{2}\right)d\xi. \qquad (7.59)$$

This expression confirms an important property of self-similar flow: as long as we keep ξ_m constant, E_m will remain a constant fraction of total energy E, given that $R^{5-k}\propto t^2$ and the integral a function only of ξ_m. A spherical volume up to fixed ξ_m will still grow in actual size over time, and the requirement that E_m is constant translates into a balance between the energy flux through the surface of this sphere during a time interval dt and the energy in the additional volume covered by the sphere during the same time. The expansion velocity v_m of the sphere is given by setting the full derivative of ξ_m with respect to time equal to zero, keeping in mind that this time we are evaluating along a path of velocity v_m rather than the fluid velocity v:

$$\frac{\partial\xi}{\partial t}+v_m\frac{\partial\xi}{\partial r}=0\Rightarrow v_m=\frac{2}{5-k}\frac{r}{t}. \qquad (7.60)$$

The energy flux through the expanding sphere over a time interval dt is given by $-4\pi r^2\left(ev+pv+\frac{1}{2}\rho v^2 v\right)dt$, acting as a loss term. The additional energy added by the increasing radius of the sphere is given by $4\pi r^2 \left(e + \frac{1}{2}\rho v^2\right) v_m dt$. The sum of the two therefore implies:

$$4\pi r^2 \left(e + \frac{1}{2}\rho v^2\right) v_m dt - 4\pi r^2 \left(ev + pv + \frac{1}{2}\rho v^2 v\right) dt = 0. \qquad (7.61)$$

If our expression for v_m is used in this equation, along with the self-similar expressions for the fluid variables, it can be shown that

$$Z = \frac{\hat{\gamma}(\hat{\gamma} - 1)(1 - V)V^2}{2(\hat{\gamma}V - 1)}. \qquad (7.62)$$

This expression has the significant advantage over Equations 7.55, 7.56 and 7.57 that it is not a differential equation.

Unfortunately, the same approach to mass does not pay the same dividends. The cumulative mass m up to some fixed radius r is defined by

$$m = 4\pi \int_0^{r_m} r^2 \rho dr = 4\pi R^3 \rho_1 \int_0^{\xi_m} \xi^2 G d\xi = M(3 - k) \int_0^{\xi_m} \xi^2 G d\xi, \qquad (7.63)$$

in terms of total swept-up mass M. Even though the cumulative mass m will remain a fixed fraction of M, the total mass M is not a constant. The mass version of Equation 7.61 would have to include a RHS side term to account for a total mass growth $\frac{m}{M}dM$ within the sphere, to ensure that $(m + dm)/(M + dM) = m/M$ and the mass fraction indeed stays fixed. We have

$$4\pi r^2 \rho v_m dt - 4\pi r^2 \rho v dt = (3 - k)\left(\int_0^{\xi_m} G\xi^2 d\xi\right) 4\pi R^2 \rho_1 U dt. \qquad (7.64)$$

Plugging in self-similar functions, it can be shown that the mass budget equation reduces to

$$G(\xi_m)\xi_m^3 (1 - V(\xi_m)) = (3 - k) \int_0^{\xi_m} G\xi^2 d\xi, \qquad (7.65)$$

which has to hold for arbitrary ξ_m. However, differentiating this result on both sides just takes us back to Equation 7.55, so its use is limited to being merely a consistency check on our work.

Going back to the energy budget evaluation and differentiating the resulting expression for Z (Equation 7.62), we find

$$\frac{d \ln Z}{d \ln \xi} = \frac{dV}{d \ln \xi}\left(\frac{-1}{1 - V} + \frac{2}{V} - \frac{\hat{\gamma}}{\hat{\gamma}V - 1}\right). \qquad (7.66)$$

This we can use to eliminate terms involving Z from Equation 7.57. After that, we can combine the result with Equation 7.55 in order to eliminate terms involving G as well. After some reorganizing, it can then eventually be shown that

$$\frac{dV}{d \ln \xi} = -\frac{V(\hat{\gamma}V - 1)(3\hat{\gamma}V - V + k - 5)}{2 + \hat{\gamma}(1 + \hat{\gamma})V^2 - 2(\hat{\gamma} + 1)V}. \qquad (7.67)$$

At this point, an expression for $\xi(V)$ can be obtained by integrating

$$-\frac{2+\hat{\gamma}(1+\hat{\gamma})V^2-2(\hat{\gamma}+1)V}{V(\hat{\gamma}V-1)(3\hat{\gamma}V-V+k-5)}dV = d\ln\xi \tag{7.68}$$

on both sides over their respective variables. We will provide the result below. However, because this result is both a messy expression and 'merely' an implicit expression for the velocity profile (i.e. ξ as a function of V, rather than the other way around), we first take a look at a special case that has a more appealing simplicity to it.

Consider the case of an ideal gas with $\hat{\gamma}=5/3$ expanding in a medium with $k=2$. Directly behind the shock front, we have $V=3/4$, according to the jump conditions from Equations 7.49. But this set of values is potentially problematic, as can be seen by taking a close look at the LHS of Equation 7.68, where this combination results in a singularity. Going back one equation further to Equation 7.67, we conclude that at the shock front $dV/d\ln\xi = 0$.

At this point, we might want to test a trial solution: what if V is actually constant *throughout*? As it turns out, this is indeed a viable solution to the $\hat{\gamma}=5/3$, $k=2$ case. It then directly follows from Equation 7.62 that Z is constant as well. G then follows from Equation 7.55, and altogether we get

$$V=\frac{3}{4}, \qquad G=4\xi, \qquad Z=\frac{5}{16}. \tag{7.69}$$

All values are consistent with the jump conditions (as they must be), and yield the following fluid profile:

$$v=\frac{1}{2}\frac{r}{t}, \qquad \rho=4\frac{r}{R}\rho_1, \qquad c_s^2=\frac{5}{36}\frac{r^2}{t^2}, \qquad p=\frac{1}{3}\frac{r^3}{Rt^2}\rho_1. \tag{7.70}$$

When the fluid profile is this simple, it becomes very straightforward to compute the total mass and energy contained in the blast wave. The mass is just what we expect and equal to the total swept-up mass M. The total energy can be used to determine the dimensionless prefactor C_{ST} in the expression for radius Equation 7.44. Setting $\xi_m = 1$ in Equation 7.59 and substituting Equation 7.44 for R, it follows that

$$C_{ST} = \left(\frac{6}{4\pi}\right)^{\frac{1}{3}} \approx 0.782 \qquad \text{(stellar wind with } \hat{\gamma}=\frac{5}{3}\text{).} \tag{7.71}$$

While this value is not equal to unity, it nevertheless is of the same order of magnitude. This is not unexpected, given that there were no unidentified properties of the system left that could have introduced a radically different length scale.

The full solution to the Sedov-Taylor problem can be obtained by the same method as used for the wind case, but now without the benefit of a simple expression for V. Without going through the details[7], it can eventually be shown that:

$$\xi^{5-k} = \left(\frac{V}{V_2}\right)^{-2} \left(\frac{5-(3\hat{\gamma}-1)V-k}{5-(3\hat{\gamma}-1)V_2-k}\right)^{v_1} \left(\frac{\hat{\gamma}V-1}{\hat{\gamma}V_2-1}\right)^{v_2}, \tag{7.72}$$

---

[7]There is a reason even Landau & Lifshitz deem these steps in the process to be 'laborious', even if 'elementary', in their analysis of the $k=0$ case ([23]).

$$G = 4\left(\frac{\hat{\gamma}V-1}{\hat{\gamma}V_2-1}\right)^{v_3}\left(\frac{5-(3\hat{\gamma}-1)V-k}{5-(3\hat{\gamma}-1)V_2-k}\right)^{v_4}\left(\frac{1-V}{1-V_2}\right)^{v_5}\left(\frac{V}{V_2}\right)^{v_6}, \quad (7.73)$$

with exponents given by

$$v_1 = -\frac{(k^2-4k+13)\hat{\gamma}^2+(k^2-6k-7)\hat{\gamma}+12-2k}{(2\hat{\gamma}+1-\hat{\gamma}k)(3\hat{\gamma}-1)}, \quad (7.74)$$

$$v_2 = \frac{(\hat{\gamma}-1)(5-k)}{(2-k)\hat{\gamma}+1}, \quad (7.75)$$

$$v_3 = \frac{3-\hat{\gamma}k}{2\hat{\gamma}+1-\hat{\gamma}k}, \quad (7.76)$$

$$v_4 = -v_1\frac{15-(2+3\hat{\gamma})k}{(6-3\hat{\gamma}-k)(5-k)}, \quad (7.77)$$

$$v_5 = -\frac{6-k-\hat{\gamma}k}{6-3\hat{\gamma}-k}, \quad (7.78)$$

$$v_6 = \frac{2k}{5-k}. \quad (7.79)$$

Together with Equation 7.62, this constitutes a full solution in terms of V to the self-similar blast wave problem. We plot the solution for $\hat{\gamma} = 7/5$ and $k = 0$ in Figure 7.7. In this case, $C_{ST} \approx 1.033$.

7.6 FIRST-ORDER FERMI ACCELERATION ACROSS STRONG SHOCKS

An appealing topic to illustrate the use of shock-jump conditions in astrophysical problems is that of diffusive shock-acceleration. The main idea is that by bouncing back-and-forth across a shock front a test particle is actually able to gain momentum and energy at a steady rate. As we will see, it is the sudden headwind experienced by the particle after crossing the shock in either direction that makes this possible. Without this abrupt change in bulk velocity of the environment net acceleration will not take place, so particles in a homogeneous environment will not begin to acquire a large kinetic energy in order to e.g. take off eventually as cosmic rays or radiate at frequencies detectable up to gamma rays. The process is called *first-order Fermi acceleration*. A first iteration of the general principle was proposed in 1949 by Enrico Fermi[8]. We shall see below that the particle momentum increase per crossing cycle is a linear ('first order') fraction of its current momentum.

Since we have not yet discussed special relativistic kinematics in detail, we will anchor our discussion of first-order Fermi acceleration in Newtonian dynamics, postponing a relativistic argument[9] to Section 8.7.

---

[8]The paper [11] discusses particle acceleration by charged particles bouncing against magnetized clouds that can either have a net motion towards or away from the local environment from which the particle originates prior to a given collision. When shock waves are involved, the process is more efficient.

[9]Alternatively, one can take a hybrid approach. For example, the argument presented by [30] mixes classical mechanics with the light speed limit in a one-dimensional model.

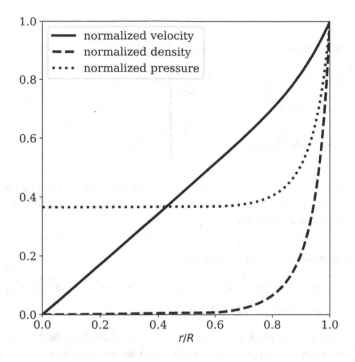

Figure 7.7 The full self-similar Sedov-Taylor profile, for $k = 0$ and $\hat{\gamma} = 7/5$. Due to the choice of $\hat{\gamma}$, this resembles the profile of a strong explosion in the earth's atmosphere.

We set the stage with a strong shock with shock velocity U moving into a medium at rest. Assume a polytropic gas with index $\hat{\gamma} = 5/3$. According to Equations 7.19 and 7.20, we will have a fluid velocity of magnitude $v = 3U/4$ when expressed in the rest frame of the upstream medium. Now consider a test particle of mass m with initial velocity magnitude u_0 in the upstream medium. By 'test particle', we refer to an individual particle whose evolution is shaped by its environment, while we can neglect its impact on the environment. Let us follow step-by-step what would happen to the particle momentum in case it repeatedly crosses the shock (see Figure 7.8 for an illustration).

- The test particle starts its cycle in the upstream fluid and has an initial momentum $\mathbf{p}_0$.
- The particle crosses the shock front at an angle θ with momentum components $p_0^x = -\cos(\theta)mu_0$ and $p_0^y = \sin(\theta)mu_0$.
- Arriving in the downstream medium, the particle has a horizontal momentum $\bar{p}^x = -\cos(\theta)mu_0 - mv$ in the frame of the downstream fluid. The extra term reflects the fact that the downstream fluid sees the upstream fluid approach it with velocity v. The vertical momentum does not appear different between the two frames, $\bar{p}^y = \sin(\theta)mu_0$. The magnitude of the momentum

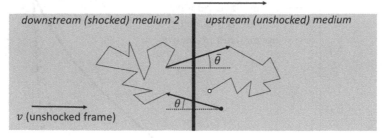

Figure 7.8 An illustration of first-order Fermi acceleration across a strong shock. In a full cycle, a test particle crosses the shock front twice and randomizes its direction through a series of elastic collisions at either side.

in the downstream frame follows from

$$\bar{\mathbf{p}}^2 = \cos^2(\theta)m^2 u_0^2 + 2\cos(\theta)mu_0 mv + m^2 v^2 + \sin^2(\theta)m^2 u_0^2,$$
$$\bar{\mathbf{p}}^2 = m^2 u_0^2 + 2\cos(\theta)mu_0 mv + m^2 v^2. \tag{7.80}$$

We assume the particle's velocity to greatly exceed the bulk motion of the fluid, such that $u_0 \gg v$. The momentum above can therefore be approximated as

$$\bar{\mathbf{p}}^2 \approx m^2 u_0^2 + 2\cos(\theta)mu_0 mv. \tag{7.81}$$

Likewise, the actual magnitude of the momentum can be approximated as

$$|\bar{\mathbf{p}}| = \sqrt{\bar{\mathbf{p}}^2} \approx mu_0 \left(1 + \cos(\theta)\frac{v}{u_0}\right), \tag{7.82}$$

using $\sqrt{1+x} \approx 1 + x/2$.

- Taking the above as representative of a process that plays out many times per particles and for many particles, we average over the angle of approach θ in a manner that also weights the *rate* at which a given angle occurs. Crossings with $\theta = \pi/2$ will not take place and should not be weighted, whereas crossings with $\theta = 0$ occur the most often. In other words, we are weighing by horizontal velocity component, which we can do with a weight factor $\cos(\theta)$. Further, we have weight factor $\sin(\theta)$ if all direction solid angles are equally likely. The normalized probability distribution P for averaging is therefore given by

$$P(\theta)\,d\theta = 2\cos(\theta)\sin(\theta)\,d\theta. \tag{7.83}$$

Replacing the magnitude of momentum with its averaged version, we get:

$$|\bar{\mathbf{p}}| \rightarrow mu_0 \left(1 + \int_{\theta=0}^{\frac{\pi}{2}} \cos(\theta)P(\theta)\,d\theta \frac{v}{u_0}\right) = mu_0 \left(1 + \frac{2}{3}\frac{v}{u_0}\right) \tag{7.84}$$

- The particle now undergoes a series of elastic collisions while in the downstream medium. These will not alter the magnitude of its momentum in this frame, but will randomize its direction.
- After some time bouncing around, the particle recrosses the shock front, with an angle $\bar{\theta}$ and momentum components

$$\bar{p}^x = mu_0 \left(1 + \tfrac{2}{3}\tfrac{v}{u_0}\right) \cos(\bar{\theta}), \tag{7.85}$$

$$\bar{p}^y = mu_0 \left(1 + \tfrac{2}{3}\tfrac{v}{u_0}\right) \sin(\bar{\theta}), \tag{7.86}$$

both still expressed in the frame of the downstream medium.
- In the frame of the upstream medium, the horizontal momentum is given by $p^x = \bar{p}^x + mv$, again accounting for the velocity difference between the frames. It is key that, due to the particle direction having been turned around by elastic collisions, once again the mv term has the same sign as the term it is being added to.
- Following the exact same procedure of elastic collisions and averaging as we did for the particle in the downstream medium, we eventually end up with a particle in the upstream medium with a momentum magnitude given by

$$|\mathbf{p}| \to mu_0 \left(1 + \frac{2}{3}\frac{v}{u_0} + \frac{2}{3}\frac{v}{u_0}\right) = mu_0 \left(1 + \frac{4}{3}\frac{v}{u_0}\right) \tag{7.87}$$

As previously, we dropped any term involving v/u_0 beyond first order terms in the process.

The last step shows why this acceleration process is called *first-order* acceleration. At the end of a cycle, the fractional gain in absolute momentum is given by,

$$\frac{\Delta|\mathbf{p}|}{|\mathbf{p}_0|} = \frac{|\mathbf{p}| - |\mathbf{p}_0|}{|\mathbf{p}_0|} = \frac{4}{3}\frac{v}{u}. \tag{7.88}$$

showing a linear growth with a first-order dependency on v/u.

As long as the particle remains near the shock front it will repeatedly cross it whenever its randomized direction of motion takes it sufficiently far in the right direction. The upstream medium is in the process of being overtaken by the shock front, so to the extent that the particle follows the bulk flow velocity its proximity to the shock is guaranteed. The downstream medium, however, moves away from the shock front and may advect the test particle with it. To compute the odds of the particle remaining near the shock in the downstream medium, we need to compare the advection velocity to the randomized motion of the particle. The flux of N particles with velocity u leaving the zone in a given direction is given by

$$F = \frac{1}{4\pi} \int_{\phi=0}^{2\pi} \int_{\theta=0}^{\frac{\pi}{2}} Nu\cos\theta \sin\theta \, d\theta \, d\phi = \frac{Nu}{4}, \tag{7.89}$$

when averaging over angles. At the same time there is a steady attrition of particles carried off by advection, given by relative advection flux $N(U - v) = NU/4$. The probability of remaining after one cycle follows from a comparison between the particle fluxes with and without accounting for the loss through advection:

$$\frac{N}{N_0} = P = \frac{\frac{Nu}{4} - \frac{NU}{4}}{\frac{Nu}{4}} = 1 - \frac{U}{u}. \tag{7.90}$$

Per cycle, the ratio between the change in particle number and particle momentum is given by

$$\frac{\ln \frac{N}{N_0}}{\ln \frac{|\mathbf{p}|}{|\mathbf{p}_0|}} = \frac{\ln\left(1 - \frac{U}{u}\right)}{\ln\left(1 + \frac{4v}{3u}\right)} \approx \frac{-\frac{U}{u}}{\frac{4v}{3u}}, \tag{7.91}$$

where we used that the fractional number loss and energy gain are very small. This ratio between the two is independent of u,

$$\frac{\ln \frac{N}{N_0}}{\ln \frac{|\mathbf{p}|}{|\mathbf{p}_0|}} \approx \frac{-U}{\frac{4v}{3}}, \tag{7.92}$$

and applies equally to any cycle in the process. Plugging in that $v = 3U/4$ and rewriting, we have

$$\frac{N}{N_0} = \left(\frac{|\mathbf{p}|}{|\mathbf{p}_0|}\right)^{-1}. \tag{7.93}$$

If we differentiate the expression above on both sides to explore this relationship across a range of absolute particle momenta (and across corresponding ranges of successful cycles), our final result is

$$\frac{dN}{N_0} = -1 \times \left(\frac{|\mathbf{p}|}{|\mathbf{p}_0|}\right)^{-2} \frac{d|\mathbf{p}|}{|\mathbf{p}_0|} \Rightarrow dN \propto |\mathbf{p}|^{-2} d|\mathbf{p}|. \tag{7.94}$$

Through a fairly generic argument, we have therefore established that shock acceleration can give rise to a *power law* distribution of particle momenta. The argument is generic in the sense that we at no point specified the exact nature of the collision process, other than that it was assumed to be elastic. Verifying the model in detail in a realistic scenario is a very challenging problem in plasma physics, attempted for example through large-scale computer simulations that include many individual interacting charged particles and their electromagnetic fields.

The power-law distribution is one of the most important examples of a nonthermal distribution of particles within a fluid, one that has been observationally inferred for a wide range of (astro-)physical phenomena that involve non-equilibrium physics. Deviations from the predicted -2 slope may result, depending on the details of the actual acceleration process being considered, when these details alter our core assumptions (e.g. about the isotropy of particle velocities, or the elasticity of collisions).

PROBLEM 7.6
A weak shock in a planetary nebula

For this question, you will need Boltzmann's constant $k_B = 1.380658 \times 10^{-16}$ erg K^{-1}, and the hydrogen mass $m_H = 1.6733 \times 10^{-24}$ g.

(i) Starting from Euler's equations in the absence of heating terms,

$$\frac{\partial \rho}{\partial t} + \nabla \cdot (\rho \mathbf{v}) = 0,$$

$$\frac{\partial \rho \mathbf{v}}{\partial t} + \nabla \cdot (\rho \mathbf{v}\mathbf{v} + p\mathbb{1}) = \mathbf{f},$$

$$\frac{\partial \mathscr{E}}{\partial t} + \nabla \cdot ([\mathscr{E} + p]\mathbf{v}) = \mathbf{v} \cdot \mathbf{f},$$

derive three distinct shock-jump conditions (focus on flow perpendicular to the shock front). All symbols have their usual meaning. Explain / justify your steps.

(ii) For a gas that could be described using adiabatic exponent $\hat{\gamma}_{ad}$, the following shock-jump condition was derived from the basic jump-conditions:

$$\frac{p_2}{p_1} = \frac{(\hat{\gamma}_{ad} - 1)p_1 + (\hat{\gamma}_{ad} + 1)p_2}{(\hat{\gamma}_{ad} + 1)p_1 + (\hat{\gamma}_{ad} - 1)p_2}.$$

Introducing the notation $\bar{p} \equiv (p_2 - p_1)/p_1, \bar{\rho} \equiv (\rho_2 - \rho_1)/\rho_1$, demonstrate that this implies

$$\bar{\rho} = \frac{\bar{p}}{\hat{\gamma}_{ad}},$$

for the case of a *weak* shock, where $\bar{p} \ll 1$.

(iii) The result of part (ii) appears to suggest that equal pressures imply equal densities and therefore no flow discontinuity at all. However...

    (a) explain that density jumps can nevertheless exist in the absence of a pressure jump,

    (b) name this type of flow discontinuity,

    (c) and provide a physical interpretation.

(iv) Two further jump-conditions are (considering the direction perpendicular to the shock front):

$$v_{2,s}^2 = \rho_2^{-2} \frac{p_2 - p_1}{\rho_1^{-1} - \rho_2^{-1}}; \qquad v_{1,s}^2 = \rho_1^{-2} \frac{p_2 - p_1}{\rho_1^{-1} - \rho_2^{-1}}.$$

Consider a planetary nebula at rest, with temperature $T_1 = 10^4$ K and density 10^3 m_H cm^{-3}. A *weak* shock runs through the nebula with shock velocity $U = 1.2 \times 10^6$ cm s^{-1}. Using the shock-jump conditions that have been provided in this question and/or those you have derived above, compute the post-shock state (ρ_2, p_2, v_2). Consider the nebula to be a single-species polytropic gas with index $\hat{\gamma}_{ad} = 5/3$.

(v) What would happen to the shock in the planetary nebula if it had a velocity $U = 10^6$ cm s^{-1} instead?

8 Fluid Dynamics in Special Relativity

Relativistic fluid dynamics is becoming increasingly relevant in astrophysics, where a growing range of phenomena require the acknowledgement of at minimum *special* relativity if they are to be modelled successfully. Astrophysical flows close to black holes, where highly energetic jets can be launched during the process of accretion (which we will discuss in more depth in Chapter 11) will even require *general* relativity to account for the distorted spacetime environment. Examples of relativistic fluid dynamics can be found across a range of scales, both in mass and velocity. On the largest scales, active galactic nuclei (AGN) containing central black holes of 10^7 solar masses and more are able to sustain jets with Lorentz factors of about 10–30 (with the Lorentz factor γ defined in the usual manner in terms of flow velocity v and light speed c, i.e. $\gamma \equiv \left(1 - v^2/c^2\right)^{-1/2}$; we will recap the key points of special relativity and 'officially' introduce the Lorentz factor shortly). When pointed directly towards the observer, the jets associated with AGN are known as blazars. More sudden releases of energy due to interaction with supermassive black holes are triggered by the tidal disruption of individual stars that venture too deep into the black hole's gravitational field. Since the discovery of tidal disruption event Swift J1644+57 in 2011, we know that in this case, too, a relativistic outflow can sometimes be launched.

On the scale of stellar mass black holes, jets produced by a sub-class of X-ray binaries labelled microquasars have Lorentz factors of a few, with $\gamma \sim 5$ already corresponding to a flow velocity of 98% of light speed. Gamma-ray bursts, produced by the collapse of highly massive stars or the merging of two neutron stars can reach Lorentz factors of well over a hundred. Pulsar wind nebulae, dilute winds of plasma generated by spinning pulsars can reach even far higher Lorentz factors due to their low mass content. Supernova explosions have been observed in cases to lead to (trans-)relativistic velocities in their ejecta, which begins to bridge the divide between supernovae from stellar collapse and the more rare cases of gamma-ray bursts from massive stars. Since the first detailed observations in 2017 of kilonova emission from a merging neutron star pair (along with a direct detection of gravitational waves) we can add kilonovae to the list of outflows capable of reaching near-light speed velocity.

8.1 CORE CONCEPTS IN SPECIAL RELATIVITY

The theory of special relativity can be derived from two postulates[1]:

---

[1] Along with some assumptions about the isotropy of empty space and a leap of faith when applying special relativity against the background of a non-empty environment.

1. The laws of physics should be independent of the inertial (i.e. 'non-accelerating') reference frame they are expressed in.
2. The speed of light in vacuum c is invariant under frame transformations between inertial frames.

We begin by quickly recapping some of the key implications of special relativity[2]. As a direct consequence of the second postulate, the classical *Galilean* velocity addition rule for non-relativistic velocities needs to be amended to ensure the speed limit is upheld for any frame transformation. Consider the velocity of object 1 relative to object 3, when expressed in terms of the velocity of object 2 (e.g. person 1 on train 2, relative to ground 3):

$$v_{1,3} = v_{1,2} + v_{2,3} \text{ (Galilean)} \rightarrow v_{1,3} = \frac{v_{1,2} + v_{2,3}}{1 + v_{1,2}v_{2,3}/c^2} \text{ (Relativistic)}. \qquad (8.1)$$

Here the second subscript indicates in whose frame of reference the quantity is expressed.

PROBLEM 8.1
Quick check velocity addition

A simple warm-up exercise: check that if frame 2 above moves with light speed relative to an observer in frame 3, $v_{1,3}$ will also be light speed.

The three big consequences of the constant speed of light and the different addition rule for velocities are *relativity of simultaneity*, *time dilation* and *Lorentz contraction*. Let us resort to the usual examples with trains and light bulbs to illustrate the three. With the finite speed of information, it is only possible to unambiguously compare two *events* if they happen to coincide in spacetime. If two events are at a spatial or temporal distance, then simultaneity in one reference frame will not imply simultaneity in all reference frames.

This is illustrated in Figure 8.1. If a light bulb hanging in the middle of a freight car moving in the positive x-direction relative to a stationary observer emits light both to the left and to the right, an observer on the train will see the light strike both ends of the car at the same time, covering the same distance $\Delta x'/2$ in both directions. On the other hand, an observer on the ground will see the light hit the left side first, then the right side.

The second implication of constant light speed is that of time dilation. This is illustrated in Figure 8.2. We now consider a light ray emitted downward, covering a distance h to the floor. In the frame of the freight car, the light will take a time $\Delta t' = h/c$ to cross this distance. On the other hand, in the frame of a stationary observer, the light ray will have travelled at an angle, albeit still in a straight line, with velocity c. It will have taken a time $\Delta t = \sqrt{h^2 + v^2(\Delta t)^2}/c$ instead, which we can solve for Δt to obtain

$$\Delta t = \frac{\Delta t'}{\sqrt{1 - \beta^2}} \equiv \gamma \Delta t', \qquad (8.2)$$

---

[2]For this purpose, we will use the same instructive train-and-light-bulb examples as [16].

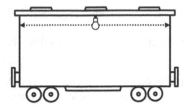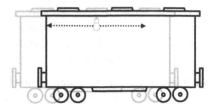

Figure 8.1 Loss of simultaneity in special relativity. On the left, the observer is comoving with the train and will conclude that light will hit both walls simultaneously when the light bulb is switched on. On the right, a stationary observer will witness the left wall getting hit first.

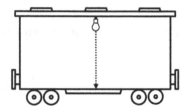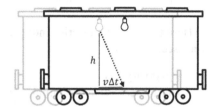

Figure 8.2 Time dilation in special relativity. In a frame comoving with the train (left) and in a stationary frame (right), the same ray of light will be seen to hit the same spot on the floor. But in the stationary frame the light will have to travel a longer distance.

introducing the Lorentz factor γ and normalized velocity $\beta \equiv v/c$. In other words, a clock on the train will be perceived to move slower by a stationary observer. Equation 8.2 represents a special case of a *Lorentz transformation* (to be defined below), for which $\Delta x' = 0$ (h is directed perpendicular to x').

The third implication, Lorentz contraction, is illustrated in Figure 8.3. Light emitted from the rear end of the car and bouncing off a mirror on the other end takes $\Delta t' = 2\Delta x'/c$ to travel back and forth according to an observer moving with the train. An observer on the ground measures the two events, hitting front end and arriving back at rear end, differently, finding respectively a longer and shorter travel time. Separating the forward and return trips of the light, the stationary observer will find $c\Delta t_1 = \Delta x + v\Delta t_1$ and $c\Delta t_2 = \Delta x - v\Delta t_2$. These can both separately be solved for their respective Δt variables, leading to $c\Delta t_1 = \Delta x/(1 - v/c)$ and $c\Delta t_2 = \Delta x/(1 + v/c)$. These time intervals add up to the total time interval back and forth for a stationary observer:

$$c\Delta t = \frac{\Delta x}{1 - v/c} + \frac{\Delta x}{1 + v/c} = \Delta x \left(\frac{1 + v/c + 1 - v/c}{1 - (v/c)^2} \right) = 2\gamma^2 \Delta x. \tag{8.3}$$

On the other hand, we have time dilation, which we can apply above, so

$$\Delta t = \gamma \Delta t' \Rightarrow c\gamma \Delta t' = 2\gamma^2 \Delta x. \tag{8.4}$$

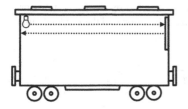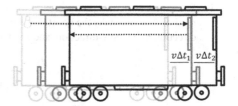

Figure 8.3 Lorentz contraction in special relativity. Light bouncing off a mirror will travel twice the same distance according to a frame comoving with the train (left). In a stationary frame not moving with the train (right), light will first travel a longer distance, then a shorter distance.

Substituting the full distance $2\Delta x'$ for $c\Delta t'$ then yields

$$\Delta x = \Delta x'/\gamma, \tag{8.5}$$

i.e. moving objects are perceived to be shorter. This too is a consequence of Lorentz transformations (albeit not simply a case of $\Delta t'$ being zero; instead $\Delta t'$ was eliminated in favour of $\Delta x'$ using the constancy of light speed).

More generally, we can define the *Lorentz transformation* (or Lorentz *boost*) from a frame moving with speed v to a stationary observer, according to

$$\Delta t = \gamma\left(\Delta t' + \frac{v}{c^2}\Delta x'\right),$$
$$\Delta x = \gamma\left(\Delta x' + v\Delta t'\right),$$
$$\Delta y = \Delta y',$$
$$\Delta z = \Delta z'. \tag{8.6}$$

Here, we chose to express the transformation using coordinate systems that had their x-directions aligned with each other, as well as with the direction of the Lorentz boost **v**. The transformation rules for the inverse transformation follow from swapping primed and unprimed quantities and adding a minus sign to the velocity:

$$\Delta t' = \gamma\left(\Delta t - \frac{v}{c^2}\Delta x\right),$$
$$\Delta x' = \gamma\left(\Delta x - v\Delta t\right),$$
$$\Delta y' = \Delta y,$$
$$\Delta z' = \Delta z. \tag{8.7}$$

PROBLEM 8.2
From Lorentz boost to velocity addition rules

i) Show that the Lorentz transformation rules from Equation 8.6 imply the relativistic velocity addition rule from Equation 8.1 for velocities along the direction of the boost.

ii) Show that velocities perpendicular to the boost are affected by the frame transformation. You may take the boost direction to be the x-direction and the perpendicular direction to be the y-direction, and show that

$$\frac{dy}{dt} = \frac{d\bar{y}/d\bar{t}}{\gamma\left(1 + \frac{v}{c^2}\frac{d\bar{x}}{d\bar{t}}\right)}.$$

The Lorentz transformation rules and the three consequences of the constant light speed postulate are all manifestations of having to account for the existence of an *invariant* interval ds. Light is postulated to obey

$$c = \frac{dr}{dt} = \frac{dr'}{dt'}, \tag{8.8}$$

with dr the differential change in position in an arbitrary direction. The first equality implies

$$cdt = dr \Rightarrow c^2(dt)^2 = (dx)^2 + (dy)^2 + (dz)^2. \tag{8.9}$$

We can define a spacetime interval ds, which for light obeys

$$(ds)^2 \equiv -c^2(dt)^2 + (dx)^2 + (dy)^2 + (dz)^2 = 0, \qquad \text{(light)}. \tag{8.10}$$

Since $c = dr'/dt'$ too, we also have

$$(ds')^2 \equiv -c^2(dt')^2 + (dx')^2 + (dy')^2 + (dz')^2 = 0, \qquad \text{(light)}. \tag{8.11}$$

Light is an extreme case, for which $ds = 0$. As far as a photon is concerned, there is zero distance between any two points in space, if using ds as a generalized distance measure. But the interval is invariant for arbitrary velocities, i.e.

$$(ds)^2 = (ds')^2, \tag{8.12}$$

even when not equal to zero, which implies

$$-(cdt)^2 + (dx)^2 + (dy)^2 + (dz)^2 = -(cdt')^2 + (dx')^2 + (dy')^2 + (dz')^2. \tag{8.13}$$

PROBLEM 8.3
A Lorentz-invariant distance measure

Demonstrate the validity of the second equality of Equation 8.12 using Lorentz transforms.

8.1.1   FOUR-DIMENSIONAL SPACETIME

The Lorentz transform rules show an inevitable mixing of space and time coordinates when changing reference frames, and through this introduce us to four-dimensional *spacetime*. To extend our understanding of coordinate transformations and three-dimensional geometry to spacetime it is helpful to continue our focus on ds, which we argued above was a measure of the distance interval experienced when undergoing a displacement. Here, the phrasing in terms of a subjective experience is intentional and allows us to e.g. fold the $ds = 0$ limit for light into our interpretation of ds.

In Cartesian coordinates, consider an infinitesimal displacement **w** cast in terms of dx, dy, dz (not necessarily of equal length, maybe we just travel along the x-axis and have $dy = dz = 0$). The interval ds is linked to the inner product of the displacement vector with itself according to

$$[dx \quad dy \quad dx] \begin{bmatrix} dx \\ dy \\ dz \end{bmatrix} = (dx)^2 + (dy)^2 + (dz)^2 = (ds)^2, \qquad (8.14)$$

in matrix notation. Recall from Section 1.1.1 that matrix notation implicitly assumes an orthogonal and normalized basis, and that the above expression corresponds to

$$(ds)^2 = \mathbf{w} \cdot \mathbf{w} = (dx\hat{\mathbf{e}}_x + dy\hat{\mathbf{e}}_y + dz\hat{\mathbf{e}}_z)^2. \qquad (8.15)$$

Expressed in the *natural basis* for spherical coordinates, the displacement would translate as

$$dx\hat{\mathbf{e}}_x + dy\hat{\mathbf{e}}_y + dz\hat{\mathbf{e}}_z = dr\mathbf{e}_r + d\theta\mathbf{e}_\theta + d\phi\mathbf{e}_\phi. \qquad (8.16)$$

This equality can be shown to hold by plugging in expressions for x, y, z in spherical coordinates and taking a displacement, $dx = d(r\sin\theta\cos\phi)$ etc. Since the natural basis for spherical coordinates is not normalized, a naive application of matrix notation would lead to an erroneous outcome. If instead the natural basis vectors are normalized, and $dr\hat{\mathbf{e}}_r + rd\theta\hat{\mathbf{e}}_\theta + r\sin\theta d\phi\hat{\mathbf{e}}_\phi$ is used, the same inner product can be expressed as

$$[dr \quad d\theta \quad d\phi] \begin{bmatrix} 1 & 0 & 0 \\ 0 & r^2 & 0 \\ 0 & 0 & r^2\sin^2\theta \end{bmatrix} \begin{bmatrix} dr \\ d\theta \\ d\phi \end{bmatrix}. \qquad (8.17)$$

In Section 1.3 we identified this matrix in the middle as the matrix notation manifestation of the *metric tensor* $g_{ij} \equiv \mathbf{e}_i \cdot \mathbf{e}_j$ for spherical coordinates in the natural basis. In index notation for the contravariant components w^i of **w**, we have

$$(ds)^2 = w^i g_{ij} w^j, \qquad (8.18)$$

which also introduces a means to define covariant components $w_i \equiv g_{ij}w^j$ and a shorter notation $(ds)^2 = w^i w_i$. In all these expressions, the Einstein summation rule is applicable.

It is this mechanism that generalizes straightforwardly to spacetime, and we can use a four-dimensional metric tensor to account for the minus sign:

$$(ds)^2 = [cdt \quad dx \quad dy \quad dz] \begin{bmatrix} -1 & 0 & 0 & 0 \\ 0 & 1 & 0 & 0 \\ 0 & 0 & 1 & 0 \\ 0 & 0 & 0 & 1 \end{bmatrix} \begin{bmatrix} cdt \\ dx \\ dy \\ dz \end{bmatrix} = W^\alpha g_{\alpha\beta} W^\beta. \qquad (8.19)$$

Note how the time dimension gets an extra factor of c to ensure all entries have the same units. In the case of four-dimensional spacetime, we use greek letters to denote indices that run from zero to three instead of 1, 2, 3 like roman letters indicating three-dimensional space. We have a set of contravariant components $[W^\alpha] = [cdt, dx, dy, dz]$ and a set of covariant components $[W_\alpha] = [-cdt, dx, dy, dz]$, via $W_\alpha = g_{\alpha\beta} W^\beta$. The metric $[g_{\alpha\beta}]$ corresponds to Cartesian space coordinates in a flat spacetime (an alternative convention for the same metric is to put the minus sign in the spatial entries instead). Again, the effect of the metric is to 'raise' or 'lower' an index of a vector or array. Since this amounts to putting a minus sign in front of the zeroth component, doing it twice cancels out its effect. We also have $g^{\alpha\beta} = g_{\alpha\beta}$, while $g^\alpha{}_\beta = \delta^\alpha{}_\beta$, i.e. a unit operator that has no effect.

The Lorentz transform can be written in terms of four-vectors as well, as demonstrated here for a transformation in the x-direction:

$$[X^\alpha] = [\Lambda^\alpha{}_\beta][(X')^\beta] \Rightarrow \begin{bmatrix} \Delta ct \\ \Delta x \\ \Delta y \\ \Delta z \end{bmatrix} = \begin{bmatrix} \gamma & \beta\gamma & 0 & 0 \\ \beta\gamma & \gamma & 0 & 0 \\ 0 & 0 & 1 & 0 \\ 0 & 0 & 0 & 1 \end{bmatrix} \begin{bmatrix} \Delta ct' \\ \Delta x' \\ \Delta y' \\ \Delta z' \end{bmatrix}. \tag{8.20}$$

Using the metric tensor, we can work out what the different variations of the Lorentz transformation matrix look like. Dropping the y and z indices for a concise presentation, this would work out to:

$$\begin{aligned} \left[\Lambda_{\alpha\beta}\right] &= \left[g_{\alpha\gamma}\Lambda^\gamma{}_\beta\right] = \begin{bmatrix} -1 & 0 \\ 0 & 1 \end{bmatrix}\begin{bmatrix} \gamma & \gamma\beta \\ \gamma\beta & \gamma \end{bmatrix} \\ &= \begin{bmatrix} -\gamma & -\gamma\beta \\ \gamma\beta & \gamma \end{bmatrix}, \end{aligned} \tag{8.21}$$

$$\begin{aligned} \left[\Lambda_\alpha{}^\beta\right] &= \left[g_{\alpha\gamma}\Lambda^\gamma{}_\delta g^{\delta\beta}\right] = \begin{bmatrix} -1 & 0 \\ 0 & 1 \end{bmatrix}\begin{bmatrix} \gamma & \gamma\beta \\ \gamma\beta & \gamma \end{bmatrix}\begin{bmatrix} -1 & 0 \\ 0 & 1 \end{bmatrix} \\ &= \begin{bmatrix} \gamma & -\gamma\beta \\ -\gamma\beta & \gamma \end{bmatrix}, \end{aligned} \tag{8.22}$$

$$\begin{aligned} \left[\Lambda^{\alpha\beta}\right] &= \left[g^{\alpha\gamma}\Lambda_\gamma{}^\beta\right] = \begin{bmatrix} -1 & 0 \\ 0 & 1 \end{bmatrix}\begin{bmatrix} \gamma & -\gamma\beta \\ -\gamma\beta & \gamma \end{bmatrix} \\ &= \begin{bmatrix} -\gamma & \gamma\beta \\ -\gamma\beta & \gamma \end{bmatrix}. \end{aligned} \tag{8.23}$$

These different versions can be used to confirm that Lorentz transformations indeed keep the invariant interval (or any other contraction of four-vectors) unchanged:

$$(X')_\alpha (X')^\alpha = \Lambda_\alpha{}^\beta X_\beta \Lambda^\alpha{}_\gamma X^\gamma = X_\alpha X^\alpha. \tag{8.24}$$

Any combination of four-vectors and matrices (i.e. the two types of tensors we typically encounter in this book) that leaves no indices remaining is a *Lorentz invariant* or Lorentz scalar. An analogue of the Lorentz transformation operator in three dimensions is the rotation operator, which leaves the inner product between two three-dimensional vectors invariant (three-dimensional rotations are a subset of Lorentz-transformations which do not affect the time dimension). Note that, for a normalized set of spherical basis vectors, a transformation from a Cartesian coordinate system to spherical coordinates (defined at a given position) is a rotation.

PROBLEM 8.4
Lorentz invariants from contraction

Confirm that Equation 8.24 holds. You may omit the y and z dimensions.

8.1.2 POINT PARTICLES

Let X^α represent the coordinates of a point particle in spacetime. The *proper* time interval $d\tau$, is always defined in a frame comoving with the particle. Its translation to a time interval dt measured by a stationary observer, is simply $dt = \gamma d\tau$ (i.e. time dilation). It can be useful to define an expression for velocity explicitly in terms of proper time no matter what frame it is observed in. This four-velocity then obeys

$$[V^\alpha] \equiv \left[\frac{dX^\alpha}{d\tau}\right] = \left[\begin{array}{c} cdt/d\tau \\ dx/d\tau \end{array}\right] = \left[\begin{array}{c} c\gamma \\ \gamma v \end{array}\right]. \tag{8.25}$$

The old definition of velocity $v \equiv dx/dt$ is still encapsulated within this generalization, but because we use $d\tau$ we do not have to worry about how dt in dX^α/dt transforms when changing coordinate frames. Instead of having to use Equation 8.1, we can transform V^α between frames through application of a Lorentz transformation like any other four-vector.

At this point you might wonder, which is the 'real' velocity? Is this v, or is this the spatial part $[V^i]$ of the four-velocity? Arguing from the four-momentum to be defined below, one could claim it is $[V^i]$, but the fact is that nobody has a stopwatch capable of measuring proper time directly if not comoving with the object experiencing the time interval, and the more accessible observable therefore is v. Ultimately, the question is subjective, and both definitions are equally valid approaches to describe the same physical behaviour.

A four-momentum vector is now readily defined as $P^\alpha \equiv mV^\alpha = [\gamma mc, \gamma mv]$. While the spatial indices show the classical momentum mv, multiplied by γ, the temporal index contains a measure of energy (modulo a factor c). To confirm the latter, it can be shown that cP^0 indeed reduces to Newtonian energy in the limit of small γ, albeit with a rest mass term included:

$$cP^0 = \gamma mc^2 = \frac{mc^2}{\sqrt{1-(v/c)^2}} \approx m^2 c^2 \left\{1 + \frac{1}{2}\left(\frac{v}{c}\right)^2\right\} = mc^2 + \frac{1}{2}mv^2. \tag{8.26}$$

In a primed frame co-moving with the particle, velocity is zero (and $\gamma = 1$), so the contraction of the momentum four-vector with itself gives

$$(P')_\alpha (P')^\alpha = -m^2 c^2. \tag{8.27}$$

But this contraction has no remaining indices and must therefore describe a Lorentz-invariant quantity. We have

$$P_\alpha P^\alpha = (P')_\alpha (P')^\alpha$$
$$\Rightarrow \quad -\gamma^2 m^2 c^2 + \gamma^2 m^2 v^2 = -m^2 c^2$$
$$\Rightarrow \quad \frac{E^2}{c^2} - P^i P_i = mc^2$$
$$\Rightarrow \quad E = \sqrt{P_i P^i c^2 + m^2 c^4} \tag{8.28}$$

which provides useful relationship between energy and momentum that is applied often in particle physics analysis. The point is here that energy and momentum are intertwined in spacetime in a manner similar to time and space. This tells us that when constructing relativistic analogues of the fluid dynamical equations, we can no longer separately write equations for energy and momentum conservation like we did in the Newtonian case.

The evolution of a collection of particles can be modelled by following their collective behaviour in *phase space*. In Section 2.4 we considered a phase space in terms of particle positions and velocities. It would have been equally possible to consider phase space in terms of particle positions and momenta instead (the difference being a fixed mass multiplication factor m per particle that can be accounted for straightforwardly). Whenever discussing relativistic phase space, we will consider particle momenta rather than velocities, which allows us to include the zero mass case where all particles have velocity c but can still be distinguished through their momentum (e.g. photons, see Section 3.4.1).

It can be demonstrated in various ways that relativistic phase space is Lorentz invariant, thus generalizing Liouville's theorem mentioned in Section 2.4. For this purpose, let us limit ourselves to Lorentz transformations along the x-axis, which is sufficient to demonstrate the principle. Consider a phase space element with a spread of momenta dP^x, dP^y, dP^z around $P^x = mV^x$, $P^y = mV^y$, $P^z = mV^z$ in a given reference frame (frame one, unprimed). We define a second reference frame (frame two, primed) which itself moves with velocity $[V^i]$ relative to the first. In this second frame, the phase space element momenta are spread around zero. If the velocity (in units of c) of frame two relative to frame one is given by β, with a corresponding Lorentz factor γ, we obtain for the inferred phase space element width $dx = dx'/\gamma$ on account of Lorentz contraction (see Equations 8.6 & 8.7):

$$dx = \gamma dx' + \beta \gamma c dt'$$
$$\Rightarrow \quad dx = \gamma dx' + \beta \gamma (\gamma dt - \beta)$$
$$\Rightarrow \quad dx(1 + \beta^2 \gamma^2) = \gamma dx'$$
$$\Rightarrow \quad dx = dx'/\gamma. \tag{8.29}$$

Here we used $dt = 0$, i.e. the observer in reference frame one counts the particles in this phase space element at a single instant in time.

For element dP^x we have $dP^x = \gamma d \left(P'\right)^x + \beta \gamma d \left(P'\right)^0$. However, for frame two, Equation 8.28 implies that

$$d \left(P'\right)^0 = \frac{\left(P^x\right)'}{\left(P'\right)^0} d \left(P^x\right)' \approx 0, \tag{8.30}$$

since in this frame nearly comoving with the particles $\left(P^x\right)' \ll \left(P'\right)^0$ because the rest mass contribution vastly exceeds the momentum term. Combining the transformations for momentum and space interval, we therefore conclude

$$dx dP^x = dx' d \left(P'\right)^x, \tag{8.31}$$

since the Lorentz factors γ cancel out. Because this also holds when going from frame two to a third reference frame we can conclude that we can transform between frame two and three via frame one without changing the phase space element. In fact, we might as well cut out the middleman and state that phase space is invariant between arbitrary frames one and three without reference to a nearly comoving frame two (which might not necessarily be easy to construct, e.g. in the case of a 'photon gas').

8.2 SPECIAL RELATIVISTIC FLUID DYNAMICAL EQUATIONS

8.2.1 THE CONTINUITY EQUATION

To derive a set of special relativistic fluid dynamics conservation laws, let us start by generalizing the continuity equation. The goal is to have equations that obey the first postulate, which states that *the laws of nature should be independent of reference frame*. The Newtonian fluid dynamical equations obey this principle under Galilean transformations, as was demonstrated back in Problem 2.8 from Chapter 2.

We find that we can generalize the continuity equation to special relativity very quickly, by merely making explicit that the density should be expressed in the same frame as the other variables occurring in the equation (the temporal and spatial derivatives). If we henceforth use ρ to refer *only* to a density measured in the frame locally comoving with the fluid parcel, i.e., a rest mass density, we will need to cast mass conservation in terms of $\gamma \rho$ instead. The factor γ is a consequence of Lorentz contraction, since the volume is contracted in one of its three spatial directions. The same amount of mass is therefore compressed into a smaller volume, increasing its density. The continuity equation then reads

$$\frac{\partial \left(\gamma \rho\right)}{\partial t} + \frac{\partial \left(\gamma \rho v^i\right)}{\partial x^i} = 0 \Rightarrow \frac{\partial \left(c \gamma \rho\right)}{\partial \left(ct\right)} + \frac{\partial \left(\gamma \rho v^i\right)}{\partial x^i} \Rightarrow \frac{\partial}{\partial X^\alpha} \left(\rho V^\alpha\right) = 0. \tag{8.32}$$

This last step demonstrates that indeed we did not need to search far in order to arrive at a Lorentz-invariant expression for conservation of mass. The expression also demonstrates the true elegance of the spacetime formalism, by showing how spatial flux terms accounting for local temporal changes are naturally embedded in our notion of spacetime.

We can also use the expected structure of scalar and vector conservation laws as a guide when proposing relativistic generalizations. For a scalar quantity such as density, we expect (in the absence of source terms)

$$\frac{\partial Z^\alpha}{\partial X^\alpha} = 0, \tag{8.33}$$

like we just saw above. Here Z^α represents a generic four-vector that includes a conserved scalar quantity Z^0 and its corresponding flux terms Z^i. This four-vector can be transformed using a Lorentz transformation, and we can demonstrate the concept using a single spatial dimension: going from a frame co-moving with the local fluid velocity, where Z^α should reduce to its non-relativistic counterpart, to a frame moving with arbitrary velocity $v = \beta c$, we have for Z^α that

$$[Z^\alpha] = \begin{bmatrix} \gamma & \beta\gamma \\ \beta\gamma & \gamma \end{bmatrix} \begin{bmatrix} \rho c \\ 0 \end{bmatrix} = \begin{bmatrix} \gamma\rho c \\ \gamma\rho v \end{bmatrix}. \tag{8.34}$$

When applied to Equation 8.33, this expression for Z^α yields the same result as we got when transforming density directly.

8.2.2 CONSERVATION OF ENERGY-MOMENTUM

Energy and momentum are combined in a single four-dimensional object, which makes it difficult to generalize the energy and momentum conservation laws separately. Furthermore, the transformation rules for quantities describing the intrinsic nature of the fluid such as p and ε are not straightforward. To come up with a proposed expression for relativistic conservation of energy and momentum density, we start by assuming that the resulting four-vector equation takes the following form

$$\frac{\partial}{\partial X^\alpha} T^{\alpha\beta}, \tag{8.35}$$

with β the remaining free index. For the sake of simplicity, we have assumed no source or sink terms (these can always be added later). The tensor $T^{\alpha\beta}$ is a generic tensor that we will soon identify as the energy-momentum tensor. Equation 8.35 expresses our expectation that the conservation law is identical across reference frames once the frame-dependent appearance of the energy-momentum density four-vector is accounted for. In the case of zero velocity the tensor components should be identical to their non-relativistic limits, both for conserved quantities and flux terms. Applying a Lorentz boost to $(T')^{\alpha\beta}$,

$$T^{\alpha\beta} = \Lambda^\alpha{}_\gamma \Lambda^\beta{}_\delta (T')^{\gamma\delta}, \tag{8.36}$$

will then suggest a formulation for the relativistic conservation laws. We demonstrate this for one spatial dimension, like we did for the scale density conservation law. In the rest-frame version $(T')^{\alpha\beta}$ we include rest mass energy in the energy component, set the momenta equal to zero (no velocity) and include the pressure term that is present in the momentum equation even at zero velocity. We take the same approach to p and ε as we did to ρ, that is, we will take them to *always* express rest frame quantities. We can switch from index notation to matrix notation as long as we properly order the matrices such that the correct row entries get summed with the correct column entries:

$$\left[T^{\alpha\beta}\right] = \begin{bmatrix} \gamma & \beta\gamma \\ \beta\gamma & \gamma \end{bmatrix} \begin{bmatrix} e+\rho c^2 & 0 \\ 0 & p \end{bmatrix} \begin{bmatrix} \gamma & \beta\gamma \\ \beta\gamma & \gamma \end{bmatrix}^T. \tag{8.37}$$

The transpose of the last matrix is redundant, given that $\Lambda^\beta{}_\delta$ happens to be symmetric, but is added for completeness. Working our way through the matrix products, we obtain

$$\begin{aligned}
\left[T^{\alpha\beta}\right] &= \begin{bmatrix} \gamma & \beta\gamma \\ \beta\gamma & \gamma \end{bmatrix} \begin{bmatrix} \gamma e+\gamma\rho c^2 & \beta\gamma e+\beta\gamma\rho c^2 \\ \beta\gamma p & \gamma p \end{bmatrix}, \\
&= \begin{bmatrix} \gamma^2 e+\gamma^2\rho c^2+\beta^2\gamma^2 p & \beta\gamma^2 e+\beta\gamma^2\rho c^2+\beta\gamma^2 p \\ \beta\gamma^2 e+\beta\gamma^2\rho c^2+\beta\gamma^2 p & \beta^2\gamma^2 e+\beta^2\gamma^2\rho c^2+\gamma^2 p \end{bmatrix}, \\
&= \begin{bmatrix} \gamma^2\left(e+\beta^2 p+\rho c^2\right) & \gamma^2\left(e+p+\rho c^2\right)\beta \\ \gamma^2\left(e+p+\rho c^2\right)\beta & \gamma^2\left(e\beta^2+p+\rho c^2\beta^2\right) \end{bmatrix}, \\
&= \begin{bmatrix} \gamma^2\left(e+p+\rho c^2\right)-p & \gamma\left(e+p+\rho c^2\right)\beta \\ \gamma^2\left(e+p+\rho c^2\right)\beta & \gamma^2\left(e+p+\rho c^2\right)\beta^2+p \end{bmatrix}.
\end{aligned} \tag{8.38}$$

In the last step we made some use of alternative formatting of combinations of Lorentz factor γ and β,

$$\gamma^2-1 = \frac{1}{1-\beta^2}-1 = \frac{1-1+\beta^2}{1-\beta^2} = \beta^2\gamma^2. \tag{8.39}$$

The result we can cast back into index notation by making use of the flat spacetime metric

$$\left[g^{\alpha\beta}\right] = \begin{bmatrix} -1 & 0 \\ 0 & 1 \end{bmatrix}, \tag{8.40}$$

and the four-velocity V^α from Equation 8.25. The resulting expression is

$$T^{\alpha\beta} \equiv \left(e+\rho c^2+p\right)\frac{V^\alpha}{c}\frac{V^\beta}{c}+pg^{\alpha\beta}. \tag{8.41}$$

It can be verified by plugging in actual values for indices α and β how this reduces to our result in matrix notation. In the energy-momentum tensor, enthalpy density ρh can be substituted for $e+p$. Text books on the subject also sometimes absorb the rest mass energy in the definition of internal energy density and enthalpy density. To establish some notation that will be useful later on, let us too define some symbols that incorporate a rest mass term. The internal energy density plus rest mass term we define by e:

$$\mathrm{e} \equiv \rho c^2+e = \rho c^2+\rho\varepsilon. \tag{8.42}$$

The corresponding enthalpy density including rest mass we define by w:

$$w \equiv e + p = \rho c^2 + e + p. \tag{8.43}$$

Using this definition of enthalpy density, we can write energy density conservation as

$$\frac{\partial}{\partial X^\alpha} T^{\alpha 0} = 0 \Rightarrow \frac{\partial}{\partial ct} \left(w\gamma^2 - p \right) + \frac{\partial}{\partial x^i} \left(w\gamma^2 \beta^i \right) = 0. \tag{8.44}$$

Momentum density conservation generalizes to

$$\frac{\partial}{\partial X^\alpha} T^{\alpha i} = 0 \Rightarrow \frac{\partial}{\partial ct} \left(w\gamma^2 \beta^i \right) + \frac{\partial}{\partial x^j} \left(w\gamma^2 \beta^j \beta^i + p\delta^{ji} \right) = 0. \tag{8.45}$$

The free index i indicates that we have three conserved quantities, one for each spatial dimension. In this equation we assume Cartesian coordinates, such that the spatial part of the metric tensor obeys $g^{ji} = \delta^{ji}$.

> **PROBLEM 8.5**
> **Small velocity limit for energy-momentum equation**
>
> Check whether Equation 8.35 reproduces results consistent with the Newtonian fluid dynamical equations in the case of small velocities, where you can truncate higher order terms of Taylor expansions in terms of v/c.

Problem 8.5 covers the limiting case of $v \ll c$ for the expression of energy-momentum conservation. The fact that we can reproduce the Newtonian version of these conservation laws in this limit (not just at $v = 0$), acts as a consistency check on our work. If we are to include the effect of external forces like gravity in our expression for energy-momentum conservation, our extra terms should again reduce to their non-relativistic counterparts in the limit of $v \ll c$. We would also need a four-vector that generalizes force and work into a single object. For this purpose, we therefore define four-force density F^α as follows:

$$[F^\alpha] \equiv \begin{bmatrix} \mathbf{f} \cdot \mathbf{v}/c \\ \mathbf{f} \end{bmatrix}. \tag{8.46}$$

The conservation law now becomes

$$\frac{\partial}{\partial X^\alpha} T^{\alpha\beta} = F^\beta. \tag{8.47}$$

8.3 THE MICROPHYSICS OF RELATIVISTIC GASES

It is important to keep in mind that merely changing the bulk flow velocity of a gas, even to relativistic values, will not alter the nature of the gas itself. A gas outflow moving with bulk velocity at nearly the speed of light will be perceived as a perfectly normal gas by an observer moving along at similar velocity. Indeed, this is one of the postulates of special relativity. On the other hand, as we have learned from our study of the microphysics of gases, the macroscopic behaviour of a fluid is ultimately dictated (through the equation of state) by what happens at the microscopic scale. Specifically, for a gas of point particles, it was the motion of these particles that generated the pressure term.

It therefore follows that there actually is a way in which a relativistic gas can behave fundamentally differently from a Newtonian gas: if the *peculiar* velocities of the constituent particles approach light speed. After all, given that each particle has its own direction and velocity, there can be no single frame transformation that transforms a gas with relativistic peculiar velocities into a frame where the velocities become non-relativistic again. A gas with peculiar velocities approaching light speed is *relativistically hot*, characterized by $p \gg \rho c^2$. In a manner, the comparison between p and ρc^2 provides us with yet another *similarity parameter* for our collection. The *relativistic temperature* Θ (we are using the term 'temperature' in a generalized sense here) is given by

$$\Theta \equiv \frac{p}{\rho c^2}. \tag{8.48}$$

When $\Theta > 1$, the fluid requires treatment using a relativistic EOS. The existence of such a transition is a consequence of our introduction of a fixed maximum light speed c that serves to break the scale-invariance of the fluid dynamics conservation laws.

It is worthwhile revisiting the case of an ideal gas of point particles. In Section 2.4.3 we introduced an EOS of the form $p = (\hat{\gamma}_{ad} - 1) e$. Comparing expressions for energy density and pressure from kinetic theory we concluded that for a non-relativistic ideal gas of point particles we have $\hat{\gamma}_{ad} = 5/3$ (we also revisited the definition of the adiabatic exponent in Section 3.4).

Generalizing to relativistic flow, we want to start with definitions for internal energy density and pressure that themselves are rooted in relativistic dynamics of point particles. By taking the relativistic four-momentum $[P^\alpha] = [\gamma mc, \gamma m\mathbf{v}]$ as a starting point, a reasonable definition of internal energy density would be as follows:

$$e \equiv \int (\tilde{\gamma} - 1) mc^2 \mathscr{F} \left(P^1, P^2, P^3\right) dP^1 dP^2 dP^3. \tag{8.49}$$

In this expression, we define internal energy density e (which is always considered in the local frame where the fluid element is at rest) in terms of 'peculiar' Lorentz factor $\tilde{\gamma}$. The total peculiar energy of the particle is $\tilde{\gamma} mc^2$, and in the above equation we have subtracted the rest mass contribution from this total. The energy contribution per particle is integrated over the momentum part of particle phase space, weighted using a distribution function $\mathscr{F}$ as usual. The switch from phase space in terms of velocity in the Newtonian case to one in terms of momenta has no impact on the result (see the discussion in Section 8.1.2). In the limit of ultra-relativistic peculiar velocities and infinite peculiar Lorentz factors, the internal energy density approaches

$$e \approx \int \tilde{\gamma} mc^2 \mathscr{F} dP^1 dP^2 dP^3, \tag{8.50}$$

dropping the reminder that the distribution function is in terms of four-momentum for conciseness.

If we assume the distribution $\mathscr{F}$ to be isotropic, we can also reformulate the definition for pressure in terms of a flux of peculiar momentum across fluid parcel boundaries. Following our Newtonian approach from Section 2.4.2, we define

$$\delta^{ij}p = \int \tilde{P}^i \tilde{u}^j \mathscr{F} dP^1 dP^2 dP^3 = \int \tilde{\gamma} m \tilde{u}^i \tilde{u}^j \mathscr{F} dP^1 dP^2 dP^3. \tag{8.51}$$

In the relativistic limit the peculiar velocities approach light speed. In this case, the equation above yields

$$3p = \int \tilde{\gamma} m c^2 \mathscr{F} dP^1 dP^2 dP^3, \tag{8.52}$$

once we take the trace on both sides. Comparing the relativistic limits for internal energy density and pressure reveals that for a relativistic hot gas of isotropically distributed point particles we have

$$p = \frac{e}{3}. \tag{8.53}$$

In other words, in this limit the adiabatic exponent of an ideal gas of point particles is no longer $\hat{\gamma}_{ad} = 5/3$, but instead given by $\hat{\gamma}_{ad} = 4/3$. This matches the results obtained for photons from Section 3.4.1.

In principle, the relation between pressure and energy density can be worked out exactly for the case of trans-relativistic point particles. This is a known analytical result, but an unwieldy one involving Bessel functions[3]:

$$\frac{w}{\rho c^2} = \frac{K_3(1/\Theta)}{K_2(1/\Theta)}. \tag{8.54}$$

In this expression K_2 and K_3 are modified Bessel functions of the second kind of orders 2 and 3 respectively. Because Equation 8.54 is computationally expensive to evaluate (something which needs doing *a lot* when solving fluid dynamics problems numerically), and because it is mathematically somewhat opaque, simplified approximations mimicking this behaviour are often used instead in numerical and theoretical astrophysics. An example[4] of such an equation of state is

$$p = \frac{\rho \varepsilon}{3} \frac{2\rho c^2 + \rho \varepsilon}{\rho c^2 + \rho \varepsilon}, \tag{8.55}$$

We can use Equations 8.54 and 8.55 in $p = (\hat{\gamma}_{ad} - 1)\rho\varepsilon$ to obtain expressions for a quantity $\hat{\gamma}_{ad,eq}$ that reduces to the correct adiabatic exponent in limiting cases of Θ. For Equation 8.55 we can find the relatively straightforward expression

$$\hat{\gamma}_{ad,eq} = \frac{4}{3} - \Theta + \sqrt{\frac{\Theta^2}{4} + \frac{1}{9}}, \tag{8.56}$$

---

[3]A famous presentation of the full expression dating back to 1957 can be found in [52], but the full expression itself dates from 1911 [19].

[4]This particular version is due to [35].

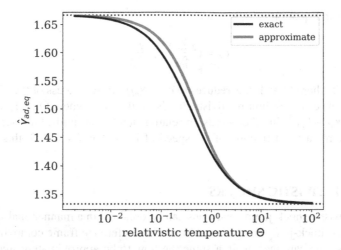

Figure 8.4 The equivalent to the adiabatic exponent for trans-relativistic point-particles as a function of relativistic temperature. Both the exact solution derived from Equation 8.54 and from the approximation of Equation 8.55 are plotted.

if we pick the positive root when solving Equation 8.55 for $\hat{\gamma}_{ad,eq}$ after substituting $e = p/(\hat{\gamma}_{ad,eq} - 1)$. While not equal to the exact solution (as shown in Figure 8.4), this describes an approximation that is always accurate up to at least 4%. An alternative version of the previous equation in terms of ρ and e rather than Θ is given by

$$\hat{\gamma}_{ad,eq} = \frac{5\rho c^2 + 4e}{3\left(\rho c^2 + e\right)}, \tag{8.57}$$

in which the limiting cases $\hat{\gamma}_{ad,eq} \to 4/3$ (if $e \gg \rho c^2$) and $\hat{\gamma}_{ad,eq} \to 5/3$ (if $e \ll \rho c^2$) can be recognized easily.

An expression for the speed of sound of an acoustic wave in relativistic flow can be obtained by linearizing the special relativistic fluid dynamics equations rather than their non-relativistic counterpart. Following this standard linearization procedure to derive wave equations, it can be shown that

$$c_s^2 = c^2 \left(\frac{\partial p}{\partial e}\right)_s, \tag{8.58}$$

at constant entropy. If we assume a *fixed* $\hat{\gamma}_{ad}$, we can obtain an expression for the sound speed as follows. First, we look at

$$\left(\frac{\partial e}{\partial p}\right)_s = c^2 \left(\frac{\partial \rho}{\partial p}\right)_s + \left(\frac{\partial e}{\partial p}\right)_s. \tag{8.59}$$

The first term on the RHS can be rewritten for polytropic gas flow obeying $p \propto \rho^{\hat{\gamma}_{ad}}$, the second using $e = p/(\hat{\gamma}_{ad} - 1)$. We obtain

$$\left(\frac{\partial e}{\partial p}\right)_s = \frac{c^2}{\hat{\gamma}_{ad}} \frac{\rho}{p} + \frac{1}{\hat{\gamma}_{ad} - 1}. \tag{8.60}$$

Using this in Equation 8.58, we then get

$$c_s^2 = c^2 \frac{\hat{\gamma}_{ad} \Theta \left(\hat{\gamma}_{ad} - 1 \right)}{\hat{\gamma}_{ad} - 1 + \hat{\gamma}_{ad} \Theta}. \tag{8.61}$$

In the limit where $\Theta \ll 1$, this reduces to $c_s^2 = \hat{\gamma}_{ad} p / \rho$, regardless of the (fixed) value of $\hat{\gamma}_{ad}$, as expected for non-relativistic fluids. In the limit where $\Theta \gg 1$, this reduces to $c_s^2 = \left(\hat{\gamma}_{ad} - 1 \right) c^2$. In other words, in relativistically hot fluids, the speed of sound tends towards a fixed fraction of the speed of light. For $\hat{\gamma}_{ad} = 4/3$, this fraction is $1/\sqrt{3}$.

8.4   RELATIVISTIC SHOCKS

The relativistic shock-jump conditions can be obtained in a manner analogous to the Newtonian shock-jump conditions. Moving to a reference frame comoving with the shock front we can once again assume the flow to be approximately steady for the brief instant it takes us to establish the jump conditions. And once again the jump conditions follow quickly from considering the flux terms in the conservation laws, in this case, Equations 8.32, 8.44 and 8.45. From mass conservation we find

$$\gamma_{2,S} \rho_2 \beta_{2,S} = \gamma_{1,S} \rho_1 \beta_{1,S}. \tag{8.62}$$

From energy conservation we find

$$w_2 \gamma_{2,S}^2 \beta_{2,S} = w_1 \gamma_{1,S}^2 \beta_{1,S}. \tag{8.63}$$

From momentum conservation we find

$$w_2 \gamma_{2,S}^2 \beta_{2,S}^2 + p_2 = w_1 \gamma_{1,S}^2 \beta_{1,S}^2 + p_1. \tag{8.64}$$

We have dropped the superscript x for β (assuming x was the direction we chose to align with the shock direction of motion), it being understood that β refers to the velocity across the shock. The subscript S both on β and γ indicates that these expressions are in the frame of the shock. A complicating factor is that the velocity addition rule for relativistic velocities is no longer linear, so the equations quickly become ugly when transforming back to an arbitrary reference frame.

As was the case for Newtonian flow, the jump conditions can be combined in various ways and expressed in forms more suitable to any particular problem at hand. We will provide a few examples with a forthcoming application to blast waves in mind. From Equation 8.64, it follows with some minor rewrites that the pressure jump across a relativistic shock obeys

$$p_2 - p_1 = w_2 \gamma_{2,S}^2 \left(1 - \frac{\beta_{2,S}}{\beta_{1,S}} \right) \beta_{1,S} \beta_{2,S} = -w_1 \gamma_{1,S}^2 \left(1 - \frac{\beta_{1,S}}{\beta_{2,S}} \right) \beta_{1,S} \beta_{2,S}. \tag{8.65}$$

Here we used Equation 8.63 to eliminate one of the enthalpies in favour of the other. As for Newtonian shocks, no pressure difference means no velocity difference. The same features of non-relativistic shocks (compressions shocks, rarefaction waves, contact discontinuities) translate straightforwardly to their relativistic counterparts.

Rewriting the product of velocity and Lorentz factor like we did in Equation 8.39, we can use Equation 8.64 to derive a jump in internal energy density (including rest mass term) $e = w - p$:

$$p_2 - p_1 = w_1 \left(\gamma_{1,s}^2 - 1 \right) - w_2 \left(\gamma_{2,s}^2 - 1 \right) \Rightarrow e_2 - e_1 = w_2 \gamma_{2,s}^2 - w_1 \gamma_{1,s}^2. \qquad (8.66)$$

In alternative form, using mass conservation, this becomes:

$$e_2 - e_1 = w_2 \gamma_{2,s}^2 \left(1 - \frac{\beta_{2,s}}{\beta_{1,s}} \right), \qquad (8.67)$$

which closely resembles the result in Equation 8.65. In fact, this allows for an expression for the velocities in terms of pressure and internal energy difference of the form

$$\beta_{1,s} \beta_{2,s} = \frac{p_2 - p_1}{e_2 - e_1}. \qquad (8.68)$$

According to Equation 8.63, the ratio of velocities follows

$$\frac{\beta_{2,s}}{\beta_{1,s}} = \frac{w_1 \gamma_{1,s}^2}{w_2 \gamma_{2,s}^2}. \qquad (8.69)$$

An alternative expression for this ratio is given by

$$\frac{\beta_{2,s}}{\beta_{1,s}} = \frac{e_1 + p_2}{e_2 + p_1}. \qquad (8.70)$$

PROBLEM 8.6
Relativistic velocity ratio jump condition derivation

Derive Equation 8.70, for example by working backwards from the result and substituting energy jump conditions for the energy terms in the numerator and denominator.

We can use the expressions for the velocity product and ratio to extract separate jump conditions for the velocities:

$$\beta_{1,s}^2 = \beta_{1,s} \beta_{2,s} \frac{\beta_{1,s}}{\beta_{2,s}} = \frac{p_2 - p_1}{e_2 - e_1} \frac{e_2 + p_1}{e_1 + p_2}, \qquad (8.71)$$

$$\beta_{2,s}^2 = \beta_{2,s} \beta_{1,s} \frac{\beta_{2,s}}{\beta_{1,s}} = \frac{p_2 - p_1}{e_2 - e_1} \frac{e_1 + p_2}{e_2 + p_1}. \qquad (8.72)$$

An expression for the velocity in a frame other than the shock frame is likely more useful, however. A natural choice of frame would be that of the unshocked fluid because in actual applications the upstream fluid state is often known. We trivially have $\beta_{1,1} = 0$. According to the velocity addition rule we have for the downstream fluid velocity that

$$\beta_{2,1} = \frac{\beta_{2,s} - \beta_{1,s}}{1 - \beta_{1,s}\beta_{2,s}} = -\beta_{1,s}\frac{1 - \beta_{2,s}/\beta_{1,s}}{1 - \beta_{1,s}\beta_{2,s}} = -\beta_{1,s}\frac{e_2 - e_1}{e_2 + p_1}, \qquad (8.73)$$

from which we can eliminate the $\beta_{1,s}$ term by squaring both sides and substituting Equation 8.71 for $\beta_{1,s}^2$:

$$\beta_{2,1}^2 = \frac{(p_2 - p_1)(e_2 - e_1)}{(e_1 + p_2)(e_2 + p_1)}. \qquad (8.74)$$

By this point we also have expressions for the shock fluid Lorentz factor $\gamma_{2,1}$ and for the Lorentz factor Γ of the shock (equivalent to $\gamma_{1,s}$):

$$
\begin{aligned}
\gamma_{2,1}^2 &= \frac{1}{1 - \beta_{2,1}^2}, \\
&= \frac{(e_1 + p_2)(e_2 + p_1)}{(e_1 + p_2)(e_2 + p_1) - (p_2 - p_1)(e_2 - e_1)}, \\
&= \frac{(e_2 + p_1)(e_1 + p_2)}{w_1 w_2}, \qquad (8.75)
\end{aligned}
$$

and

$$\Gamma^2 = \frac{1}{1 - \beta_{1,s}^2} = \frac{(e_2 - e_1)(e_1 + p_2)}{(e_2 - e_1)(e_1 + p_2) - (p_2 - p_1)(e_2 + p_1)}. \qquad (8.76)$$

Although these expressions for Γ and $\gamma_{2,1}$ do not show it clearly, we know that both Lorentz factors need to be of the same order in any shock realization: as is the case for non-relativistic shocks, sound waves triggered downstream must be able to reach to the shock front to maintain causal contact between the shock front and its energy source. The shock velocity $\beta_{S,1}$ must therefore lie in between that of the shocked medium ($\beta_{2,1}$) and that of a sound wave originating behind the shock ($\beta_{sound,1}$):

$$\beta_{2,1} < \beta_{S,1} < \beta_{sound,1}. \qquad (8.77)$$

We can estimate an upper limit to $\beta_{sound,1}$ and its corresponding Lorentz factor $\gamma_{sound,1}$ by assuming the maximum possible sound speed ($c_s = 1/\sqrt{3}$, assuming $\hat{\gamma}_{ad} = 4/3$). Transforming the sound speed into the frame of the unshocked medium, we get

$$\beta_{sound,1} \leq \frac{1/\sqrt{3} + \beta_{2,1}}{1 + \beta_{2,1}/\sqrt{3}}$$

$$\Rightarrow \beta_{sound,1} \leq \frac{1 + \sqrt{3}\beta_{2,1}}{\sqrt{3} + \beta_{2,1}}$$

$$\Rightarrow \beta_{sound,1}^2 \leq \frac{1 + 2\sqrt{3}\beta_{2,1}3\beta_{2,1}^2}{3 + 2\sqrt{3}\beta_{2,1} + \beta_{2,1}^2}$$

$$\Rightarrow \quad 1 - \beta_{sound,1}^2 \geq \frac{3 + 2\sqrt{3}\beta_{2,1} + \beta_{2,1}^2 - 1 - 2\sqrt{3}\beta_{2,1} - 3\beta_{2,1}^2}{3 + 2\sqrt{3}\beta_{2,1} + \beta_{2,1}^2}$$

$$\Rightarrow \quad 1 - \beta_{sound,1}^2 \geq \left(1 - \beta_{2,1}^2\right) \frac{2}{3 + 2\sqrt{3}\beta_{2,1} + \beta_{2,1}^2}$$

$$\Rightarrow \quad \gamma_{sound,1}^2 \leq \gamma_{2,1}^2 \frac{3 + 2\sqrt{3}\beta_{2,1} + \beta_{2,1}^2}{2}. \tag{8.78}$$

If we take $\beta_{2,1} \to 1$ to get the largest possible margin, we therefore find that the shock Lorentz factor Γ obeys

$$\gamma_{2,1}^2 < \Gamma^2 < \gamma_{2,1}^2 \left(2 + \sqrt{3}\right), \tag{8.79}$$

confirming that Γ and $\gamma_{2,1}$ are of comparable order.

So far we have not yet included many variations on the mass conservation jump condition. However, using the velocity ratio Equation 8.70, one alternative jump condition can quickly be found to be

$$\frac{\rho_2^2}{\rho_1^2} = \frac{\gamma_{1,s}^2 \beta_{1,s}^2}{\gamma_{2,s}^2 \beta_{2,s}^2} = \frac{w_2}{w_1} \frac{e_2 + p_1}{e_1 + p_2}. \tag{8.80}$$

Alternative expressions for the density ratio can be formed by using the result for $\gamma_{2,1}^2$ in the previous equation:

$$\frac{\rho_2^2}{\rho_1^2} = \frac{w_2}{w_1} (e_2 + p_1) \frac{e_2 + p_1}{w_1 w_2} \frac{1}{\gamma_{2,1}^2} = \frac{(e_2 + p_1)^2}{w_1^2 \gamma_{2,1}^2}, \tag{8.81}$$

and

$$\frac{\rho_2^2}{\rho_1^2} = \frac{w_2}{w_1} \frac{1}{(e_1 + p_2)} \frac{w_1 w_2 \gamma_{2,1}^2}{(e_1 + p_2)} = \frac{w_2^2 \gamma_{2,1}^2}{(e_1 + p_2)^2}. \tag{8.82}$$

8.4.1 STRONG SHOCKS

For the Newtonian case we defined strong shocks as obeying $p_2 \gg p_1$. In view of the moderate jump in density even for strong shocks (at least in the adiabatic case), we could also have defined strong shocks as obeying $p_2/\rho_2 \gg p_1/\rho_1$. This is equivalent to requiring a strong adiabatic shock to lead to a significant heating of the gas. It is this definition that we carry over to relativistic strong shocks, where we will focus on adiabatic shocks. Using the results from previous sections in this chapter, we can formulate shock jump conditions for strong shocks that allow for relativistic blast wave velocities while still covering the non-relativistic limit(s) for shock velocity and/or fluid temperature(s).

We will find that strong shocks running into a relativistically cold medium (i.e. with $p_1 \ll \rho_1 c^2$) can have a range of Lorentz factors Γ from non-relativistic to ultra-relativistic. Perhaps surprisingly, however, strong shocks running into a relativistically hot medium will *always* have $\Gamma \gg 1$. This can be seen from Equation

8.76, where taking the limits $e \rightarrow e$ and $e_2, p_2 \gg e_1, p_1$ results in Γ tending towards infinity.

The relativistic definition of a strong shock encapsulates $p_2 \gg p_1$, since $p_2/\rho_2 \gg p_1/\rho_1 \Rightarrow p_2/\rho_1 \gg p_1/\rho_1$ (for shocks, $\rho_2 > \rho_1$, after all). Expecting pressure and energy density to be linked by an EOS that renders them of at least somewhat comparable order, it follows that $e_2 \gg p_1$, and thus $e_2 = \rho_2 c^2 + e_2 \gg p_1$. This holds for strong shocks regardless of whether the shock velocity or any of the temperatures are relativistic. For strong shocks, Equation 8.81 thus reduces to

$$\frac{\rho_2^2}{\rho_1^2} = \frac{e_2^2}{w_1^2 \gamma_{2,1}^2} \Rightarrow \frac{e_2}{p_2} = \gamma_{2,1} \frac{w_1}{\rho_1} \qquad \text{(strong shock)}. \qquad (8.83)$$

This is getting close to a statement about the downstream conditions purely in terms of the upstream state and the strength of the shock, assuming we can also find a way to cast $\gamma_{2,1}$ purely in terms of the latter. Already, the jump condition tells us a few things. For example, if the unshocked medium is cold (in the relativistic sense), $w_1 \sim \rho_1 c^2$, and we find that

$$\frac{e_2}{\rho_2 c^2} = \gamma_{2,1} - 1. \qquad \text{(strong shock, cold upstream medium)} \qquad (8.84)$$

From this we learn that a shock wave that is sufficiently relativistic to accelerate the downstream medium to relativistic velocity (i.e. $\gamma_{2,1} \gg 1$), is capable of shock-heating even a cold fluid to relativistic temperatures.

Another limiting case is where the unshocked medium is already relativistically hot with $p_1 \gg \rho_1 c^2$. We can now no longer ignore pressure and internal energy density in the upstream medium, and Equation 8.83 indicates that

$$\frac{e_2}{\rho_2 c^2} = \gamma_{2,1} \frac{\rho_1 c^2 + p_1 + e_1}{\rho_1 c^2} \qquad \text{(strong shock)}$$

$$\Rightarrow \quad 1 + \frac{e_2}{\rho_2 c^2} = \gamma_{2,1} \left(1 + \frac{p_1}{\rho_1 c^2} + \frac{e_1}{\rho_1 c^2} \right). \qquad \text{(strong shock)} \qquad (8.85)$$

In the limiting case of relativistically hot fluids both upstream and downstream, this equation implies

$$\frac{e_2}{\rho_2 c^2} \sim \gamma_{2,1} \frac{p_1}{\rho_1 c^2}, \qquad \text{(strong shock, hot up- and downstream media)} \qquad (8.86)$$

since p_1 and e_1 are of comparable order. We conclude

$$\gamma_{2,1} \sim \frac{p_2}{\rho_2} \frac{\rho_1}{p_1} \gg 1, \qquad \text{(strong shock, hot up- and downstream media)} \qquad (8.87)$$

on account of our definition of a strong shock.

Further jump conditions can be derived if we specialize to the case of adiabatic shocks and perfect gases, obeying $p = (\hat{\gamma}_{ad} - 1) e$. Because the adiabatic exponent will depend on how relativistically hot the gas is, we will also need to apply a subscript to it, indicating whether it is the upstream or downstream exponent. For a strong shock, it can be shown that (see Problem 8.7):

$$\frac{\rho_2}{\rho_1} = (\gamma_{2,1} - 1)\frac{\rho_2 c^2}{p_2} + \frac{\hat{\gamma}_{ad,2}}{\hat{\gamma}_{ad,2} - 1}\gamma_{2,1}. \qquad \text{(strong shock)} \qquad (8.88)$$

For a cold upstream medium, this reduces to

$$\frac{\rho_2}{\rho_1} = \frac{\hat{\gamma}_{ad,2}\gamma_{2,1} + 1}{\hat{\gamma}_{ad,2} - 1}. \qquad \text{(strong shock, cold upstream medium)} \qquad (8.89)$$

These expressions reduce to the expected minor increase in density in the non-relativistic limit, e.g. a factor four if $\hat{\gamma}_{ad,2} = 5/3$ and $\gamma_{2,1} \sim 1$. But they also show that relativistic shocks are able to arbitrarily increase the density jump across a shock once their Lorentz factors become appreciable.

An expression for the shock Lorentz factor can also be derived (again, see Problem 8.7):

$$\Gamma^2 = \frac{(\gamma_{2,1} + 1)\left\{\hat{\gamma}_{ad,2}(\gamma_{2,1} - 1) + 1\right\}^2}{\hat{\gamma}_{ad,2}(2 - \hat{\gamma}_{ad,2})(\gamma_{2,1} - 1) + 2}. \qquad \text{(strong shock, cold upst. med.)} \qquad (8.90)$$

This equation is somewhat opaque in this generic form, but for example reduces to

$$\Gamma^2 = 2\gamma_{2,1}^2, \qquad (8.91)$$

in the relativistic limit where $\gamma_{2,1} \gg 1$ and $\hat{\gamma}_{ad,2} = 4/3$. A relativistic shock will indeed accelerate the fluid in its wake to relativistic velocity, while still outrunning it. Note also how Equation 8.91 is consistent with our earlier estimate for the range of Γ based on the sound speed (Equation 8.79).

PROBLEM 8.7
Relativistic jump conditions for adiabatic strong shocks

i) Derive Equation 8.89. You might want to start from Equation 8.82 and plug in the equation of state for an adiabatic gas.
ii) Derive Equation 8.90. You might want to start from Equation 8.76 and/or work partially backwards from Equation 8.90.

8.5 RELATIVISTIC BLAST WAVES

We can construct a an approximate homogeneous shell model for relativistic spherical blast waves in a manner completely similar to the derivation of the non-relativistic shell model from Section 7.4. Such shell models are used to model gamma-ray bursts as well as more recently discovered (semi-)relativistic transients like the kilonovae mentioned in the beginning of this chapter[5].

Let us assume the medium into which the blast wave evolves to be relativistically cold, with $p_1 \ll \rho_1 c^2$. Further, let us use the approximate equation of state from Equation 8.55. It turns out that under these condition various jump conditions take on surprisingly simple form. First, we show this for the density jump, starting with substituting $\hat{\gamma}_{ad}$ according to Equation 8.57 into Equation 8.89:

$$\frac{\rho_2}{\rho_1} = \frac{\hat{\gamma}_{ad,2}\gamma_{2,1} + 1}{\hat{\gamma}_{ad,2} - 1} = \frac{(5\rho_2 c^2 + 4e_2)\,\gamma_{2,1} + 3\,(\rho_2 c^2 + e_2)}{(5\rho_2 c^2 + 4e_2) - 3\,(\rho_2 c^2 + e_2)}. \tag{8.92}$$

This expression we can simplify further by using Equation 8.84:

$$
\begin{aligned}
\frac{\rho_2}{\rho_1} &= \frac{\left(5 + 4\frac{e_2}{\rho_2 c^2}\right)\gamma_{2,1} + 3\left(1 + \frac{e_2}{\rho_2 c^2}\right)}{2 + \frac{e_2}{\rho_2 c^2}}, \\
&= \frac{(5 + 4\gamma_{2,1} - 4)\,\gamma_{2,1} + 3\,(1 + \gamma_{2,1} - 1)}{2 + \gamma_{2,1} - 1}, \\
&= \frac{4\,(\gamma_{2,1} + 1)\,\gamma_{2,1}}{\gamma_{2,1} + 1},
\end{aligned}
\tag{8.93}
$$

and ultimately,

$$\rho_2 = 4\gamma_{2,1}\rho_1. \tag{8.94}$$

This shows that $\rho_2/\gamma_{2,1}$ will *always* jump by a factor four relative to the upstream density, which also includes the non-relativistic limit where this factor is as expected. Using this result in Equation 8.84 quickly shows that the jump in energy simply becomes

$$e_2 = 4\gamma_{2,1}\,(\gamma_{2,1} - 1)\,\rho_1 c^2. \tag{8.95}$$

A few straightforward manipulations in the same vein as the previous show that the pressure jump will be given by

$$p_2 = \frac{4}{3}\,(\gamma_{2,1}^2 - 1)\,\rho_1 c^2. \tag{8.96}$$

The procedure for obtaining the shock Lorentz factor Γ in terms of the shocked fluid Lorentz factor $\gamma_{2,1}$ is no different, and when applied to Equation 8.76, leads to

$$\Gamma^2 = \frac{\left(4\gamma_{2,1}^2 - 1\right)^2}{8\gamma_{2,1}^2 + 1}. \tag{8.97}$$

The next step is to consider global conservation of energy and mass in the shell model, just like we did for the non-relativistic version of Section 7.4. We will again

---

[5]The specific trans-relativistic shell model discussed here was first presented independently in [54] and [38]. A nice shell model for kilonovae (although not a relativistic one) is presented in [31].

assume a homogeneous medium. We should however be careful to be consistent with the reference frame that the various quantities are expressed in. In the lab frame at rest relative to the unshocked medium, we compare the swept-up mass within a sphere of radius R to the mass of the shell with width ΔR:

$$M = \frac{4\pi}{3}\rho_1 R^3 = 4\pi\left(\rho_2\gamma_{2,1}\right)R^2\Delta R. \tag{8.98}$$

Using the jump condition for mass Equation 8.94, the width of the shell is found to be

$$\Delta R = \frac{R}{12\gamma_{2,1}^2}, \tag{8.99}$$

throughout the evolution of the blast wave. Where non-relativistic blast waves led to thin shells, relativistic blast waves flatten the swept-up mass profile even more so. For highly relativistic blast waves (e.g. $\gamma_{2,1} > 100$, as in the early stages of gamma-ray bursts), the difference between the shell width ΔR and the size of the system R will exceed five orders in magnitude. This range in scales, further exacerbated by the many orders in magnitude that the blast wave covers from relativistic to non-relativistic evolution, poses a challenge to numerically simulating such systems.

The total energy E of the explosion must be contained in the shell. Expressing the volume V_{shell} as total mass divided by mass density, and using that the 00-component of the energy-momentum tensor represents energy density in the lab frame, we have

$$E = \left(T^{00} - \gamma_{2,1}\rho_2 c^2\right)V_{shell} = \left(w_2\gamma_{2,1}^2 - p_2 - \gamma_{2,1}\rho_2 c^2\right)\frac{M}{\rho_2\gamma_{2,1}}. \tag{8.100}$$

Here we subtracted the rest-mass energy from the energy total, since this was provided by the swept-up mass rather than the initial explosion. Using the jump conditions to simplify this result, it takes a few steps to show

$$E = \frac{1}{3}\left(4\gamma_{2,1}^2 - 1\right)M\beta_{2,1}^2 c^2. \tag{8.101}$$

Here we can recognize the non-relativistic shell model limit ($\gamma_{2,1} \to 1$) where the energy $E = Mv_2^2$ is twice the kinetic energy of the system. In this limit all other conclusions from Section 7.4 apply as well, such as $R \propto t^{2/5}$. On the other hand, in the ultra-relativistic limit ($\beta_{2,1} \to 1$), we have $R \sim ct$. Therefore $M \propto t^3$ and the energy balance equation implies

$$\gamma_{2,1}^2 \propto M^{-1} \propto t^{-3}. \tag{8.102}$$

What is interesting is that relativistic spherical shell models can even be useful for obviously jetted outflow like gamma-ray bursts. The reason for this is that when the jet is ultra-relativistic, most parts of the blast wave will not yet be 'aware' of the existence of an edge to the jet since it takes time for a sound wave to communicate this fact between the edges of the jet and the tip. In the lab frame, these sound waves end up suppressed by Lorentz factor Γ (see Problem 8.9 for details). A collimated ultra-relativistic jet will therefore pretty much proceed radially outward in the form of an expanding conic wedge as long as the sound waves occupy just a small patch near the edge. During this stage, the shell model can be applied up to a truncation angle θ_0 that describes the opening angle of the jet. Mass M and energy E are then to be understood as not describing the actual total of the conic wedge, but rather the values the blast wave would have had if it were spherical. In the literature, these are deemed the 'isotropic equivalent' mass and energy.

Eventually a jetted outflow will spread out sideways. Ultimately, the blast wave will indeed become spherical or at least quasi-spherical. At this stage, the shell model can be applied again without caveat, and E represents the actual energy content of the jet. Note that the latter will be far smaller than the isotropic equivalent value inferred from the same blast wave in case it started out tightly collimated. Between early ultra-relativistic stage and late non-relativistic stage, the motion of the blast wave is not merely radially outward and the shell model is of limited value (but can be tweaked to include sideways spreading). The relativistic shell model can be expanded in a similar manner as its non-relativistic counterpart, to include e.g. ongoing injection of energy into the system or an upstream medium profile that is inhomogeneous.

PROBLEM 8.8
A spherical gamma-ray burst

We explore the actual scales of gamma-ray burst afterglows. We set the physical parameters to values typical for a *long* gamma-ray burst, where a relativistic jet is launched into the interstellar medium that surrounds a collapsing massive star (this stage is known as the 'afterglow' stage, since the jet will radiate at wavelengths progressively longer than produced in the initial burst of gamma rays). We take a total explosion energy $E = 10^{53}$ erg, and a surrounding medium number density $n = 1$ cm^{-3}, with the medium consisting of particles with mass $m_p = 1.6726231 \times 10^{-24}$ g. You will further need to make use of the speed of light $c = 2.99792458 \times 10^{10}$ cm s^{-1}.

i) Assuming the spherical model to apply, show that the radius of the sphere at $\gamma_{2,1} = 100$ is about $R = 1.06 \times 10^{17}$ cm.

ii) Show that the corresponding width of the blast wave shell at this radius is about $\Delta R = 8.83 \times 10^{11}$ cm.

iii) Show that the radius of the sphere at $\beta_{2,1} = 0.01$ (that is, down to 1% of light speed), is about $R = 5.4 \times 10^{19}$ cm.

iv) Show that the transition between relativistic and non-relativistic evolution occurs around $R = 2.39 \times 10^{18}$ cm, where we take $\beta\gamma \sim 1$ to mark this turning point.

PROBLEM 8.9
A gamma-ray burst jet

We consider a jetted outflow with an opening angle θ_0. This opening angle indicates the maximum angle θ associated with an outflow aligned with the z-axis (with $\theta = 0$ the tip of the jet). As long as causal contact between the tip and the edges of the jet has not yet been achieved, the dynamics of the spherical shell model are roughly applicable. In this exercise, we calculate the time it takes for the presence of an outer edge at $\theta = \theta_0$ to become apparent at $\theta = 0$. The speed of communication is given by the sound speed, and throughout this exercise we assume the jet to move relativistically fast with shock front Lorentz factor $\Gamma \gg 1$ and to be relativistically hot with $p \gg \rho c^2$. We compare two reference frames, the lab frame in which the origin of the explosion and the surrounding medium are at rest and the frame comoving with the shocked fluid (indicated by a bar).

i) The sound wave running inwards along the shock front needs to keep up with the radial motion of the shock front. Show that in the frame of the shocked fluid, the radial signal velocity $\bar{\beta}_s^r$ can be approximated as $\bar{\beta}_s^r \sim 1/3$.

ii) Using that $(\beta_s^r)^2 + (\beta_s^\theta)^2$ equals the square of the relativistic sound speed, show that in the lab frame, the signal speed along the shock front $\beta_s^\theta \sim \sqrt{2}/(4\gamma)$. The more relativistic the jet, the slower a sound wave is perceived to move along the shock front in the lab frame.

iii) Show that a sound wave departing from the edge of the jet at the point of explosion will reach the tip of the jet once the shock fluid decelerates to a Lorentz factor γ given by

$$\gamma = \frac{\sqrt{2}}{2} \frac{1}{3\theta_0}.$$

The narrower the jet, the earlier the *jet break*, the transition point from radial flow to a spreading jet stage.

iv) For a typical gamma-ray burst jet with opening angle of $\theta_0 = 0.1$ rad, what is the difference between the isotropic-equivalent energy E_{iso} that dictates the dynamics before the jet break and the actual energy E_{jet} in the jet? Gamma-ray burst outflows come in the form of *bipolar* jets, so an individual jet will have half the energy of the entire burst outflow.

8.6 SELF-SIMILAR RELATIVISTIC EXPLOSIONS

At first glance it seems unlikely that a self-similar solution exists for blast waves in the relativistic limit like it does for the non-relativistic point explosion. After all, in Section 7.5, we argued for self-similarity from dimensional analysis. In relativistic fluid dynamics, the speed of light c adds more options for combining variables and parameters of the explosion E, ρ_1, r, t and c. For example, the radius of a relativistic blast wave evolves at least to first order approximately as $R \sim ct$, featuring c rather than a combination of E and ρ_1.

However, the argument from dimensional analysis always was a sufficient means to establish self-similarity rather than a necessary one. It turns out that an approximately self-similar solution to the relativistic blast wave can still be formulated. This *Blandford-McKee* solution[6] to the relativistic point explosion problem is approximately self-similar wherever the Lorentz factor of the outflow obeys $\gamma^{-2} \ll 1$, a constraint not too dissimilar from the implicit assumption that $v \ll c$ in the Sedov-Taylor blast wave solution.

---

[6]This self-similar solution was first presented in 1976 [1] and shot up in popularity with the discovery of gamma-ray burst afterglows, which turned out to be a far more direct application of the theory than the active galactic nuclei referred to in the original paper.

We take our cue from the relativistic shell model of the previous section when setting up the self-similar point explosion solution. In the relativistic limit, we found that $\gamma_{2,1} \propto t^{-3/2}$ (see Equation 8.102), while $\Gamma \propto \gamma_{2,1}$ according to Equation 8.91. We also have an expression for the velocity of the blast wave in the relativistic limit:

$$\Gamma^2 = \frac{1}{1 - (U/c)^2} \Rightarrow \frac{U}{c} \approx 1 - \frac{1}{2\Gamma^2},$$

(8.103)

where we truncate after first order in Γ^{-2}. Integrating this expression over time while making use of $\Gamma \propto t^{-3/2}$, we obtain an expression for radius R in the form

$$R = ct\left(1 - \frac{1}{8\Gamma^2}\right).$$

(8.104)

We now define a dimensionless self-similarity variable χ according to

$$\chi = \left(1 + 8\Gamma^2\right)\left(1 - \frac{r}{ct}\right).$$

(8.105)

What exact definition to use for a self-similarity variable is always a matter of convenience (in the same way that for the non-relativistic case any function of ξ, e.g. ξ^5, would also have been valid choice). The simple form of the self-similar solution when expressed in terms of χ will provide a justification after the fact for this particular choice. The variable χ starts at 1 at $r = R$ (up to order Γ^{-2}, that is) and increases further downstream.

In the ultra-relativistic limit, the shock jump conditions become even simpler than they do for the trans-relativistic shell model (with the exception of the density jump condition, which is already as simple as can be). Writing $D \equiv \rho\gamma$ for Lorentz-contracted density in the lab frame, Equation 8.94 becomes

$$D_2 = 4\gamma_{2,1}^1\rho_1 = 2\Gamma^2\rho_1,$$

(8.106)

where we already used the relativistic limit $\Gamma^2 = 2\gamma_{2,1}^2$. At $\hat{\gamma}_{ad,2} = 4/3$ and $\gamma_{2,1} \gg 1$, pressure and energy become

$$p_2 = \frac{1}{3}e_2 = \frac{2}{3}\Gamma^2 w_1.$$

(8.107)

This suggests we express pressure, Lorentz factor and lab frame density in terms of self-similar functions as follows:

$$p = \tfrac{2}{3}w_1\Gamma^2 f(\chi),$$ (8.108)

$$\gamma_{2,1}^2 = \tfrac{1}{2}\Gamma^2 g(\chi),$$ (8.109)

$$D = 2D_1 h(\chi).$$ (8.110)

The jump conditions imply that $f(1) = g(1) = h(1) = 1$ by construction. These self-similar forms for the fluid variables now need to be substituted in the relativistic fluid dynamics equations in spherical symmetry, which in turn need to be written not in terms of r and t, but in terms of χ and Γ, keeping only lower order terms in Γ^{-2}. The spherically symmetric equations are

$$\frac{\partial D}{\partial ct} + \frac{1}{r^2}\frac{\partial}{\partial r}\left(r^2 D\beta\right) = 0, \tag{8.111}$$

$$\frac{\partial}{\partial ct}\left(w\gamma^2 - p\right) + \frac{1}{r^2}\frac{\partial}{\partial r}\left(r^2 w\gamma^2\beta\right) = 0, \tag{8.112}$$

$$\frac{\partial}{\partial ct}\left(w\gamma^2\beta\right) + \frac{1}{r^2}\frac{\partial}{\partial r}\left(r^2 w\gamma^2\beta^2\right) + \frac{\partial p}{\partial r} = 0. \tag{8.113}$$

The relevant derivatives to the required accuracy can be written as

$$t\frac{\partial}{\partial t} = -3\frac{\partial}{\partial \ln\Gamma^2} + \left(4\left(2\Gamma^2 - \chi\right) + 1\right)\frac{\partial}{\partial \chi}, \tag{8.114}$$

$$t\frac{\partial}{\partial r} = -\left(1 + 8\Gamma^2\right)\frac{\partial}{\partial \chi}, \tag{8.115}$$

$$t\frac{d}{dt} = -3\frac{\partial}{\partial \ln\Gamma^2} + 4\left(\frac{2}{g} - \chi\right)\frac{\partial}{\partial \chi}. \tag{8.116}$$

The extra factor t is easily included in the original conservation equations by just multiplying them with t on both sides. After all substitutions are applied to the conservation equations the results can be reorganized to isolate the derivatives of the functions f, g and h. This eventually leads to

$$\frac{d\ln f}{d\chi} = g\frac{16 + g\chi}{4\left(4 - 8g\chi + g^2\chi^2\right)}, \tag{8.117}$$

$$\frac{d\ln g}{d\chi} = g\frac{17 - 5g\chi}{4\left(4 - 8g\chi + g^2\chi^2\right)}, \tag{8.118}$$

$$\frac{d\ln h}{d\chi} = g\frac{38 - 18g\chi + g^2\chi^2}{4\left(4 - 8g\chi + g^2\chi^2\right)\left(2 - g\chi\right)}. \tag{8.119}$$

When combined with boundary conditions at the shock front provided by the jump conditions, the solution to these equations turns out to be quite simple:

$$f = \chi^{-17/12}, \qquad g = \chi^{-1}, \qquad h = \chi^{-7/4}. \tag{8.120}$$

While some of the other algebra of this section can be tedious if redone in full, this at least can be confirmed quickly by substituting the solutions back into the self-similar differential equations. It is also possible to obtain $g(\chi)$ quickly by following the same approach to energy conservation up to arbitrary radius as was used in Section 7.5 for the non-relativistic case (see Problem 8.10). We show the shock profile for a self-similar blast wave in Figure 8.5.

PROBLEM 8.10
Energy conservation within a relativistic point explosion

i) Show that the velocity (in units of c) β_m of a sphere anchored at fixed χ is given by

$$\beta_m = 1 - \frac{\chi}{2\Gamma^2},$$

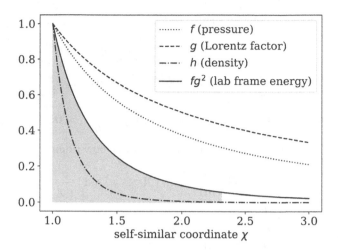

Figure 8.5 The Blandford-McKee self-similar profile for a point explosion. The area up to the back of the shock at $\chi = 7/3$ according to a homogeneous shell estimate is indicated in grey.

up to order Γ^{-2}. Be careful not to use Equation 8.116, where the recasting in terms of partial derivatives assumes the velocity dr/dt to be equal to the fluid flow velocity rather than that of a shell at fixed χ.

ii) Derive $g(\chi) = 1/\chi$ for the self-similar relativistic point explosion by assuming that the energy up to constant χ remains fixed for any χ, not just $\chi = 1$.

The solution in Equation 8.120 does not apply all the way to the origin of the explosion. At $r = 0$, we have $\chi = 1 + 8\Gamma^2$. At this point,

$$\gamma_{2,1}^{1} = \frac{\Gamma^2}{2(1 + 8\Gamma^2)} < 1, \tag{8.121}$$

which is clearly unphysical. According to the self-similar solution, the Lorentz factor will drop below unity already at $\chi = \Gamma^2/2$ (compare Equations 8.109 and 8.120). Once we start moving away from the shock front, neglecting terms of order γ^{-2} and above becomes less and less accurate. Luckily, within the region of the blast wave that contains the vast majority of the explosion energy and swept-up mass, the solution is still very accurate. If we take the estimate of the shock width from the homogeneous shell model, $\Delta R \sim R/(6\Gamma^2)$, which is still valid as an order-of-magnitude estimate, we can show that:

$$\chi_{back} = (1 + 8\Gamma^2)\left(1 - \frac{r_{back}}{R}\frac{R}{t}\right),$$

$$\chi_{back} = (1 + 8\Gamma^2)\left(1 - \left(1 - \frac{1}{6\Gamma^2}\right)\left(1 - \frac{1}{8\Gamma^2}\right)\right),$$

$$\chi_{back} = 1 + \frac{4}{3} + \frac{1}{8\Gamma^2} + \frac{1}{6\Gamma^2}. \tag{8.122}$$

The difference between this coordinate $\chi_{back} = 7/3$ for the 'back' of the blast wave and the point $\chi = \Gamma^2/2$ (where $\gamma \sim 1$), scales with Γ^2 as expected. Figure 8.5 includes the estimate of the characteristic width of the shell profile in grey.

With the self-similar solution in hand, the explosion energy E can be compared to the spatially integrated energy density. For the non-relativistic self-similar point explosion we used this to determine the pre-factor C_{ST} setting the radius of the blast wave. Here we can use this to set the scale of the Lorentz factor of the blast wave Γ instead:

$$E = 8\pi w_1 c^3 t^3 \Gamma^2/17. \tag{8.123}$$

8.7   FIRST-ORDER FERMI SHOCK-ACCELERATION REVISITED

It is not uncommon for particles in a plasma to achieve relativistic velocities when accelerated across shocks in the manner described in Section 7.6. At minimum, this requires an analysis in terms of relativistic kinematics for the particles, if not for both particles and shocks. Below, we briefly revisit first-order Fermi shock-acceleration for relativistic particle velocities crossing non-relativistic shocks[7]. Again, we will find a slope of -2, now best expressed in terms of particle energies. The case where the shocks themselves become relativistic lies beyond the scope of the book and is an open area of research. Somewhat steeper slopes of about -2.2 are typically inferred in this case.

The main difference between non-relativistic and relativistic particles is that in the latter case all particles have approximate velocity c and an energy greatly exceeding their rest mass energy. The relation between particle energy and momentum becomes $E \approx \sqrt{P_i P^i} c$ (see Equation 8.28), so at this point a power law in magnitude of momentum implies the same power law in energy.

Following the steps from Section 7.6, we can no longer apply Galilean velocity addition rules. Instead, we apply a Lorentz boost in the horizontal direction when switching frames between the upstream and downstream fluid. We have

$$\bar{E} = \gamma E + \gamma v P^x \approx E + v P^x, \tag{8.124}$$

where γ the Lorentz factor corresponding to fluid velocity v. Because the fluid is not assumed to move relativistically, $\gamma \sim 1$. The same fortunate alignment of fluid velocity and particle momentum takes place as in the non-relativistic case. Either both are negative (going from right to left into the downstream medium in the depiction of Figure 7.8), or both are positive (going back into the upstream medium). Other than the frame transformation rule, the argument for the increase in momentum and energy in relativistic particles runs analogous to the non-relativistic case. It therefore follows that

$$\frac{\Delta E}{E} = \frac{4v}{3c}, \tag{8.125}$$

---

[7] We now fully follow the approach by [27], which we already used in part to set up the steps of Section 7.6.

where we already assume that $u \sim c$. Similarly plugging in $u \sim c$ in the derivation of the particle number attrition rate that can also be found in Section 7.6 and combining results for particle number and energy, we eventually find

$$dN \propto E^{-2}dE. \qquad (8.126)$$

This power law distribution is a ubiquitous ingredient in applications of non-thermal emission models describing astrophysical phenomena, in particular, synchrotron emission and Compton scattering.

9 Viscous Flow

Up to now, we have limited our treatment to *inviscid* fluids. These lack the internal friction that occurs when different parts of the fluids move relative to one another and the fluid parcels of an inviscid fluid offer no resistance to deformation. By contrast, the fluid parcels of viscous fluids *do* resist deformation. The stronger the viscosity of the fluid, the more momentum is converted to heating and transferred between fluid parcels due to shearing and compression.

Under realistic conditions, viscosity can often not be ignored and Euler's equations need to be augmented with terms describing its impact. The most common generalization of the non-relativistic conservation laws of fluid dynamics to include viscosity are called the *Navier-Stokes* equations, which are used in a wide range of applications of fluid dynamics, from water flows and weather prediction to medical fluid dynamics. Even in astrophysics, viscosity can play an important role. Examples of such include viscous accretion discs (we discuss accretion flow in Chapter 11) and shock waves. Regarding the latter, recall from Section 7.3 that shocks inevitably lead to an increase in entropy. Although in Chapter 7 we sidestepped a discussion of the length scales at which this effect occurs (on the order of the mean free path in a gas) and focused on the larger scale implications of shocks, the conditions within shocks are typically such that viscosity is important.

9.1 THE NAVIER-STOKES EQUATION

The continuity equation is not affected by fluid viscosity. Viscosity may alter the flow velocity of a fluid, but it cannot make mass disappear. Our starting point for including viscosity will therefore be to generalize the momentum equation

$$\frac{\partial \rho \mathbf{v}}{\partial t} + \nabla \cdot (\rho \mathbf{v} \mathbf{v} + p \mathbb{1}) = \mathbf{f}, \tag{9.1}$$

which we first derived in Section 2.1.3, starting from

$$\int_{V_{fixed}} \frac{\partial (\rho \mathbf{v})}{\partial t} dV_{fixed} = -\int_{V_{fixed}} \nabla \cdot (\rho \mathbf{v} \mathbf{v}) dV_{fixed} + \int_{V_{fixed}} \mathbf{f} dV_{fixed} +$$
$$\oint_{A_{fixed}} \mathbf{t} dS_{fixed}. \tag{9.2}$$

The pressure term arose from surface forces acting on fluid parcel surface S, and we introduced the stress tensor T to describe these surface forces:

$$\mathbf{t} dS = \mathbf{T} \cdot d\mathbf{S}; \quad \mathbf{T} = -p \mathbb{1} = \begin{pmatrix} -p & 0 & 0 \\ 0 & -p & 0 \\ 0 & 0 & -p \end{pmatrix}. \tag{9.3}$$

DOI: 10.1201/9781003095088-9

Our approach will be to generalize T to include terms accounting for viscous effects[1]. Viscosity is a consequence of friction, in turn produced by local velocity differences within the fluid. It is therefore a reasonable assumption to take the viscosity terms to depend on dv^i/dx^j. We add a *viscous stress tensor* σ describing velocity to T that is a function of these velocity differences:

$$T^{ij} = -p\delta^{ij} + \sigma^{ij}, \qquad \sigma = \sigma\left(\frac{\partial \mathbf{v}}{\partial \mathbf{x}}\right). \tag{9.4}$$

At this point σ can still be any function of the velocity differences, something that in theory could then be expressed as a Taylor series

$$\sigma = C_1\left(\frac{\partial \mathbf{v}}{\partial \mathbf{x}}\right) + C_2\left(\left(\frac{\partial \mathbf{v}}{\partial \mathbf{x}}\right)^2\right) + C_3\left(\left(\frac{\partial \mathbf{v}}{\partial \mathbf{x}}\right)^3\right) + \dots \tag{9.5}$$

In practice, it is normally sufficient to consider the first order terms only. If σ depends on linear space derivatives of velocity only, the fluid is called a *Newtonian* fluid. Fluids that depend quadratically on velocity gradients are called *Stokesian* fluids. The first-order terms can be decomposed into two effects, *bulk* viscosity and *shear* viscosity. Before we go into detail on these manifestations of viscosity, it is worthwhile to establish that T^{ij} (and thus σ^{ij}) needs to be symmetric to ensure that the angular momentum of fluid parcels is conserved in the absence of forces acting on the parcels, as expected for Newtonian fluids.

9.1.1   ANGULAR MOMENTUM AND THE STRESS TENSOR

To demonstrate that the stress tensor is symmetric we will at some point make use of the permutation symbol ε^{ijk} introduced in Section 1.2.1. Readers who are satisfied to take the statement that T must be symmetric on account of angular momentum conservation at face value can safely skip this section and pick up again at Section 9.1.2.

The statement that the angular momentum of a fluid parcel can only change as a consequence of forces acting on the parcel can be expressed mathematically as follows:

$$\frac{d}{dt}\int \rho\,(\mathbf{r}\times\mathbf{v})\,dV = \int (\mathbf{r}\times\mathbf{f})\,dV + \oint (\mathbf{r}\times\mathbf{t})\,dS, \tag{9.6}$$

where $\mathbf{f}$ and $\mathbf{t}$ the same volume and surface force densities as previously and $\mathbf{r}$ the position vector with components x^i. The LHS of this equation can be rewritten by applying the product rule,

$$\frac{d}{dt}\int \rho\,(\mathbf{r}\times\mathbf{v})\,dV = \int (\mathbf{r}\times\mathbf{v})\frac{d}{dt}(\rho dV) + \int \left(\frac{d\mathbf{r}}{dt}\times\mathbf{v}\right)\rho dV$$

$$+ \int \left(\mathbf{r}\times\frac{d\mathbf{v}}{dt}\right)\rho dV. \tag{9.7}$$

---

[1]Do not confuse T with the energy-momentum tensor from Chapter 8.

Of the three resulting terms, the first one is zero on account of mass conservation. The second term is zero too, because the time derivative of the position vector is just the flow velocity **v** and the cross product between identical vectors is zero. We are therefore left with

$$\frac{d}{dt} \int \rho \, (\mathbf{r} \times \mathbf{v}) \, dV = \int \left(\mathbf{r} \times \frac{d\mathbf{v}}{dt} \right) \rho \, dV. \tag{9.8}$$

Now let us examine the rightmost term in Equation 9.6, which contains the surface forces. At this point, we switch to index notation and start using the permutation symbol, in order to avoid getting lost completely in sorting our what contracts and multiplies with what. We can work our way towards an application of the divergence theorem:

$$\oint \varepsilon^i{}_{jk} x^j t^k dS = \oint \varepsilon^i{}_{jk} x^j T^{km} dS_m = \int \frac{\partial}{\partial x^m} \left(\varepsilon^i{}_{jk} x^j T^{km} \right) dV. \tag{9.9}$$

Using this result and the previous result in Equation 9.6 and realising that it has to hold for arbitrary sized fluid parcels, we arrive at

$$\varepsilon^i{}_{jk} x^j \frac{dv^k}{dt} \rho - \varepsilon^i{}_{jk} x^j f^k - \frac{\partial}{\partial x^m} \left(\varepsilon^i{}_{jk} x^j T^{km} \right) = 0. \tag{9.10}$$

We can use Euler's equation (see e.g. Equation 3.17) to replace the time derivative of the velocity in the first term:

$$\left(\varepsilon^i{}_{jk} x^j \frac{\partial}{\partial x^m} T^{mk} + \varepsilon^i{}_{jk} x^j f^k \right) - \varepsilon^i{}_{jk} x^j f^k - \frac{\partial}{\partial x^m} \left(\varepsilon^i{}_{jk} x^j T^{km} \right) = 0. \tag{9.11}$$

Cleaning up some cancelling terms and applying the product rule then gets us to

$$\varepsilon^i{}_{jk} x^j \frac{\partial}{\partial x^m} T^{mk} - \varepsilon^i{}_{jk} \frac{\partial x^j}{\partial x^m} T^{km} - \varepsilon^i{}_{jk} x^j \frac{\partial}{\partial x^m} T^{km} = 0. \tag{9.12}$$

This expression has to hold for arbitrary values of x^i, which means that the central term on the LHS needs to be independently equal to zero. We have $\partial x^j / \partial x^m = \delta^j_m$, so we are looking at terms $\varepsilon^i{}_{jk} T^{kj}$ that need to be equal to zero for each entry i. Using the properties of the permutation symbol, we have for the different entries:

$$\varepsilon^1{}_{jk} T^{kj} = \varepsilon^1{}_{23} T^{32} + \varepsilon^1{}_{32} T^{23} = T^{32} - T^{23}, \tag{9.13}$$

$$\varepsilon^2{}_{jk} T^{kj} = \varepsilon^2{}_{13} T^{31} + \varepsilon^2{}_{31} T^{13} = -T^{31} + T^{13}, \tag{9.14}$$

$$\varepsilon^3{}_{jk} T^{kj} = \varepsilon^3{}_{12} T^{21} + \varepsilon^3{}_{21} T^{12} = T^{21} - T^{12}, \tag{9.15}$$

Since all these entries must equal zero, we have therefore established that T must be symmetric. Once the symmetry of T is established, the outer terms on the LHS of Equation 9.12 can be seen to cancel as well.

9.1.2 BULK VISCOSITY

Bulk viscosity, associated with compression ('pinching') or expansion ('stretching') of the fluid, depends on $\dot{\mathscr{V}}/\mathscr{V}$ (that is, the compression or expansion rate of the specific volume). This is in line with our previous assertion that viscous effects are related to velocity gradients, as can be seen from the Lagrangian version of the continuity equation:

$$\frac{d\rho}{dt} + \rho\left(\nabla \cdot \mathbf{v}\right) = 0 \Rightarrow \nabla \cdot \mathbf{v} = -1\frac{1}{\rho}\frac{d\rho}{dt} = -\mathscr{V}\frac{d\mathscr{V}^{-1}}{dt} = \frac{\dot{\mathscr{V}}}{\mathscr{V}}. \qquad (9.16)$$

In general form, velocity gradients leading to bulk viscosity can be included in the momentum equation as follows:

$$\frac{\partial \rho v^i}{\partial t} + \frac{\partial}{\partial x^j}\left(\rho v^j v^i + p\delta^{ji} - \zeta^{ji}\left(\nabla \cdot \mathbf{v}\right)\right) = f^i. \qquad (9.17)$$

In this expression, ζ^{ji} provides a measure of the strength of bulk viscosity and is itself a function of fluid temperature and density in general. For the bulk viscosity we limit ourselves to viscous stress forces that act perpendicularly to the fluid parcel surface orientations, which means that $\zeta^{ij} = 0$ if $i \neq j$. We will revisit whatever viscous effects we miss out on as a consequence of the latter restriction once we discuss shear viscosity. We then assume the viscosity to be independent on the orientation of the reference frame, requiring $\zeta^{xx} = \zeta^{yy} = \zeta^{zz}$. We take $\zeta^{ij} = \zeta\delta^{ij}$, with equal diagonal entries ζ:

$$\frac{\partial \rho v^i}{\partial t} + \frac{\partial}{\partial x^j}\left(\rho v^j v^i + p\delta^{ji} - \zeta\delta^{ji}\left(\nabla \cdot \mathbf{v}\right)\right) = f^i. \qquad (9.18)$$

We take the coefficient of bulk viscosity ζ to be positive by construction. The minus sign in front of ζ is then chosen to ensure that viscous effects lead to dissipation of kinetic energy into heating the fluid, rather than the other way around. For (as of yet undefined) shear viscosity, this is demonstrated in Section 9.2, while the demonstration of this effect for bulk viscosity is delegated to Problem 9.2

9.1.3 SHEAR VISCOSITY

If the viscosity tensor $\sigma^{ij} = \sigma_{bulk}^{ij} + \sigma_{shear}^{ij}$ and $\sigma_{bulk}^{ij} = \zeta\delta^{ij}\left(\nabla \cdot \mathbf{v}\right)$ as per the discussion of the preceding section, this leaves a shear viscosity term σ_{shear}^{ij} to be included in Equation 9.4. The purpose of σ_{shear}^{ij} is to capture the viscosity produced by the shearing of two fluid layers due to a difference in their parallel velocities (i.e. 'rubbing', rather than 'pinching'). We have some basic considerations that help constrain the form of σ_{shear}^{ij}:

1. For uniform linear velocity of the fluid, $\sigma_{shear} = 0$, since there are no velocity differences causing friction. This we have imposed from the outset both for bulk and shear viscosity, given that they were taken to depend on *gradients* of **v**.

2. For uniform angular rotation ('rigid rotation'), the same should hold, because there is no slippage between rotating annuli in this case.

3. $\sigma_{shear}^{ij} = \sigma_{shear}^{ji}$, as demonstrated in Section 9.1.1.

4. σ_{shear}^{ij} should not double-count bulk viscosity effects that are already dealt with using the ζ term.

We take a closer look at condition two. A fluid in rigid rotation obeys $\mathbf{v} = \mathbf{\Omega} \times \mathbf{r}$. For example, a rotation along the x-axis would lead to a fluid velocity profile of the form

$$
V = \begin{bmatrix} \Omega \\ 0 \\ 0 \end{bmatrix} \times \begin{bmatrix} x \\ y \\ z \end{bmatrix} = \begin{bmatrix} 0 \\ -\Omega z \\ \Omega y \end{bmatrix},
\tag{9.19}
$$

where V the velocity vector in linear algebra notation. From this fluid profile it can easily be checked that the following relations hold:

$$
\frac{\partial v^x}{\partial y} + \frac{\partial v^y}{\partial x} = 0; \qquad \frac{\partial v^x}{\partial z} + \frac{\partial v^z}{\partial x} = 0; \qquad \frac{\partial v^y}{\partial z} + \frac{\partial v^z}{\partial y} = 0.
\tag{9.20}
$$

As a matter of fact,

$$
\frac{\partial v^i}{\partial x^j} + \frac{\partial v^j}{\partial x^i} = 0, \qquad i \neq j,
\tag{9.21}
$$

and

$$
\frac{\partial v^i}{\partial x^i} = 0 \quad \text{(no summation over } i\text{)},
\tag{9.22}
$$

hold for *any* rigid rotation $\mathbf{\Omega}$, no matter its rotation angle. This then suggests we try

$$
\sigma_{shear,trial}^{ij} \equiv \eta \left(\frac{\partial v^i}{\partial x_j} + \frac{\partial v^j}{\partial x_i} \right),
\tag{9.23}
$$

to capture the effect of shear viscosity, because it nicely collapses to zero in the case of rigid rotation. Here, we introduce η as the coefficient of shear viscosity. This trial version also obeys conditions 1 and 3 from our list above. Condition 4 refers to leaving out the diagonal terms of the viscosity stress tensor already captured by $\delta^{ij}\zeta$. A solution obeying $\sigma_{shear}^{ij} = 0$ along the diagonal for $\frac{\partial v^x}{\partial x} = \frac{\partial v^y}{\partial y} = \frac{\partial v^z}{\partial z}$ (isotropic stretching or shrinking), is given by

$$
\sigma_{shear}^{ij} = \eta \left(\frac{\partial v^i}{\partial x_j} + \frac{\partial v^j}{\partial x_i} - \frac{2}{3} \delta^{ij} \frac{\partial v^k}{\partial x^k} \right),
\tag{9.24}
$$

as can be verified from writing out the diagonal terms.

Collecting our results for bulk and shear viscosity, we conclude that the momentum conservation law can be amended to

$$
\frac{\partial \rho v^i}{\partial t} + \frac{\partial}{\partial x^j} \left(\rho v^j v^i + p \delta^{ji} - \eta \left(\frac{\partial v^i}{\partial x_j} + \frac{\partial v^j}{\partial x_i} - \delta^{ji} \frac{2}{3} \frac{\partial v^k}{\partial x^k} \right) - \zeta \delta^{ji} \frac{\partial v^k}{\partial x^k} \right) = f^i.
\tag{9.25}
$$

As a reminder: there is no difference between upper and lower indices in Newtonian Cartesian geometry in e.g. x^i and x_i, and the choice of upper of lower placement is only to benefit from the summation convention. In vector notation, the equation becomes

$$\frac{\partial \rho \mathbf{v}}{\partial t} + \nabla \cdot (\rho \mathbf{vv} + p\mathbb{1} - \sigma_{shear} - \zeta\mathbb{1}(\nabla \cdot \mathbf{v})) = \mathbf{f}, \qquad (9.26)$$

or more concisely

$$\frac{\partial \rho \mathbf{v}}{\partial t} + \nabla \cdot (\rho \mathbf{vv} - \mathrm{T}) = \mathbf{f}, \qquad (9.27)$$

albeit with a more general form of T than we have used for inviscid fluid dynamics.

The previous discussion demonstrated that in practice implementing viscosity in the momentum equation was achieved by the substitution $-p\mathbb{1} \to -p\mathbb{1} + \sigma$. We can generalize the velocity equation (see e.g. 2.17) and the kinetic energy density equation (e.g. Equation 2.24) in the same manner by applying this substitution. For the velocity equation we get

$$\frac{\partial v^i}{\partial t} = -v^j \frac{\partial}{\partial x^j} v^i - \frac{1}{\rho}\frac{\partial}{\partial x^j}\left(p\delta^{ji} - \sigma^{ji}\right) + \frac{f^i}{\rho}. \qquad (9.28)$$

Once viscosity is accounted for in the manner described above, the equation

$$\rho\frac{d\mathbf{v}}{dt} = \mathbf{f} + T, \qquad (9.29)$$

for the evolution of the velocity components is known as the *Navier-Stokes* equation. If T is left unspecified, the equation is known in the literature as the more general *Cauchy's equation of motion*.

The kinetic energy density equation becomes

$$\frac{\partial}{\partial t}\left(\frac{1}{2}\rho v^2\right) = -\frac{\partial}{\partial x^i}\left(\frac{1}{2}\rho v^2 v^i\right) - v_i\frac{\partial}{\partial x^j}\left(p\delta^{ji} - \sigma^{ji}\right) + v_i f^i. \qquad (9.30)$$

More generally, conservation of (total) energy density takes the form

$$\frac{\partial \mathscr{E}}{\partial t} + \frac{\partial}{\partial x^i}\left(\mathscr{E}v^i - T^{ji}v_j\right) = v_i f^i. \qquad (9.31)$$

to which heat dissipation and conduction terms can be added if needed.

In general, viscosity is not of importance in areas of astrophysical modelling where the gases are dilute, although there are notable exceptions (such as the afore-mentioned accretion discs where viscosity is generated by turbulence). Astrophysics aside, for a perfect monatomic gas it can be shown that $\zeta = 0$, while incompressible fluids by definition imply that $\zeta(\nabla \cdot \mathbf{v}) = 0$. In the remainder of this chapter, we limit ourselves to incompressible fluids and thus to the effect of shear viscosity.

9.2   VISCOSITY AND DISSIPATION

As was discussed previously in Section 2.1.4 on conservation laws, the pressure term
in the kinetic energy density equation above provides a means to transfer energy
between internal and kinetic energy and vice versa. The same term but with different
sign appears in the internal energy density equation. Given how closely the viscosity
tensor σ is linked to the pressure tensor $p\mathbb{1}$, it stands to reason that viscosity plays a
very similar role. But there is one important difference, which is that in the case of
viscosity this connection is a one-way street. As a consequence of the second law of
thermodynamics, it is only possible to dissipate kinetic energy into heat on account
of viscosity and not the other way around. This can be shown by considering the
effect of viscosity on kinetic energy in a larger (fixed) volume. Starting from the
conservation law for kinetic energy density, we can apply the product rule in reverse
and the divergence theorem. We get (leaving out most of the non-viscous terms from
Equation 9.30 for the sake of brevity):

$$\frac{\partial E_{kin}}{\partial t} = \int_V \frac{\partial}{\partial t}\left(\frac{1}{2}\rho v^2\right) dV = \ldots + \int_V v_i \frac{\partial}{\partial x_j}\sigma^{ji} dV,$$

$$\frac{\partial E_{kin}}{\partial t} = \ldots + \int_V \left(\frac{\partial}{\partial x^j}\left(v_i \sigma^{ji}\right) - \sigma^{ji}\frac{\partial}{\partial x^j}v_i\right) dV,$$

$$\frac{\partial E_{kin}}{\partial t} = \ldots + \oint_S v_i \sigma^{ji} dS_j - \int_V \sigma^{ji}\frac{\partial v_i}{\partial x_j} dV.$$

The surface fluxes have been taken to be zero, considering either an enclosed fluid
or a sufficiently large volume that $\mathbf{v} \to 0$. The resulting expression can be rewritten
using the identity $\frac{\partial v^i}{\partial x^j} = \frac{1}{2}\left(\frac{\partial v^i}{\partial x^j} + \frac{\partial v^j}{\partial x^j}\right)$ in order to obtain

$$\frac{\partial E_{kin}}{\partial t} = \ldots + 0 - \int_V \frac{1}{2}\sigma^{ji}\left(\frac{\partial v_i}{\partial x_j} + \frac{\partial v_j}{\partial x_i}\right) dV,$$

$$\frac{\partial E_{kin}}{\partial t} = \ldots + 0 - \int_V \frac{1}{2}\frac{1}{\eta}\sigma \cdot \sigma dV. \tag{9.32}$$

In the last step we used the definition of σ for the case $\nabla \cdot \mathbf{v} = 0$. The thing to note
here is that the final result for the viscosity term is *always* negative given that it
contains a positive coefficient η and the square of σ, preceded by a minus sign. This
confirms that the effect of viscosity is to *always* reduce the kinetic energy density no
matter what the fluid flow velocity profile.

9.3   PHYSICAL INTERPRETATION OF SHEAR VISCOSITY

Ultimately, viscosity has its roots in the microphysical behaviour of the fluid. Like
an equation of state, its mathematical expression cannot be derived from the conservation laws alone and the viscosity will take on different form for different types
of fluids (e.g. liquids versus gases). With ζ typically either negligible or irrelevant
on account of the assumption of incompressibility, we concentrate on estimating the

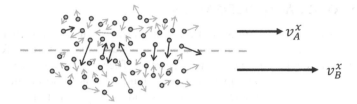

Figure 9.1   The layers of fluid showing shear flow. Individual constituent particles are shared between layers on account of their random (thermal) motions.

strength of η from microphysics considerations. We start with considering a gas, where the fluid particle mean free paths λ comfortably exceed the effective particle radius l_D. This way we do not need to worry about any influences gas particles exert on each other: direct collisions aside, the particles are assumed to be free to move around. We compare two layers with a difference in bulk velocity v^x (see Figure 9.1) but otherwise identical. For the average peculiar velocity $\langle \tilde{u} \rangle$ of the constituent particles, we have $m \langle \tilde{u} \rangle^2 \sim k_B T$, as per the connection between fluid kinetic temperature and particle kinetic energies.

Normally the microscopic fluxes $\left| \rho v_A^x \tilde{u}^y \right|$ and $\left| \rho v_B^x \tilde{u}^y \right|$ are in equilibrium, with layer A losing on average as much momentum to layer B as it gains from that layer. However, once $v_A^x \neq v_B^x$, this equilibrium is violated. The difference in bulk velocity density between the two layers can be taken to be small at small distances δy and approximated as

$$v_B^x \approx v_A^x + \Delta v^x \approx v_A^x + \frac{\partial v^x}{\partial y} \Delta y. \tag{9.33}$$

Plugging in our estimate for the average magnitude of the peculiar velocity, this implies a net momentum flux on average of

$$\sigma^{xy} \sim \left| \rho \Delta v \tilde{u}^y \right| \sim \left| \rho \frac{\partial v_x}{\partial y} \Delta y \sqrt{\frac{k_B T}{m}} \right|. \tag{9.34}$$

A reasonable distance scale for Δy is the mean free path λ

$$\Delta y \sim \lambda = \frac{1}{\pi l_D^2 n} = \frac{m}{\pi l_D^2 \rho}, \tag{9.35}$$

where l_D the classical radius of the particles and n the number density as usual. If we generalize these considerations to all directions, we estimate

$$\sigma^{ij} \sim \left| \rho \left(\frac{\partial v^i}{\partial x_j} + \frac{\partial v^j}{\partial x_i} \right) \frac{m}{\pi l_D^2 \rho} \sqrt{\frac{k_B T}{m}} \right| \rightarrow \eta \sim \frac{1}{\pi l_D^2} \sqrt{m k_B T}. \tag{9.36}$$

Indeed, for a dilute gas of hard spheres it can formally be shown that $\eta = 5\sqrt{\pi} \sqrt{m k_B T} / (64 \pi l_D^2)$, but what matters more here is not the numerical pre-factor per se, but the dependence of η on the physics parameters. The density can be seen to have dropped out. Even though one might intuitively expect the viscosity to increase with increasing density, this effect is balanced by the reduced mean free path available to the particles. A (weak) dependence on temperature T remains. If an isothermal fluid is assumed or if the temperature gradients are assumed to be small, η can be taken outside of the differentiation, simplifying the equations. A quantity closely related to the viscosity coefficient η is a quantity ν that measures η per unit mass (i.e. *specific*) known as the *kinematic viscosity coefficient*. According to the above, we have $\nu = \eta/\rho \Rightarrow \nu \sim \lambda \langle \bar{u} \rangle$, that is, the product of the average thermal velocity and the mean free path.

This analysis applies specifically to dilute gases and it bears repeating that the resulting expression for viscosity is by no means universal. Physical chemists, for example, are more likely to encounter viscosity in the context of the *Arrhenius equation*. This is an empirical relation between reaction rate and temperature that accounts for the local potential field experienced by particles in the fluid. If each particle sits in a potential well of energy E_{local}, increasing the temperature will at first *decrease* the viscosity of the fluid by making it easier for particles to escape their local environment. In terms of Arrhenius behaviour, we can posit for this situation that

$$\eta \propto \exp\left(\frac{E_{local}}{k_B T}\right). \tag{9.37}$$

The larger the temperature of the liquid, the larger the random thermal motions of the particles. Therefore, the interpretation here is that the larger the temperature of the liquid, the deeper the local potential wells need to be to keep the particles confined to their local area. At high temperature, the potential field that follows from material properties of the liquid ceases to impede particle motions and the viscosity for the liquid drops as a result.

9.4 FLOW THROUGH A PIPE

As a practical example, we apply the Navier-Stokes equation to the flow through a pipe (see Figure 9.2). If we assume steady state, no external force, constant η and

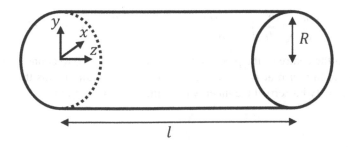

Figure 9.2 Viscous flow through a pipe of length l and with radius R.

incompressible flow, the Navier-Stokes equation (Equation 9.28) reduces to

$$v^j \frac{\partial}{\partial x^j} v^i = -\frac{1}{\rho} \frac{\partial}{\partial x^j} p \delta^{ij} + \frac{\eta}{\rho} \frac{\partial}{\partial x^j} \left(\frac{\partial v^j}{\partial x_i} + \frac{\partial v^i}{\partial x_j} \right). \tag{9.38}$$

In vector form this equation can be written as

$$\mathbf{v} \cdot \nabla \mathbf{v} = -\frac{\nabla p}{\rho} + \frac{\eta}{\rho} \left(\nabla \left(\nabla \cdot \mathbf{v} \right) + \nabla^2 \mathbf{v} \right) = -\frac{\nabla p}{\rho} + \frac{\eta}{\rho} \left(\nabla^2 \mathbf{v} \right), \tag{9.39}$$

revealing a further simplification on account of the incompressibility assumption. For the pipe flow we take v^x and v^y to be zero everywhere and v^z to be zero at the inner edge of the pipe[2] where $r = R$ (taking r to be the distance from the z-axis). Constant ingoing mass flow rate $\dot{M}$ and constant ingoing pressure (neither a function of x, y), together with $v^x = v^y = 0$, means that there will also be no further x, y gradients in pressure and density further into the pipe. Along streamlines we have constant density ($d\rho/dt = 0$), and a pressure profile that is a function of z only, $p = p(z)$. Constant density implies that v^z remains fixed along a streamline, but can still vary between streamlines to be a function of r, i.e., $v^z = v^z(x,y)$. The term $\mathbf{v} \cdot \nabla \mathbf{v}$ reduces to zero:

$$\begin{bmatrix} 0 \\ 0 \\ v^x \frac{\partial v^z}{\partial x} + v^y \frac{\partial v^z}{\partial y} + v^z \frac{\partial v^z}{\partial z} \end{bmatrix} = \begin{bmatrix} 0 \\ 0 \\ 0 \end{bmatrix}. \tag{9.40}$$

This leaves the Navier-Stokes equation as

$$\nabla^2 \mathbf{v} = \frac{1}{\eta} \nabla p \Rightarrow \frac{\partial^2 v^z}{\partial x^2} + \frac{\partial^2 v^z}{\partial y^2} = \frac{1}{\eta} \frac{\partial p}{\partial z}. \tag{9.41}$$

The result can be integrated along each streamline to yield

$$(z - z_0) \left(\frac{\partial^2 v^z}{\partial x^2} + \frac{\partial^2 v^z}{\partial y^2} \right) = \frac{p - p_0}{\eta} \rightarrow l \left(\frac{\partial^2 v^z}{\partial x^2} + \frac{\partial^2 v^z}{\partial y^2} \right) = -\frac{\Delta p}{\eta}. \tag{9.42}$$

Here we anticipated a decreasing pressure further in the pipe along a pipe length $l \equiv z - z_0$ and defined a positive $\Delta p \equiv -(p - p_0)$. We rewrite this expression in cylindrical coordinates (i.e. polar coordinates for x, y) and solve it:

$$\frac{1}{r} \frac{d}{dr} \left(r \frac{dv^z}{dr} \right) = \frac{-\Delta p}{\eta l} \Rightarrow v^z(r) = -\frac{\Delta p}{4\eta l} r^2 + C_1 \ln r + C_2. \tag{9.43}$$

The flow at the centre of the pipe should be finite, so $C_1 = 0$ to counteract the divergence of the $\ln r$ term at $r = 0$. The outer boundary condition was the requirement that the velocity be zero at the inner wall of the pipe. As a result

$$v^z = \frac{\Delta p}{4\eta l} \left(R^2 - r^2 \right), \tag{9.44}$$

---

[2]This is known as the *no slip* condition, a reasonable approximation assuming a limiting case where the adhesion to the wall strongly exceeds the cohesion of the fluid.

describing a parabolic velocity profile. The mass flow rate is then

$$\dot{M} = 2\pi\rho \int_0^R r v^z dr = \frac{\pi}{8} \frac{\Delta p}{l} \frac{\rho}{\eta} R^4. \tag{9.45}$$

What makes this result notable is the extremely strong dependence of the mass flow rate on the radius R of the pipe. This is also why clogged arteries are so dangerous: to compensate for a change in radius, the pressure needs to be increased by a fourth power.

PROBLEM 9.1
Eddies in viscous flow

(i) The Navier-Stokes equation

$$\frac{\partial v^i}{\partial t} + v^j \frac{\partial}{\partial x^j} v^i = -\frac{1}{\rho} \frac{\partial}{\partial x^j} p\delta^{ji} + \frac{1}{\rho} \frac{\partial}{\partial x^j} \sigma^{ji} + \frac{f^i}{\rho},$$

can be written in vector-form as

$$\frac{\partial \mathbf{v}}{\partial t} + \mathbf{v} \cdot \nabla \mathbf{v} = -\frac{1}{\rho} \nabla p + \frac{\eta}{\rho} \left(\frac{1}{3} \nabla (\nabla \cdot \mathbf{v}) + \nabla^2 \mathbf{v} \right) - \nabla \Psi,$$

when we assume no bulk viscosity and a conservative force density $\mathbf{f} = -\nabla\Psi$. Show, in particular, that $\frac{\partial}{\partial x^j} \sigma^{ji}$ evaluates to

$$\nabla \cdot \sigma = \eta \left(\frac{1}{3} \nabla (\nabla \cdot \mathbf{v}) + \nabla^2 \mathbf{v} \right),$$

in vector form.

(ii) Recall the definition of vorticity $\omega \equiv \nabla \times \mathbf{v}$. Taking the curl of the Navier-Stokes equation therefore leads to an expression for the evolution of the vorticity:

$$\frac{\partial \omega}{\partial t} + \nabla \times (\mathbf{v} \cdot \nabla \mathbf{v}) = \nabla \times \left(-\frac{1}{\rho} \nabla p + \frac{\eta}{\rho} \left(\frac{1}{3} \nabla (\nabla \cdot \mathbf{v}) + \nabla^2 \mathbf{v} \right) - \nabla \Psi \right).$$

We consider a *barotropic* fluid, for which p is a function *only* of ρ. Show that in this case

$$\nabla \times \left(\frac{1}{\rho} \nabla p \right) = 0.$$

(iii) Using the vector identity

$$\mathbf{v} \cdot \nabla \mathbf{v} = \frac{1}{2} \nabla v^2 - \mathbf{v} \times (\nabla \times \mathbf{v}),$$

derive the expression

$$\frac{\partial \omega}{\partial t} = \nabla \times (\mathbf{v} \times \omega) + \frac{\eta}{\rho} \nabla^2 \omega + \nabla \left(\frac{\eta}{\rho} \right) \times \nabla^2 \mathbf{v}.$$

(iv) Contrast the above result to the inviscid (i.e. $\eta = 0$ fluid case). How does viscosity impact Kelvin's theorem for circulation and what is the physical interpretation?

PROBLEM 9.2
Dissipation through bulk viscosity

Show that, like shear viscosity, bulk viscosity also leads to dissipation of kinetic energy into heat and never the other way around. Use a similar line of reasoning as was applied to the shear viscosity case, by considering the kinetic energy in a fixed volume. Note that a flux term through the boundary of this volume has no impact in terms of dissipation: what is lost to one region is gained by another.

9.5   TWO EXAMPLE SIMILARITY PARAMETERS IN VISCOUS FLOWS

9.5.1   THE REYNOLD'S NUMBER

Viscosity introduces a whole new range of possible similarity parameters. Recall that in the case of incompressible flow (again we take a simplified case to make a more general point), we have

$$\frac{\partial v^i}{\partial t} = -v^j \frac{\partial}{\partial x^j} v^i - \frac{1}{\rho} \frac{\partial}{\partial x^j} \sigma^{ji} + \frac{f^i}{\rho}, \qquad \sigma^{ij} = \eta \left(\frac{\partial v^i}{\partial x_j} + \frac{\partial v^j}{\partial x_i} \right). \qquad (9.46)$$

Using kinematic viscosity coefficient ν, the *Reynold's number* Re is now defined according to

$$\mathrm{Re} \equiv \frac{VL}{\nu}. \qquad (9.47)$$

Here V is a characteristic velocity scale and L a characteristic length scale. If $\nu < VL$ then Re > 1 and the fluid is not viscous ('inviscid'), whereas Re < 1 describes viscous flow. A concrete example for L, V would be the diameter R and flow velocity v of the pipe flow problem discussed in Section 9.4.

Interestingly, once Re $\gg 1$, the question whether ν is sufficient to render the flow viscous becomes moot. It is known from experiments that at Re ~ 3000 the flow is susceptible to turbulence, which again will introduce an effective viscosity (the fluid bulk kinetic energy can get captured by a cascade of motions at smaller and smaller scales). In astrophysics, accretion discs are a well-known example of this effect.

Note that the Reynold's number can be interpreted as a balance between forces:

$$\mathrm{Re} = \frac{VL}{\nu} = \frac{\rho VL}{\eta} = \frac{\rho VV}{\eta V/L} = \frac{\text{'inertial force'}}{\text{'viscous force'}}. \qquad (9.48)$$

9.5.2   THE PRANDTL NUMBER

If both viscosity and heat conduction play a (potential) role in the flow, we can use the *Prandtl number* Pr to compare the two effects:

$$\mathrm{Pr} \equiv \frac{c_p \eta}{\mathscr{K}} = \frac{\eta/\rho}{\mathscr{K}/c_p \rho} = \frac{\text{'kinematic viscosity'}}{\text{'thermal diffusivity'}}. \qquad (9.49)$$

Here c_p is the specific heat capacity at constant pressure, $\mathscr{K}$ the coefficient of thermal conductivity as defined in Section 2.3. The larger the heat capacity, the more heat per Kelvin is needed to raise the temperature of the fluid. Therefore both small $\mathscr{K}$ and large c_p obstruct the flow of heat through the fluid. An example of a fluid with a very large Prandtl number is oil, which is why you would not want to put out an oil fire with water. It will only spread the burning oil around, but the lack of thermal diffusivity of oil will limit the ability of water to carry away the heat of the oil. Water itself has a Prandtl number Pr ~ 8 at $17°$, whereas air has Pr ~ 0.7.

PROBLEM 9.3
Transition to turbulence and a household tap

 (i) Assume a fully opened household tap to have a flow rate of 5 litres of water per minute. The radius of the open tap is $1/4$ inch (0.635 cm). What is the velocity of the water in km / h?

(ii) Opening a tap wide usually makes the water comes out out in a turbulent flow. If the tap is opened at $1/4$ inch radius, will the water flow be turbulent? You can assume the kinematic viscosity of water to be $v \sim 10^{-2}$ cm^2 s^{-1}.

(iii) Close the tap halfway, to a radius of $1/8$ inch. Is the flow still turbulent? Assume the water pressure to stay the same, but the throughput $\dot{M}$ to be affected as computed in the pipe flow section.

10 Fluid Instabilities

Fluid flow patterns are not always *stable*. Under many circumstances small perturbations to a fluid state will not merely propagate as sound waves but will also grow in size over time, eventually disrupting the fluid state completely. Instabilities in fluid dynamics can take many forms and can be driven by a wide range of physical phenomena such as gravity, magnetic forces or temperature. It would be beyond the scope of this book to provide an exhaustive review of the topic, so we will limit ourselves to cherry-picking a few interesting examples[1]. When studying a system that is potentially unstable, there are various questions to consider. First, of course, is whether the conditions for an instability to grow are indeed met. A strongly related follow-up question to this is what the actual growth rate of such an instability would be and whether this can be quantified in a practically useful manner. Further, it might be possible to identify factors that can potentially mitigate an instability, such as for example viscosity, and under which conditions these play a role. Eventually, when a fluid instability develops and grows, it will lead to clear dynamical and observable consequences. Ultimately, a state of fully realized *turbulence* can be reached. Turbulence is a difficult subject for which no universal theory (yet) exists. A big part of the problem is that turbulence acts across scales, from the largest down to the microscopic. Although distribution of turbulence across this spectrum of length scales itself can be modelled, it does mean that the boundaries between macroscopic and microscopic theory become blurry.

Although we will touch briefly on turbulence at the end of this chapter, in the following sections we will mostly concern ourselves with stability analysis, that is, trying to assess whether the conditions for given instability are met.

10.1 CONVECTION AND STABILITY

Convection occurs in a fluid when the positions of fluid parcels are not stable under minor perturbation. To explore the conditions under which convection takes place, we therefore need to introduce a small perturbation in a fluid parcel position and check whether the fluid parcel responds by moving back to its initial position or not. The method of introducing small perturbations in a fluid state that we will use repeatedly throughout this chapter is very similar to what we did to derive the behaviour of sound waves in Chapter 5.

We will focus on a polytropic gas in an atmosphere (where gravity acts as an external force counterbalancing pressure). If we start from the polytropic gas law and take the derivative with respect to density on both sides of the equation, we find

---

[1]The extensive monograph by Subrahmanyan Chandrasekhar first published in 1961 looms large in the literature on the topic [6].

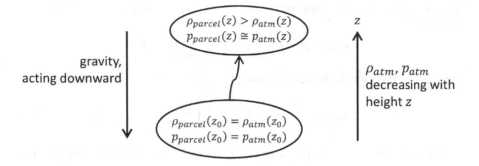

Figure 10.1   A fluid parcel in an atmosphere, displaced upward from position 0.

that

$$\frac{p}{p_0} = \left(\frac{\rho}{\rho_0}\right)^{\hat{\gamma}} \Rightarrow \frac{1}{p_0}\frac{dp}{d\rho} = \hat{\gamma}\left(\frac{\rho}{\rho_0}\right)^{\hat{\gamma}}\frac{1}{\rho} \Rightarrow \frac{dp_{parcel}}{d\rho_{parcel}} = \hat{\gamma}\frac{p_{parcel}}{\rho_{parcel}}. \tag{10.1}$$

Remember that the full derivative implies a trajectory of motion, and that we by default take this to be defined by the local fluid flow velocity $\mathbf{v}$.

Now consider a fluid parcel initially at height z_0 and at equilibrium with its environment, such that $\rho_{parcel}(z_0) = \rho_{atm}(z_0)$ and $p_{parcel}(z_0) = p_{atm}(z_0)$. Gravity acts downward and pressure and density decrease upward with height (see Figure 10.1). We now slightly perturb the position of the fluid parcel, setting the stage for a linearized perturbation analysis. If the parcel is displaced upward (we could run a similar argument for a downward perturbation) it will end up going from $\rho_{parcel}(z_0)$ to $\rho_{parcel}(z)$, and $p_{parcel}(z_0)$ to $p_{parcel}(z)$. We are considering a perturbation of the parcel, rather than a shock. The velocity acquired by the parcel will be far smaller than the sound speed, in the same way the velocity perturbation $v_1 \ll c_s$ in our analysis of sound waves in Section 5.1. Sound waves can rapidly diffuse an energy imbalance by propagating from fluid parcel to fluid parcel at the speed of sound, but it takes significantly longer for a density imbalance to be removed because this requires an actual reorganization of the fluid parcel masses themselves. After a displacement we can therefore assume that $p_{parcel}(z) \approx p_{atm}(z)$, while initially $\rho_{parcel}(z) \neq \rho_{atm}(z)$ in general. If the atmosphere is to be stable under this perturbation, it needs to satisfy the *stability criterion* $\rho_{parcel}(z) > \rho_{atm}(z)$, causing the parcel to sink back to its original position (following Archimedes' principle).

Any mathematical function can be expanded in a Taylor series for the purpose of comparing its values at two nearby points and density ρ is no different:

$$\rho(z) \approx \rho(z_0) + \left(\frac{d\rho}{dz}\right)_{z=z_0} dz. \tag{10.2}$$

If any pressure differences between parcel and environment are indeed removed rapidly by sound waves, it will be worthwhile to rewrite the previous expression such that it involves a (known) pressure difference across heights rather than a mass difference:

$$\rho(z) \approx \rho(z_0) + \left(\frac{d\rho}{dp}\right)_{z=z_0} \left(\frac{dp}{dz}\right)_{z=z_0} dz,$$

For a fluid parcel that is displaced polytropically, this expression becomes

$$\rho_{parcel}(z) \approx \rho(z_0) + \frac{\rho(z_0)}{p(z_0)\hat{\gamma}} \left(\frac{dp}{dz}\right)_{z=z_0} dz. \tag{10.3}$$

Note that the different atmospheric layers are not necessarily connected by fluid parcel flow lines in general, and that this equation therefore really is specific to our perturbed fluid parcel on account of us having used the polytropic gas law that acts along a flow line. Rewritten, the equation states

$$\frac{\rho_{parcel}(z) - \rho(z_0)}{dz} \approx \frac{d\rho_{parcel}}{dz} \approx \frac{\rho(z_0)}{p(z_0)\hat{\gamma}} \left(\frac{dp}{dz}\right)_{z=z_0}. \tag{10.4}$$

Stability therefore requires

$$\frac{\rho}{p\hat{\gamma}} \frac{dp}{dz} > \frac{d\rho_{atm}}{dz}, \tag{10.5}$$

for each arbitrary starting point z_0 (i.e. throughout the atmosphere). On the LHS, we have the change in density associated with an upwards displacement of a fluid parcel. On the RHS we have the general change in density associated with the atmosphere. If the LHS is larger, then the fluid parcel will sink back. If it is smaller, then the fluid parcel will continue to rise.

   This stability criterion is typically expressed in terms of the atmospheric temperature profile. We can do so by reminding ourselves of the ideal gas law in its various forms:

$$p = \frac{\rho k_B T}{m} \Rightarrow \rho = \frac{pm}{k_B T} \Rightarrow \frac{m}{k_B T} = \frac{\rho}{p}, \text{ etc.} \tag{10.6}$$

Applied to the stability criterion, we can derive:

$$\frac{\rho}{\hat{\gamma} p} \frac{dp}{dz} > \frac{d}{dz}\left(\frac{m}{k_B} \frac{p}{T}\right)$$

$$\Rightarrow \quad \frac{\rho}{\hat{\gamma} p} \frac{dp}{dz} > \frac{m}{k_B} \frac{1}{T} \frac{dp}{dz} - \frac{m}{k_B} \frac{p}{T^2} \frac{dT}{dz}$$

$$\Rightarrow \quad \frac{\rho}{\hat{\gamma} p} \frac{dp}{dz} > \frac{\rho}{p} \frac{dp}{dz} - \frac{\rho}{T} \frac{dT}{dz}$$

$$\Rightarrow \quad \frac{1}{\hat{\gamma}} \frac{dp}{dz} > \frac{dp}{dz} - \frac{p}{T} \frac{dT}{dz}$$

$$\Rightarrow \quad \frac{dT}{dz} > \frac{T}{p}\left(1 - \frac{1}{\hat{\gamma}}\right) \frac{dp}{dz}. \tag{10.7}$$

Since temperature and pressure both decrease with height in an atmosphere, we therefore require

$$\left| \frac{dT}{dz} \right| < \frac{T}{p} \left(1 - \frac{1}{\hat{\gamma}} \right) \left| \frac{dp}{dz} \right|, \tag{10.8}$$

for the atmosphere to be convectionally stable. If this condition is not fulfilled, convection will occur and fluid parcels will begin exhibiting random motion directed upward. An example of this in daily life is the rising of smoke above a fire (with the smoke tracing the fluid parcels of gas) as the fire introduces a rapidly changing temperature profile. Another example is the boiling of water, although here the effect is eclipsed by the phase transitions occurring in the bubbles.

PROBLEM 10.1
Deriving Archimedes' principle

Using Euler's equation and perturbation theory, derive Archimedes' principle against a hydrostatic background.

10.2   THE RAYLEIGH-TAYLOR INSTABILITY

The Rayleigh-Taylor ('RT') instability is an example of a instability that occurs when the criterion for convective stability is not satisfied. This stability is well known and results when a heavy fluid lies on top of a light fluid, e.g. water on oil. Over time, the light fluid begins to seep upwards across the contact discontinuity separating the layers and through the heavy fluid and vice versa, resulting in intricate patterns of 'fingers' across the heavy/light divide. A typical mushroom-like shape emerges, spawning smaller shapes and eddies (Figure 10.2 shows an example in two dimensions). The 'mushroom' cloud from nuclear bomb explosions, too, is an example of a RT instability. Other examples may be found in astrophysics, in particular in supernova blast waves and remnants, the Crab nebula being a particularly striking example. In these cases, as well as in their (trans-)relativistic counterparts from e.g. gamma-ray bursts and kilonovae, the deceleration of the blast wave creates conditions equivalent to gravitational acceleration when considered in the reference frame of the contact discontinuity between layers of the outflow, illustrating how the RT instability can occur more generally than under the actual influence of gravity.

In order to establish whether the RT instability occurs, we can perform a *linear perturbation analysis* that is quite similar to the perturbation analysis we used to introduce sound waves in Section 5.1. For the sake of simplicity, and to get the general flavour of this type of analysis across with a minimum of clutter, we focus on inviscid and incompressible fluids. As with sound waves, we consider baseline unperturbed states that are motionless, homogeneous and have pressure and density p_0 and ρ_0 respectively. We introduce a small perturbation across the boundary between two such states with different unperturbed values. The perturbation will have the usual small velocity $\mathbf{v}_1$, added pressure p_1 and added density ρ_1. The velocity equation (recall

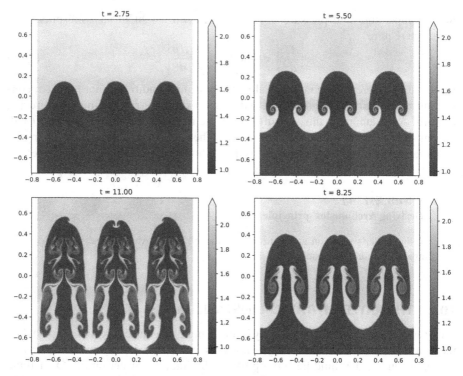

Figure 10.2 Rayleigh-Taylor instability in two dimensions, computed using numerical methods from Chapter 13. Time increases in the clockwise direction.

e.g. Equation 2.17), multiplied by density, reads

$$\rho \frac{\partial \mathbf{v}}{\partial t} + \rho \mathbf{v} \cdot \nabla \mathbf{v} = -\nabla p + \rho \mathbf{g}, \tag{10.9}$$

where $\mathbf{g}$ the gravitational acceleration. For the unperturbed state, this implies

$$\rho_0 \mathbf{g} = \nabla p_0, \tag{10.10}$$

while adding the perturbation and dropping all higher order terms beyond linear perturbation leads to

$$\rho_0 \frac{\partial \mathbf{v}_1}{\partial t} = -\nabla p_1 + \rho_1 \mathbf{g}. \tag{10.11}$$

To arrive at the latter expression, we subtracted the unperturbed state result. For an incompressible flow, we have

$$\frac{d\rho}{dt} = 0 \Rightarrow \frac{\partial \rho}{\partial t} + \mathbf{v} \cdot \nabla \rho = 0. \tag{10.12}$$

In view of the continuity equation, we know that this implies

$$\nabla \cdot \mathbf{v} = 0. \tag{10.13}$$

Let us assume that the unperturbed state is a steady state solution, varying only in the z-direction. Keeping linear perturbation terms at most, the previous two expressions then respectively imply

$$\frac{\partial \rho_1}{\partial t} + v_1^z \frac{\partial \rho_0}{\partial z} = 0, \tag{10.14}$$

and

$$\frac{\partial v_1^x}{\partial x} + \frac{\partial v_1^y}{\partial y} = -\frac{\partial v_1^z}{\partial z}. \tag{10.15}$$

For the purpose of establishing whether the system is unstable under the proposed perturbation, it is sufficient to look at any *normal mode* of the perturbation, with the understanding that a complete sum of normal modes adds up to the full perturbation. Fourier decomposition into monochromatic sound waves is based on the same principle. Recall our perturbation solution from Section 5.2, and let us take a mode

$$\mathbf{v}_1 \equiv \bar{\mathbf{v}}_1(z) \exp\left[i\left(\omega t - k^x x - k^y y\right)\right], \tag{10.16}$$

$$\rho_1 \equiv \bar{\rho}_1(z) \exp\left[i\left(\omega t - k^x x - k^y y\right)\right], \tag{10.17}$$

$$p_1 \equiv \bar{p}_1(z) \exp\left[i\left(\omega t - k^x x - k^y y\right)\right]. \tag{10.18}$$

The pre-factors with a bar set the weights of the normal mode and its dependence on z. The full expressions are very much akin to sound wave solutions in the $x - y$ plane, with $k^2 = (k^x)^2 + (k^y)^2$, and where ω is allowed to be imaginary. Without specifying k^x, k^y or ω, these expressions are generic and can serve as a stand-in for any Fourier term of the full perturbation (the exercise can be repeated for perturbations running in the other direction by taking $k \rightarrow -k$). For real ω, the perturbation behaves like a sound wave, but a complex-valued ω may lead to an exponential growth of the perturbation.

Plugging the normal mode into the previous expressions for the perturbed fluid state, we find

$$\rho_0 i \omega v_1^x = ik^x p_1, \tag{10.19}$$

$$\rho_0 i \omega v_1^y = ik^y p_1, \tag{10.20}$$

$$\rho_0 i \omega v_1^z = -\frac{\partial p_1}{\partial z} - \rho_1 g, \tag{10.21}$$

from Equation 10.11,

$$i\omega \rho_1 = -v_1^z \frac{\partial \rho_0}{\partial z}, \tag{10.22}$$

from Equation 10.14, and

$$-ik^x v_1^x - ik^y v_1^y = -\frac{\partial v_1^z}{\partial z}. \tag{10.23}$$

from Equation 10.15. Here g is the magnitude of the gravitational acceleration, which we assume is pointing downward. We can multiply Equations 10.19 and 10.20 by $-ik^x$ and $-ik^y$ respectively to obtain

$$k^x \rho_0 \omega v_1^x + k^y \rho_0 \omega v_1^y = k^2 p_1. \tag{10.24}$$

Together with Equation 10.23, this leads to

$$k^2 p_1 = -i\rho_0 \omega \frac{\partial v_1^z}{\partial z},$$

(10.25)

which we store for further use while en-route to finding some expression for ω that tells us whether complex values for ω can occur after filling in the details of a given system. Meanwhile, Equations 10.21 and 10.22 can be combined to show

$$\frac{\partial p_1}{\partial z} = -i\rho_0 v_1^z \omega - \rho_1 g = -i\rho_0 v_1^z \omega - i\frac{g}{\omega} v_1^z \frac{\partial \rho_0}{\partial z}.$$

(10.26)

Differentiating Equation 10.25 with respect to z, multiplying Equation 10.26 by k^2 and subtracting the two in order to eliminate the shared term involving p_1 then ultimately gets us to

$$\frac{\partial}{\partial z}\left(\rho_0 \frac{\partial v_1^z}{\partial z}\right) - \rho_0 k^2 v_1^z = \frac{k^2}{\omega^2} g \frac{\partial \rho_0}{\partial z} v_1^z.$$

(10.27)

As a first application of this result, we can consider the simple case of two homogeneous inviscid fluid states separated at $z = 0$. The fluid on top (T) may or may not be heavier than the fluid on the bottom (B). We expect water on top of oil to be RT unstable, but oil on water to be stable. We have set up a contact discontinuity, where $\mathbf{v}$ and p will be continuous but the density, as stated, may differ. Within a homogeneous layer of given density ρ_0, Equation 10.27 reduces to

$$\left(\frac{\partial^2}{\partial z^2} - k^2\right) v_1^z = 0.$$

(10.28)

Any perturbation at the contact discontinuity will need to be consistent with this expression further away from the CD and diminish with increasing distance. Therefore, the generic solution to Equation 10.28 that is continuous in v_1^z across the divide is

$$v_1^z = \begin{cases} A\exp[-kz], & \text{for } z > 0, \\ A\exp[+kz], & \text{for } z < 0, \end{cases}$$

(10.29)

where A an arbitrary constant. Exactly *at* the contact discontinuity ρ_0 and $\partial v_1^z/\partial z$ are undetermined, but here we can deploy the same approach that we used to establish the shock-jump conditions in Section 7.1 and integrate Equation 10.27 over a tiny interval across the discontinuity. Because v_1^z is continuous across $z = 0$, we have

$$\lim_{h\downarrow 0} \int_{-h}^{h} \frac{\partial \rho_0}{\partial z} v_1^z dz \approx v_1^z(z=0) \lim_{h\downarrow 0} \int_{-h}^{h} \frac{\partial \rho_0}{\partial z} dz \approx A\left(\rho_{0,T} - \rho_{0,B}\right).$$

(10.30)

Integrating all of Equation 10.27 therefore leads to

$$-\rho_{0,T} Ak - \rho_{0,B} Ak = \frac{k^2}{\omega^2} gA\left(\rho_{0,T} - \rho_{0,B}\right),$$

(10.31)

which we can rewrite to find a constraint on ω:

$$\omega^2 = kg\frac{\rho_{0,B} - \rho_{0,T}}{\rho_{0,B} + \rho_{0,T}}. \tag{10.32}$$

This allows for solutions for ω of the form

$$\omega = \pm\sqrt{gk\frac{\rho_{0,B} - \rho_{0,T}}{\rho_{0,B} + \rho_{0,T}}}. \tag{10.33}$$

If the top fluid is lighter than the bottom fluid, these roots are real and perturbations of the type $\propto \exp[i\omega t]$ will oscillate, but not grow, over time. However, if the top fluid is heavier, the roots will be complex and include a perturbation that does increase over time and leads to an exponential growth of the instability.

The two-dimensional illustration shown in Figure 10.2 is for an ideal inviscid gas with $\hat{\gamma} = 1.4$. Although an ideal gas is not officially incompressible, our stability analysis applies qualitatively. The top layer was given a density $\rho = 2$ in code units[2], twice as large as the bottom layer density. The pressure was set to decrease continuously according to $p = 2.5 - 0.1\rho z$ (again in code units). The horizontal (the 'x-direction') velocity was set to zero, but a small perturbation was added in the vertical direction, $v_1^z = 0.025[1 + \cos(4\pi x)] \times [1 + \cos(3\pi z)]$ (code units). At first the instabilities grow where the fluid is perturbed the strongest. It is interesting to see how eventually the symmetry between the mushroom shapes breaks down by the appearance of further smaller features that follow from numerical noise in the simulation. This illustrates an common dilemma for numericists: while the instabilities are real and should indeed appear over time, one should remain careful to separate the physical (the appearance of the instability), from the numerical (the time and manner of appearance).

10.3 THE KELVIN-HELMHOLTZ INSTABILITY

Another famous fluid instability is the Kelvin-Helmholtz ('KH') instability, which occurs along the interface of two layers of gas in relative motion. Figure 10.3 shows an illustration of this effect, again using an inviscid gas with $\hat{\gamma} = 1.4$ and the same numerical methods as used in the RT simulation. The KH instability can sometimes be seen along the edges of clouds, and occurs in diverse astrophysical settings including planetary atmospheres (e.g. Jupiter's red spot) and along the edges of jets. We shall see that the KH instability is very easy to trigger, pretty much happening whenever shearing flows occur, even if certain modes of the instability can be suppressed by viscosity, surface tension and/or gravity. Like the RT instability, the KH instability folds and wrinkles the contact surface between fluid layers. Unlike the RT instability or thermal convection, the KH instability is not a buoyancy-driven instability. The

---

[2]The simulation was performed on a grid of 1000×1000 zones using a finite volume HLLC method, a third order Runge-Kutta time step and second order spatial reconstruction using a 'minmod' slope limiter (the technical terms here are explained in Chapter 13).

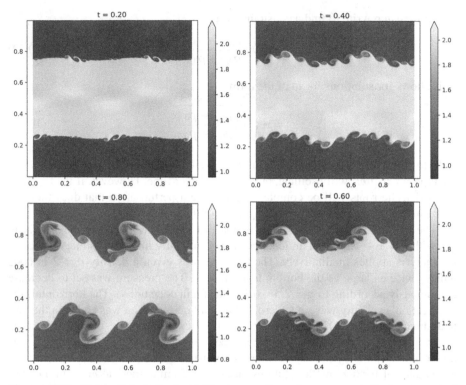

Figure 10.3 Kelvin-Helmholtz instability in two dimensions, computed using numerical methods from Chapter 13. Time increases in the clockwise direction.

KH instability therefore does not require the presence of gravity or other external acceleration field, but can certainly occur when these are present, creating a situation where KH unstable flow creates RT unstable flow patterns and vice versa.

The stability analysis for the KH case proceeds largely similar to that of the RT case from the previous section. Again, consider two layers of incompressible fluid varying at most in the z-direction, a top layer (T) and a bottom layer (B). This time, however, we need to account for the presence of a non-zero flow velocity $\mathbf{v}_0$. Retracing our steps from the RT analysis (but ignoring gravity for the sake of simplicity), we therefore get

$$\rho_0 \frac{\partial v_1^x}{\partial t} + \rho_0 v_0^x \frac{\partial v_1^x}{\partial x} + \rho_0 v_1^z \frac{\partial v_0^x}{\partial z} = -\frac{\partial p_1}{\partial x}, \tag{10.34}$$

$$\rho_0 \frac{\partial v_1^y}{\partial t} + \rho_0 v_0^x \frac{\partial v_1^y}{\partial x} = -\frac{\partial p_1}{\partial y}, \tag{10.35}$$

$$\rho_0 \frac{\partial v_1^z}{\partial t} + \rho_0 v_0^x \frac{\partial v_1^z}{\partial x} = -\frac{\partial p_1}{\partial z}, \tag{10.36}$$

$$\frac{\partial \rho_1}{\partial t} + v_0^x \frac{\partial \rho_1}{\partial x} = -v_1^z \frac{\partial \rho_0}{\partial z}. \tag{10.37}$$

Here, we assume the non-zero flow $\mathbf{v}_0$ to be aimed along the x-direction. For a mode with the same structure as for the RT case, proportional to $\exp[i(\omega t - k^x x - k^y y)]$, we find

$$\rho_0 i \omega v_1^x - \rho_0 v_0^x i k^x v_1^x + \rho_0 v_1^z \frac{\partial v_0^x}{\partial z} = i k^x p_1, \tag{10.38}$$

$$\rho_0 i \omega v_1^y - \rho_0 v_0^x i k^x v_1^y = i k^y p_1, \tag{10.39}$$

$$\rho_0 i \omega v_1^z - \rho_0 v_0^x i k^x v_1^z = -\frac{\partial p_1}{\partial z}, \tag{10.40}$$

$$i \omega \rho_1 - v_0^x i k^x \rho_1 = -v_1^z \frac{\partial \rho_0}{\partial z}. \tag{10.41}$$

Assuming incompressible flow leads to a direct repeat of Equation 10.23, now for the KH case:

$$-i k^x v_1^x - i k^y v_1^y = -\frac{\partial v_1^z}{\partial z}. \tag{10.42}$$

If we take Equations 10.38 and 10.39 and follow along the same path as in our analysis of the RT instability, we arrive at the analog of Equation 10.25, now including terms that account for nonzero velocity component v_0:

$$k^2 p_1 = i \rho_0 \left(v_0^x k^x - \omega \right) \frac{\partial v_1^z}{\partial z} - i \rho_x k^x \frac{\partial v_0^z}{\partial z}. \tag{10.43}$$

This result can be combined with Equation 10.40 to form the analog of Equation 10.27, minus gravity but plus the non-zero $\mathbf{v}_0$ contribution:

$$\frac{\partial}{\partial z} \left\{ \rho_0 \left(\omega - v_0^x k^x \right) \frac{\partial v_1^z}{\partial z} + \rho_0 k^x \frac{\partial v_0^z}{\partial z} v_1^z \right\} - \rho_0 k^2 \left(\omega - v_0^x k^x \right) v_1^z = 0. \tag{10.44}$$

Because the KH instability is a shear instability rather than a buoyancy-driven instability, we need to look at a system involving shear flows in order to explore the implications of Equation 10.44. We consider a setup of two layers of gas flowing past each other with a relative bulk velocity along their contact surface. The layers are homogeneous and we leave open the possibility that their densities differ too across the contact surface. Using Equation 10.44 it can be shown that for an ideal gas (and in the absence of mitigating factors such as surface tension) the KH instability will inevitably develop at some point once the velocities of the layers differ, no matter how small the difference. Our setup resembles the top half of the one shown in Figure 10.3, although there a perturbation is injected manually to speed up the process.

Within a given layer the unperturbed state of the fluid is homogeneous and Equation 10.44 reduces to

$$\rho_0 \left(\omega - v_0^x k^x \right) \frac{\partial^2 v_1^z}{\partial z^2} - \rho_0 k^2 \left(\omega - v_0^x k^x \right) v_1^z = 0, \tag{10.45}$$

which implies that solutions to v_1^z obey

$$\frac{\partial^2 v_1^z}{\partial z^2} - k^2 v_1^z = 0, \tag{10.46}$$

familiar from Equation 10.28. Solutions for v_1^z in the top and bottom layer obey

$$v_1^z = \begin{cases} A_T \exp[-kz], & \text{for } z > 0, \\ A_B \exp[+kz], & \text{for } z < 0, \end{cases} \tag{10.47}$$

if the perturbation is to decrease away from the contact surface. Note that we do *not* assume that $A_T = A_B$. Our set-up differs from the RT case on account of the presence of discontinuous bulk flow velocity v_0^x even in the unperturbed state, and it is not a given that v^z is continuous in this situation (in our previous discussions about shock fronts and contact discontinuities we have always allowed the shock front to advect with the flow in the direction along the front, but here there is no such possibility because there is no unique flow velocity in this direction).

Instead of assuming v_1^z to be continuous, we will need to take a look at the actual deformation of the contact surface between the fluid layers. Because for any consistent solution, both layers should be in agreement about how they wrinkle the surface layer and the surface layer should remain connected to the adjacent fluid. To account for the deformation of the contact surface between the fluid layers, we follow the z-coordinate at the perturbation point over time:

$$\frac{d}{dt} z_{1,s} = v_{1,s}^z \Rightarrow (v_0^x + v_1^x) \frac{\partial}{\partial x} z_{1,s} + \frac{\partial}{\partial t} z_{1,s} = v_{1,s}^z. \tag{10.48}$$

We attach a little label s to indicate that this expression refers to the contact surface specifically. For the generic mode under consideration, we obtain to lowest order that

$$z_{1,s} = \frac{v_{1,s}^z}{i(\omega - v_0^x k^x)}. \tag{10.49}$$

The RHS should be identical between an approach from below and an approach from the top. We therefore modify Equation 10.47 to:

$$v_1^z = \begin{cases} A\left(\omega - v_{0,T}^x k^x\right) \exp[-kz], & \text{for } z > 0, \\ A\left(\omega - v_{0,B}^x k^x\right) \exp[+kz]. & \text{for } z < 0, \end{cases} \tag{10.50}$$

The solution for v_1^z will come in useful once we integrate Equation 10.44 across the contact surface. Taking the tiny integration interval approach, we have a jump condition

$$\rho_{0,T}\left(\omega - v_{0,T}^x k^x\right) \frac{\partial v_{1,T}^z}{\partial z} = \rho_{0,B}\left(\omega - v_{0,B}^x k^x\right) \frac{\partial v_{1,B}^z}{\partial z}. \tag{10.51}$$

If we indeed apply the solution for v_1^z, this becomes

$$-\rho_{0,T}\left(\omega - v_{0,T}^x k^x\right)^2 = \rho_{0,B}\left(\omega - v_{0,B}^x k^x\right)^2. \tag{10.52}$$

This is a quadratic expression for ω that we can solve:

$$\omega = \frac{2k^x\left(\rho_{0,T} v_{0,T}^x + \rho_{0,B} v_{0,B}^x\right) \pm 2k^x \sqrt{-\rho_{0,T}\rho_{0,B}\left(v_{0,B}^x - v_{0,T}^x\right)^2}}{2\left(\rho_{0,T} + \rho_{0,B}\right)}. \tag{10.53}$$

It is the square-root term that gives us the stability information that we need. Whenever a non-zero difference between $v_{0,B}^x$ and $v_{0,T}^x$ occurs, ω will be in part imaginary, which can lead to an exponentially growing instability. Viscosity and surface tension can act to mitigate this instability, while including a gravitational acceleration g in our example can be shown to suppress the instability at low wave numbers, leading to a minimum unstable wave number

$$k_{min} = \frac{g\left(\rho_{0,B} - \rho_{0,T}\right)}{\rho_{0,B}\rho_{0,T}\left(v_{0,B}^x - v_{0,T}^x\right)^2}. \tag{10.54}$$

This can be shown by tracing the steps of the derivation for ω in this section without the $g = 0$ assumption (see also Section 10.2 to see how the gravity term propagates through the derivation).

10.4 GRAVITATIONAL INSTABILITY

The stability analysis of the previous sections showed a procedure akin to our derivation of the wave equation for sound waves, albeit with the twist that the resulting waves solutions under unstable conditions could contain a growth factor that would eventually lead to a disruption of the fluid state rather than a mere perturbation. The Rayleigh-Taylor instability achieved the disruption under the influence of an *external* gravity field, but there exists another well-known fluid instability that involves *self-gravitating* systems. This instability is called the *gravitational instability*, or *Jeans instability*, after the scientist who formulated it first.

In the case of self-gravity, we can use the potential formulation from Section 4.2.1, where we first introduced a gravitational potential Ψ in terms of local density and gravitational constant G. We arrived at Poisson's equation $\nabla^2\Psi = 4\pi G\rho$, which tells us that a perturbation in density inevitably comes with a perturbation in the gravitational field. The latter is of course completely obvious, even without Poisson's equation. What is perhaps less obvious is whether there exist conditions somewhere in the universe where such a perturbation can trigger unstable growth behaviour.

Our usual method of taking a motionless unperturbed base state and a linear perturbation on top, suggests as before a velocity $\mathbf{v} = 0 + \mathbf{v}_1$, a pressure $p = p_0 + p_1$, a density $\rho_0 + \rho_1$ and now a gravitational potential $\Psi = \Psi_0 + \Psi_1$. The perturbation of the potential obeys Poisson's equation as well, which can quickly be seen to be $\nabla^2\Psi_1 = 4\pi G\rho_1$. Taking the velocity of the base state to be zero, the velocity equation

$$\frac{\partial \mathbf{v}}{\partial t} + \mathbf{v} \cdot \nabla \mathbf{v} = -\frac{1}{\rho}\nabla p - \nabla\Psi, \tag{10.55}$$

reduces to

$$\frac{1}{\rho_0}\nabla p_0 = -\nabla\Psi_0, \tag{10.56}$$

prior to perturbation. Taking our base state to be completely homogeneous, like we did when deriving the wave equation for sound waves, would therefore imply a constant Ψ_0 as well. But Poisson's equation tells us that this in turn requires ρ_0 to be zero for consistency: either there is no mass at all in the base state, which leaves us empty-handed, or the amount of mass that is present is too small to be gravitationally relevant (i.e. the RHS of Poisson's equation is approximately zero), at which point we would have to concede that a minor perturbation ρ_1 on top of this mass is not going to make a noticeable difference either by construction. Absent the requirement that the base state be homogeneous, the first-order perturbation expressions for mass conservation and the velocity equation take the forms

$$\frac{\partial \rho_1}{\partial t} + \rho \nabla \cdot \mathbf{v}_1 + \mathbf{v}_1 \cdot \nabla \rho_0 = 0, \tag{10.57}$$

and

$$\frac{\partial \mathbf{v}_1}{\partial t} = -\frac{1}{\rho_0} \nabla p_1 + \frac{\rho_1}{\rho_0^2} \nabla p_0 - \nabla \Psi_1, \tag{10.58}$$

respectively. A comparison with the derivation of the wave equation from Section 5.1 shows which terms still clutter these expressions if the base state is not assumed homogeneous, and which therefore disrupt our derivation of a wave equation even if the gravitational potential term were absent (these would be the terms involving ∇p_0 and $\nabla \rho_0$). To get around this, we will proceed as follows: we just 'pretend' these terms can be ignored and proceed as if they did not exist. This is akin to assuming a homogeneous base state and ignoring the resulting inconsistency in Poisson's equation. We are in good company when doing so, as this follows in the footsteps of the original derivation by Jeans, and became known as the 'Jeans swindle'. We then revert to standard procedure towards a wave equation and obtain

$$\frac{\partial^2 \rho_1}{\partial t^2} + \rho_0 \nabla \cdot \frac{\partial \mathbf{v}_1}{\partial t} = -0, \tag{10.59}$$

and

$$\rho_0 \nabla \cdot \frac{\partial \mathbf{v}_1}{\partial t} = -\nabla^2 p_1 - \rho_0 \nabla^2 \Psi_1. \tag{10.60}$$

These two equations can be combined to eliminate the term involving the velocity perturbation that they have in common. An application of Poisson's equation and the definition of the sound speed c_s for an isentropic perturbation then ultimately yields

$$\frac{\partial^2 \rho_1}{\partial t^2} - c_s^2 \nabla \rho_1 - \rho_0 4\pi G \rho_1 = 0. \tag{10.61}$$

We assume ρ_1 to follow the standard form of a perturbation wave, and posit a trial solution

$$\rho_1 = \bar{\rho}_1 \exp\left(i\omega t - i\mathbf{k} \cdot \mathbf{r}\right), \tag{10.62}$$

in line with our trial solutions from the preceding sections. Using this in Equation 10.61, it follows that

$$\omega^2 = c_s^2 \left(k^2 - \frac{4\pi G \rho_0}{c_s^2} \right) = 0. \tag{10.63}$$

In the absence of gravity, we are back to a pure wave solution. But in the presence of gravity, we will have an imaginary solution for ω (and thus the potential for a growing perturbation amplitude) for wave numbers that are smaller than a limiting value

$$k_j = \sqrt{\frac{4\pi G \rho_0}{c_s^2}}. \tag{10.64}$$

Before we explore the implications of this statement, a quick remark on the 'Jeans swindle' we made use of above. Really, of course, Jeans was a gambler rather than a swindler, and if we were to repeat the argument more carefully for a consistent base state (maybe a self-gravitating disc, if exploring the instability in the context of galaxy formation), we would find a very similar outcome that would confirm that Jeans' gamble paid off in that it delivered a qualitatively correct result.

The spatial scales associated with k_j are vast for reasonable values of density and sound speed. Consider the ISM in a galaxy, at $\rho \sim m_p$ cm^{-3}, with a sound speed of, say, 10^6 cm s^{-1} (reflecting the temperature of the ISM). The wavelength $\lambda_j = 2\pi/k_j$ (the *Jeans length*) is on the order of kiloparsecs:

$$\lambda_j = 1.7 \left(\frac{\rho_0}{m_p \text{ cm}^{-3}}\right)^{-\frac{1}{2}} \left(\frac{c_s}{10^6 \text{ cm s}^{-1}}\right) \text{ kiloparsec.} \tag{10.65}$$

This is indeed comparable to the size scale of galaxies, lending plausibility to the notion that galaxies can be formed through the Jeans instability from a background that can be reasonably homogeneous. The initial mass densities were presumably smaller to begin with, but then, the initial distance scale would have been larger. The wavelength scale also shows that a semi-homogeneous object of galaxy scale is not expected to be gravitationally unstable under new perturbations, since it would be too small to contain such a perturbation.

Of course, this is mostly a back-of-the-envelope argument for the scale size of galaxies rather than a rigorous examination. The latter is typically done using massive hydrodynamics simulations these days, simulations that include non-baryonic components such as dark matter that actually make up the bulk of the mass content of the universe[3]. In the context of cosmology, a more complete argument would have to also include the expansion of the universe, which would significantly affect the analysis.

We wrap up by mentioning the concept of the *Jeans mass* M_j. Assuming indeed a roughly homogeneous environment, the total mass contained within a volume of size λ_j^3 is given by $M_j \sim \rho_0 \lambda_j^3$, or

$$M_j \sim 1.3 \times 10^8 \left(\frac{\rho_0}{m_p \text{ cm}^{-\frac{1}{2}}}\right)^{-\frac{1}{2}} \left(\frac{c_s}{10^6 \text{ cm s}^{-1}}\right)^3 M_\odot. \tag{10.66}$$

The Jeans mass can be used to demonstrate that the sound crossing time for a distance λ_j is comparable to the free-fall timescale of a volume λ_j^3 containing M_j. The free-fall velocity is given by $v \sim \sqrt{GM/\lambda_j}$ (from converting gravitational potential into

---

[3]Famous examples of cosmological simulations include the *Millennium* [51] and *IllustrisTNG* [41] simulations.

kinetic energy across a distance λ_j), and we have

$$v \sim \sqrt{G\rho_0 \lambda_j^2} = G^{\frac{1}{2}} \rho_0^{\frac{1}{2}} \pi^{\frac{1}{2}} c_s G^{-\frac{1}{2}} \rho_0^{-\frac{1}{2}} \sim c_s. \quad (10.67)$$

This helps with our interpretation of the Jeans length as indicating a tipping point beyond which sound waves would not be able to dissipate perturbations fast enough to prevent the fluid within the volume from collapsing gravitationally.

> **PROBLEM 10.2**
> **Star-formation in cores of molecular clouds**
>
> Consider a star-forming core of a giant molecular cloud. Assume a total mass $M \sim 10^3 M_\odot$ and a diameter $D \sim 1$ parsec. Assume the core to be homogeneous and predominantly filled with neutral hydrogen gas atoms with mass $m_H = 1.67 \times 10^{-24}$ g. Assume star-formation takes place because the mass locally contracts under the influence of the Jeans instability. If this process occurs, what will the temperature of the core be at most?

10.5   THERMAL INSTABILITY

Often in an astrophysical fluid, heating and cooling terms are in balance. Consider the energy equation,

$$\frac{\partial \mathcal{E}}{\partial t} + \nabla \cdot ([\mathcal{E} + p] \mathbf{v}) = -\rho \dot{q}_{cool} + \mathbf{v} \cdot \mathbf{f}, \quad (10.68)$$

or any generalization thereof (including viscosity, magnetic fields, relativity, etc.) with a net cooling term $\dot{q}_{cool}$ defined per unit mass[4]. In the case of balance, we have $\dot{q}_{cool} = 0$. But the specific net cooling rate $\dot{q}_{cool}$ will, in general, be a function of temperature. Isobaric perturbations of the cooling rate can be expressed in terms of a Taylor series about the equilibrium value:

$$\dot{q}_{cool} = \dot{q}_{cool}|_0 + \left(\frac{\partial \dot{q}_{cool}}{\partial T}\right)_p \Delta T + O\left(\Delta T^2\right). \quad (10.69)$$

This suggests an instability criterion known as *Field's criterion* of the form

$$\left(\frac{\partial \dot{q}_{cool}}{\partial T}\right)_p < 0, \text{ for an unstable fluid.} \quad (10.70)$$

Because if this derivative is smaller than zero a positive change in temperature $\Delta T > 0$ would lead to a *decreasing* rate of cooling according to the series expansion and thus to further heating. Similarly, a negative change in temperature would lead to further cooling since it would *increase* the rate of cooling.

---

[4]Note that we defined this as a *cooling* term rather than a *heating* term, due to our use of a minus sign. This is mostly for historical reasons, following the original approach to the Field criterion for thermal instability [12].

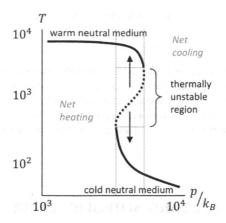

Figure 10.4 A schematic plot of the thermal instability in the interstellar medium.

The neutral gas in the interstellar medium predominantly consists of atomic hydrogen ('H I'). Such H I clouds form a good example of the impact of the thermal instability in an astrophysical setting. H I clouds can be heated in various ways, through ionization by impinging cosmic rays, photoionization of the various constituents (dust, H, He, and other elements) by light from stars or X-ray sources and through shock-heating and magnetohydrodynamical processes. The clouds cool through a range of line-emission processes where photons are released by temperature-dependent rates of deexcitation (in particular, the [C II] 158 μm and [O I] 63 μm fine structure lines; Lyman α cooling becomes dominant only at temperature $T \gtrsim 10^4$ K). The interplay of heating and cooling processes allows for a continuum of equilibrium states. This is illustrated in Figure 10.4, plotting temperature versus pressure $p = nk_BT$. At low pressure, we have an unambiguous equilibrium at high temperature, describing the *warm neutral medium* (WNM), while high pressure conditions favour the *cold neutral medium* (CNM) state. In between WNM and CNM however, roughly between 3200 cm$^{-3}K \lesssim p/k_B \lesssim 4400$ cm^{-3}K, the $p - T$ curve is multivalued. The WNM and CNM can co-exist at the same pressure level (and do so in practice). Perturbations away from the intermediate unstable equilibrium lead to either further cooling or further heating as indicated in the figure, driving the gas respectively towards a CNM or WNM state.

PROBLEM 10.3
Cooling due to thermal Bremsstrahlung

Bremsstrahlung is an example of a radiative process driven by encounters between charged particles, for example free electrons and positively charged ions. It can act as a cooling mechanism where kinetic energy from charged particles is converted into photons that subsequently leave the system.

i) Assume a constant specific heating rate $\mathscr{H}$ where energy is delivered to the system by an external source (e.g. cosmic rays or stellar radiation). We parametrize the net specific cooling rate around the equilibrium value according to

$$\dot{q}_{cool} = C\rho T^{\alpha} - \mathscr{H}, \tag{10.71}$$

where C and α two constants that are left unspecified for now. Motivate why this expression can be expected to apply to collisional processes like Bremsstrahlung.

ii) Thermal Bremsstrahlung, produced by charged particles that are in Maxwellian distributions, will produce a total power output that is proportional to $\sqrt{T}$. Show that this leads to a thermally unstable plasma. You may assume the ideal gas law to apply.

10.6 HOMOGENEOUS AND ISOTROPIC TURBULENCE

It was mentioned at the beginning of this chapter that turbulence is a complex subject acting across a wide range of scales within a fluid. A full universal theory for turbulence does not exist, even if there is a vast literature available on the subject. Nevertheless, whether we like it or not, many if not most phenomena within astrophysics to some degree display turbulence, be it accretion flows, the interstellar medium, stellar structures, galaxies or anything else. Let us therefore devote some space to at least make some observations about the subject without delving too deeply into how turbulence really works and without even attempting to rigorously define the concept.

In colloquial terms, turbulence appears to be a natural outcome of unstable behaviour in a fluid. The demonstrations of growing instabilities in Figs. 10.2 and 10.3 show how growing instability on one scale both cascades downwards to smaller scales and upwards to larger scales within the system, leading to a disruption of the original configuration by complicated flow patterns rapidly varying with distance.

Any first introduction to turbulence relies on the fundamental work done by Russian academics Andrey Kolmogorov and Alexander Obukhov in the mid-twentieth century, in particular, Kolomogorov's law about the spectrum of turbulent energy distribution across distance scales in fully developed turbulence. The interstellar medium is a striking example obeying this law across a wide range of scales, showing that there is some insight in actual astrophysical phenomena to be gathered from considering turbulence in the limit where it has become homogeneous, isotropic and has achieved a steady-state.

For the sake of simplicity, we assume the fluid to be incompressible (you might recall we also reached for this simplifying assumption in our analyses of the Rayleigh-Taylor and Kelvin-Helmholtz instabilities in Sections 10.2 and 10.3 respectively). An incompressible fluid strictly speaking does not allow for sound waves, as mentioned in closing in Section 5.1. More generally, any perturbation of an incompressible fluid will not include longitudinal disturbances, but rather be limited to transversal motion relative to the wave vector of the perturbation. For such flow $\nabla \times \mathbf{v} \neq 0$, and we might as well consider turbulent flow in terms of *eddies*, a picture at least visually supported by the KH and RT images from this chapter.

In fully developed turbulence, eddies occur at all viable spatial scales, the largest spatial scale L being that of the hydrodynamical system under investigation. In turbulence that is both homogeneous and isotropic, no other scales occur in the system other than the velocity scale V of the system (or, equivalently, a characteristic time scale T related to changes in the flow over a length scale L according to $V \sim L/T$). We define the rate ε_{rate} at which specific energy is transported across scales in the flow. This drives the generation of smaller and larger eddies out of some initial perturbation. Dimensional analysis suggests

$$\varepsilon_{rate} = \frac{[erg]}{[g]} \frac{1}{[s]} \propto V^2 \frac{1}{T} \propto \frac{V^3}{L}. \tag{10.72}$$

In a steady state, ε_{rate} must also equal the energy passing through eddies at some smaller scale L' and with associated velocity V', in order to prevent energy from accumulating at an intermediate scale. In the absence of any other scales to the problem (physical mechanisms, boundary conditions, etc.) that would introduce a pre-factor radically different from unity, *Kolmogorov's law* follows from dimensional analysis:

$$\varepsilon_{rate} \sim \frac{V^3}{L} \sim \frac{(V')^3}{L'} \Rightarrow V' \sim V \left(\frac{L'}{L} \right)^{\frac{1}{3}}. \tag{10.73}$$

According to this expression, larger eddies will have more velocity and thus turbulent energy, while the smaller eddies have higher vorticity $\omega \sim V'/L'$.

The process of energy transfer to smaller scales cannot continue indefinitely. Either we end up below length scales at which the system can still be treated as a fluid or at length scales where the transfer rate of turbulent energy starts to lose out against the viscous dissipation rate. It turns out that the latter limit is encountered before the former, as we now proceed to show. Equation 9.32 suggests a viscous dissipation rate per unit mass $\rho^{-1}\eta(V/L)^2 \sim \nu(V/L)^2$ for a given velocity scale V, length scale L and kinematic viscosity coefficient ν. In Section 9.3, it was argued that shear viscosity introduced at the microphysical level obeyed $\nu \sim \lambda \tilde{u}$, with λ the mean free path length scale and $\tilde{u}$ the velocity scale for thermal motion of individual particles in the fluid. Using all this when equating the viscous dissipation rate and the turbulent energy transfer rate to see at which length scale the two become comparable, we find

$$\nu \left(\frac{V_0}{L_0} \right)^2 \sim \frac{V_0^3}{L_0} \Rightarrow \tilde{u}\lambda \sim V_0 L_0, \tag{10.74}$$

as one would expect on dimensional grounds. If V_0 is connected to fluid flow on larger scales according to Kolmogorov's law, this implies that

$$\tilde{u}\lambda \sim V \left(\frac{L_0}{L} \right)^{\frac{1}{3}} L_0. \tag{10.75}$$

At normal values for the bulk flow velocity V, we expect $V \ll \tilde{u}$, certainly if V is to remain subsonic to ensure that the assumption of incompressibility applies at least approximately. Velocities V' at smaller scales down to a velocity scale V_0 will be increasingly smaller than V, as reflected in the cubic root term on the RHS obeying $(L_0/L)^{1/3} \ll 1$. It therefore follows that we must have $L_0 \gg \lambda$, in order to satisfy the approximate equality in Equation 10.75.

The connection between different turbulence length scales can be cast in terms of the Reynold's number $\mathrm{Re} \equiv VL/v$:

$$v \sim V\left(\frac{L_0}{L}\right)^{\frac{1}{3}} L_0 \Rightarrow L \sim \frac{VL}{v}\left(\frac{L_0}{L}\right)^{\frac{1}{3}} L_0 \Rightarrow L \sim \mathrm{Re}^{3/4}L_0. \qquad (10.76)$$

If the Reynold's number is about 1 or below, the flow is not only viscous but also leaves no range in length scales between which a turbulent cascade can develop, as the smallest allowed turbulent length scale is similar to that of the size of the system. On the other hand, if $\mathrm{Re} \gg 1$, room exists for turbulence to develop. In practice, this indeed happens once $\mathrm{Re} \sim 3000$.

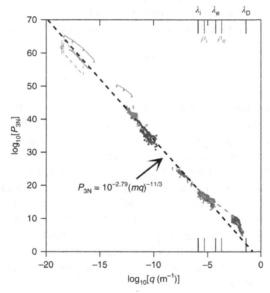

Figure 10.5 Turbulence in the interstellar medium, including measurements using Voyager 1, as indicated by a power spectrum describing density fluctuations as a function of inverse wavelength. The larger the inverse wavelength, the smaller the corresponding spatial scale. Voyager 1 data is shown alongside data from other remote sensing methods (two sets of fainter data points in the top left; Voyager data covers the data at power spectrum values below 45, in units of the figure.)

At the start of this section, we mentioned the turbulent nature of the interstellar medium. This is strikingly illustrated in Figure 10.5, which includes measurements obtained using the Voyager 1 Space Probe[5] up to 2019. Voyager 1 was launched back

---

[5]The figure was taken from [24]. A similar figure from [7] can be found showcased in a text book on the physics of the interstellar medium [10].

in 1977 and by now has moved more than 142 AU away from the sun, which means that it has crossed the heliopause (which marks the point where the pressure from the solar wind and the interstellar medium balance out). Voyager 1 can be used as a probe of the local interstellar medium and this information, along with data about the ISM obtained by different measures, is shown in the figure. On the vertical axis is a measure of the *power spectrum* of the density fluctuations in the medium. On the horizontal axis is the inverse wavelength $1/\lambda$, so the larger λ^{-1}, the smaller the corresponding spatial scale at which the fluctuation is measured. The figure demontrates a remarkable adherence to the predictions from dimensional analysis for a Kolmogorov spectrum across a vast range of scales (the rise at the lower right is argued to be due to a local change in pre-factor, rather than a change in power law for the spectrum). The density fluctuations power spectrum slope of $(1/\lambda)^{-11/3}$ can be shown to follow from the same arguments used to formulate Kolmogorov's law (Equation 10.73), assuming that the density fluctuations follow the local velocity field. The $-11/3$ slope and more famous $-5/3$ slope for turbulent spectra are derived in Problem 10.4.

PROBLEM 10.4
The power laws of Kolmogorov spectra

i) Consider the total specific energy E_{spec} in the fluid. This energy can be decomposed in energy E_k per wavenumber, with each wavenumber $k = 2\pi/L$ describing the energy at an inverse length scale L^{-1}:

$$E_{spec} \equiv \int E_k dk.$$

Using dimensional analysis, demonstrate that the power spectrum in energy obeys the famous relation

$$E_k' = E_k \left(\frac{k'}{k}\right)^{-5/3}.$$

ii) The above is a one-dimensional approach to the power spectrum. The power spectrum can also be computed in each direction:

$$E_{spec} \equiv \int E_{kkk} dk^x dk^y dk^z,$$

where the wavenumbers in each direction all scale with the inverse scale of interest. Show that, in this case, we have

$$E_{kkk}' = E_{kkk} \left(\frac{k'}{k}\right)^{-11/3},$$

where k the wavenumber in any direction (they all are taken to have the same order of magnitude at the scale of interest, so we can drop the superscript).

iii) The average density fluctuation $\langle (\Delta n)^2 \rangle$ can also be expressed using a power spectrum:

$$\langle (\Delta n)^2 \rangle = \int P_{3N} dk^x dk^y dk^z,$$

where we borrow some notation from Figure 10.5 for denoting the power term in the integrand. Under the assumption that local density fluctuations are completely dictated by the velocity field v, such that $\Delta n \propto v$, show that we also have

$$P'_{3N} = P_{3N} \left(\frac{k'}{k} \right)^{-11/3} .$$

Note that this result does not purely follow from dimensional analysis, but indeed unavoidably requires the assumption of the density fluctuation being set completely by the velocity field.

11 Accretion Flow

In astrophysics, one of the more important examples of fluid flow is that of *accretion*. Accretion describes the process by which a massive object grows its mass by gravitationally attracting matter from its environment. During accretion, gravitational potential energy of the infalling matter can be released in the form of radiation. As we shall soon see, this is the most important driver of high-energy radiation in the universe, produced by sources with masses ranging from stellar mass-sized to billions of solar masses at the centres of galaxies. Whenever we have a massive source exerting gravitational pull and a source of fuel, accretion will take place. The fuel, be it interstellar gas or plasma syphoned off directly from a donor star, will be pulled towards the accreting object, which can be a black hole or (compact) star. The accreting matter will heat up while en-route, losing energy in the form of radiation and ultimately release all or a fraction of its energy upon impact. Since the source of this energy is ultimately the gravitational potential energy of the infalling matter, the closer this matter can get to the accreting source before releasing its kinetic energy, the more extreme the amount of radiation that can be is produced. In this chapter, we describe some key features of accretion flow, both spherical and occurring within *accretion discs*[1].

11.1 ACCRETION AS A SOURCE OF ENERGY IN ASTROPHYSICS

The energy release from nuclear fusion converting hydrogen into helium (the dominant fusion process in astrophysics) is about $0.007mc^2$, or 6×10^{18} erg per gram. This is an enormous amount of energy[2], and it is perhaps counter-intuitive to expect the 'mere' infall of gas from large distance to be able to overtake this. For accretion onto an object of radius $R_\star$ and mass M, the release of energy is given by

$$\frac{dE_{acc}}{dm} = \frac{GM}{R_\star},$$

(11.1)

where G the gravitational constant, and assuming the accreted mass initially came from infinity or at least a radius $r \gg R_\star$. Table 11.1 shows example energy releases per gram for a few different stellar-type objects. While for e.g. the Sun or white dwarf stars (that have masses comparable to the Sun's mass), the energy release through accretion will always remain below that through fusion, the balance shifts for neutron stars (that have masses of a few times the Sun's mass) and black holes, suggesting that unobstructed accretion might be the most powerful source of energy for the latter. Black holes can have all sorts of masses, and range from solar mass

---

[1]A full treatise on accretion flow would take us beyond the scope of this book. The book *Accretion power in astrophysics* [14] is recommended for further reading.

[2]For comparison, the energy release associated with the nuclear bomb on Hiroshima is about 630×10^{18} erg (through fission of uranium-235, not through fusion of hydrogen).

DOI: 10.1201/9781003095088-11

Table 11.1

Energy Release Per Gram Through Accretion. The Masses of the Stellar Objects are All Taken to be 1 $M_\odot$, Although for the Black Hole This Does Not Matter. The Black Hole Has No Fixed Surface, and an Efficiency Factor η is Added (Typically $\eta \sim 0.1$). The Energy Release from Fusion of Hydrogen into Helium is About 6×10^{18} erg g^{-1}.

Object	Representative Radius	Energy Release Per Gram (erg g^{-1})	Energy Release (Relative to Fusion)
Sun	$R_\odot \approx 7 \times 10^{10}$ cm	1.9×10^{15}	3×10^{-4}
White Dwarf	10^9 cm	1.3×10^{17}	2×10^{-2}
Neutron Star	10^6 cm	1.3×10^{20}	21
Black Hole	3×10^5 cm	$\eta 4.4 \times 10^{20}$	$\eta 70$

black holes to supermassive black holes at the centres of galaxies. The Schwarzschild radius $R_{sch} = 2GM/c^2$ is typically used to denote the 'effective' radius of a black hole, but since matter can in principle disappear through this radius without leaving a trace, we have to account for this with an uncertainty factor $\eta \sim 0.1$ for the actual released energy. Because the Schwarzschild radius is linearly proportional to the mass, the energy release per gram from Equation 11.1 is independent of black hole size. A solar-mass black hole yields the same result as, e.g. M87, which has a mass $M = 6.6 \times 10^9 M_\odot$. On the other hand, an active galactic nucleus (AGN) such as M87, located in the dense centre of a galaxy, will have a lot more gas available in its environment to accrete, and this is what makes AGN the most powerful continuous sources of energy release in the universe.

11.2 BONDI ACCRETION

As one might expect, the simplest dynamical models of accretion flow assume spherical symmetry. In fact, merely inverting the velocity direction in the simplistic stellar wind model from Section 3.2.1, immediately sets up a simple accretion toy model. In this model, the fluid velocity was not only assumed to be steady but also assumed to be independent of radial distance. The latter assumption however, precludes us from modelling flow that undergoes gravitational acceleration as it gets pulled further into the gravitational potential well of the accreting object. In this section we therefore first turn to an astrophysical example of fluid flow that is in part *subsonic* (moving with velocity $v < c_s$), and in part *supersonic* (moving with $v > c_s$), with the infalling velocity presumably diminishing with increasing distance. The Bondi accretion model combines steady flow (Chapter 2), polytropic gases (Section 3.4) and sub-/supersonic flow (Chapter 5). The main dynamical quantities of interest to compute within a given accretion model are the *accretion rate* and the *radial fluid profile*, both dictated by the properties of the accreting object and its surrounding gas.

11.2.1   THE SONIC POINT

If we let $\dot{M}$ be the accretion rate onto the star and we assume the accretion to be steady, we have at each radius a flux

$$\rho v = -\frac{\dot{M}}{4\pi r^2},\tag{11.2}$$

with $v < 0$ for accretion. While for the stellar wind example we assumed a fixed v, we now relax this condition and allow v to be a function of radius (but not time). In spherical symmetry, we can rewrite the Euler equation (Equation 2.17) as

$$v\frac{dv}{dr} + \frac{1}{\rho}\frac{dp}{dr} = \frac{f}{\rho} = -\frac{GM}{r^2},\tag{11.3}$$

with M the total mass of the accreting object. We only have the variable r left, so the distinction between full and partial derivative disappears. We may also safely assume that M remains unaffected by $\dot{M}$ at the timescales of interest (i.e. $M \gg \dot{M}\Delta t$).

Equation 11.3 is going to be our key to unraveling the infalling gas profile surrounding an accreting cosmic object. First, we are going to check which family of solutions to this equation describes Bondi accretion and whether the infalling gas will at some point cross the sound barrier, that is, whether we have $v = c_s$ at some radius. For this purpose, we can use the continuity equation for steady flow in spherical coordinates and the definition for sound speed $c_s^2 \equiv dp/d\rho$, to rewrite Equation 11.3 as[3]

$$\frac{1}{2}\left(1 - \frac{c_s^2}{v^2}\right)\frac{d}{dr}\left(v^2\right) = -\frac{GM}{r^2}\left[1 - \left(\frac{2c_s^2 r}{GM}\right)\right].\tag{11.4}$$

> **PROBLEM 11.1**
> **Bondi accretion, intermediate step**
>
> Derive Equation 11.4 from Equation 11.3.

Equations 11.4 admits six types of solutions, depending on the behaviour of the flow at large and small radii and on the nature of r_s, which is defined as the point where the term in square brackets on the RHS evaluates to zero. According to this definition, r_s is given by

$$r_s \equiv \frac{GM}{2c_s^2(r_s)}.\tag{11.5}$$

We discuss the six solutions below, leaving for last the solution that is actually realized in the case of Bondi accretion. The solutions are illustrated[4] in Figure 11.1.

---

[3]This equation is known as the Parker wind equation, named after its discoverer [39], who first wrote down this expression in the context of solar winds (another illustration of the similarity between accretion and stellar wind models).

[4]This figure has been adapted from [18]. Versions of it can be found in various text books covering accretion, such as [14].

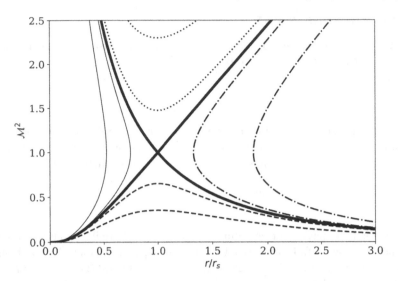

Figure 11.1 Different velocity profiles in terms of their Mach number $\mathcal{M}$. For this figure, $\hat{\gamma} = 4/3$ has been used. Different line styles indicate different solution families, as described in the main text.

Case 1: (*thin solid curves*) $d(v^2)/dr = \infty$ at $v^2 = c_s^2(r_s)$, and $r < r_s$. The flow velocity is multi-valued at radii below r_s and the solution does not cover $r > r_s$. Either shocks are present (which allow for discontinuities in the flow), or this solution is not realized. Bondi accretion does not cover (standing) shocks, so this option does not qualify as such.

Case 2: (*dash-dotted curves*) $d(v^2)/dr = \infty$ at $v^2 = c_s^2(r_s)$ and $r > r_s$. Similarly to case 1, this solution does not apply.

Case 3: (*dotted curves*) $v^2 > c_s^2$ at all points in the flow. $d(v^2)/dr = 0$ at r_s. If the flow is always supersonic, this will also hold at very large distances that are beyond the reach of the star, where the fluid should be at rest. For this reason, this solution is also not realized.

Case 4: (*dashed curves*) $v^2 < c_s^2$ at all points in the flow. $d(v^2)/dr = 0$ at r_s. The flow is always subsonic. This makes sense at large radii, where the velocity goes to zero, reflecting that these regions lie beyond the sphere of influence of the star. On the other hand, below r_s, $d(v^2)/dr$ must be a positive and v^2 a monotonically increasing function of radius, if the signs on both sides of Equation 11.4 are to remain equal at $r < r_s$. The closer we get to to the star, the smaller v^2 becomes. If $v < 0$ and the flow moves inwards, this solution describes a slowly sinking atmosphere. If $v > 0$, we have a stellar 'breeze', or mild type of stellar wind. Either way, this is not the powerful energy release process of Bondi accretion.

Case 5: (*thick solid curve*) $v^2 = c_s^2$ at $r = r_s$, and $v^2 = 0$ for $r \downarrow 0$. In this case, r_s is indeed the *sonic point*, where the flow velocity transitions from subsonic to supersonic. Case 5 suffers from the same problem as case 3. At very large radii the velocity blows up and we have the star exerting an unreasonable influence at large distance. This solution is not realized.

Case 6: (*thick solid curve*) $v^2 = c_s^2$ at $r = r_s$, and $v^2 = 0$ for $r \uparrow \infty$. *This* is Bondi accretion, and the gas strikes the stellar surface at supersonic velocity. Again, r_s is the sonic point. At radii smaller than r_s, the gas is effectively in free fall.

11.2.2 THE BONDI ACCRETION RATE

Let us now proceed to actually solve Equation 11.3, in order to find an accretion rate $\dot{M}$. We will use the polytropic gas equation $p = C\rho^{\hat{\gamma}}$ relating pressure and density through some (unspecified) constant of proportionality C. Rewriting the gravitational force term as a radial derivative, we obtain

$$v\frac{dv}{dr} + \frac{1}{\rho}\frac{dp}{dr} - GM\frac{dr^{-1}}{dr} = 0. \tag{11.6}$$

If we multiply all terms by dr and proceed to integrate with respect to their respective differentials dv, dp and dr^{-1}, we get

$$\frac{1}{2}v^2 + \frac{C\hat{\gamma}}{\hat{\gamma}-1}\rho^{\hat{\gamma}-1} - \frac{GM}{r} = \text{constant}. \tag{11.7}$$

For the second term, we expressed our result in terms of density using the polytropic gas law. For the integration we used the assumption that $\hat{\gamma} \neq 1$, which would lead to a logarithmic term in the solution instead. If we now use the continuity equation (Equation 11.2) to eliminate v in favour of $\dot{M}$ and the definition of the speed of sound for a polytropic gas to clean up the second term, we arrive at

$$\frac{1}{2}\left(\frac{\dot{M}}{4\pi r^2 \rho}\right)^2 + \frac{c_s^2}{\hat{\gamma}-1} - \frac{GM}{r} = \frac{c_s^2(\infty)}{\hat{\gamma}-1}, \tag{11.8}$$

That the constant on the RHS is equal to the sound speed at infinity divided by $(\hat{\gamma}-1)$ follows from considering the solution at $r \to \infty$. Another point of interest is r_s, where $|v| = c_s$, and where we can plug in Equation 11.5. As a result,

$$\frac{c_s^2(r_s)}{2} + \frac{c_s^2(r_s)}{\hat{\gamma}-1} - 2c_s^2(r_s) = \frac{c_s^2(\infty)}{\hat{\gamma}-1} \Rightarrow c_s(r_s) = c_s(\infty)\left(\frac{2}{5-3\hat{\gamma}}\right)^{1/2}. \tag{11.9}$$

The speed of sound at infinite radius is a known quantity, set by the temperature and density of the interstellar medium (ISM) or other environment in which the accreting object resides. Using r_s in the expression for $\dot{M}$ from Equation 11.2, we get

$$\dot{M} = 4\pi r_s^2 \rho(r_s) c_s(r_s). \tag{11.10}$$

If we manage to eliminate r_s as well as $\rho(r_s)$, $c_s(r_s)$ in favour of their values at infinity, we have what we need: an expression for the accretion rate in terms of the properties of the star (in particular, its mass) and the properties of the surrounding medium. Using the properties of the sonic point and a rewritten version of the polytropic gas equation,

$$p = C\rho^{\hat{\gamma}}, \quad c_s^2 = \hat{\gamma}\frac{p}{\rho} \quad \Rightarrow \quad \frac{c_s^2(r_s)}{c_s^2(\infty)} = \left(\frac{\rho(r_s)}{\rho(\infty)}\right)^{\hat{\gamma}-1}, \quad (11.11)$$

we can indeed obtain

$$\dot{M} = \pi G^2 M^2 \frac{\rho(\infty)}{c_s^3(\infty)} \left(\frac{2}{5-3\hat{\gamma}}\right)^{\frac{5-3\hat{\gamma}}{2\hat{\gamma}-2}}. \quad (11.12)$$

PROBLEM 11.2
Bondi accretion, mass flux

Derive Equation 11.12 from Equation 11.10.

Because $\hat{\gamma}$ potentially introduces fractional powers of v and r, it is not generally possible to solve Equation 11.7 for v analytically. The Bondi accretion profile therefore typically needs to be determined numerically. Examples of such solutions are illustrated in Figure 11.2, for two values of $\hat{\gamma}$.

You might have spotted in Equations 11.9 and 11.12 that $\hat{\gamma} = 5/3$ is a special case. Equation 11.9 then implies an infinite sound speed at the sonic point, which is actually not an issue given that r_s is zero for this case. As shown in the figure, the inflow velocity quickly (but asymptotically) approaches the speed of sound for $\hat{\gamma} = 5/3$. Since it never crosses the sound speed barrier, it also never makes it to a stage where fluid considerations no longer matter and the material is freely falling inwards. By contrast, this free-fall velocity is achieved for $\hat{\gamma} = 4/3$ already at an appreciably distance from the source (also shown in the figure). For $\hat{\gamma} = 5/3$, the $\hat{\gamma}$-dependent factor of Equation 11.12 reduces to 1, as can be shown by carefully taking the limit $\hat{\gamma} \rightarrow 5/3$. The polytropic index $5/3$ that applies to classical ideal adiabatic gas is rarely achieved in nature. As gas gets accreted it heats up significantly and will radiate as a result, violating the adiabatic assumption. In practice, the system is then modelled using a smaller value for $\hat{\gamma}$.

As an example application, our analysis of Bondi accretions tells us that accreting from the ISM is unlikely to produce a detectable amount of radiation. The cold ISM roughly has a number density $n \sim 1$ proton cm^{-3}, and a temperature on the order of 10 K, leading to a sound speed on the order of 10 km s^{-1}, and therefore an accretion rate

$$\dot{M} \approx 1.4 \times 10^{11} \left(\frac{M}{M_\odot}\right)^2 \left(\frac{\rho(\infty)}{10^{-24}\text{ gram}}\right) \left(\frac{c_s(\infty)}{10\text{ km s}^{-1}}\right)^{-3} \text{ g s}^{-1}. \quad (11.13)$$

Multiplied by the energy release values from Table 11.1, we get luminosities of around 10^{31} erg s^{-1} at most (neutron stars). For comparison, the Sun's luminosity is 3.826×10^{33} erg s^{-1}. The sun is both a hundred times more luminous as well as 10^5 times closer than the nearest neutron star (PSR J0108-1431, at 130 parsec). It

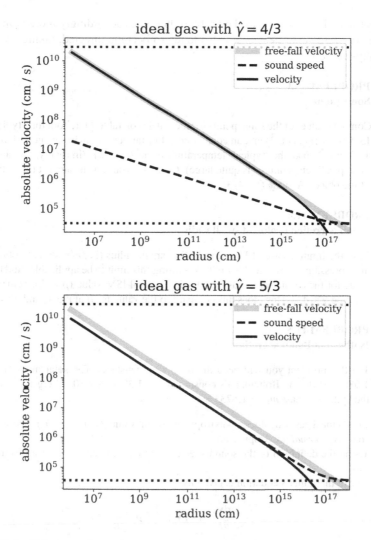

Figure 11.2 Velocity and sound speed profiles for Bondi accretion onto a neutron star of mass $M = 1.4 M_\odot$, from a cold ISM at temperature $T = 10$ K with particle number density $n_0 = 1$ cm^{-3}. The bottom dashed line shows the sound speed at infinity, the top dashed line the speed of light. Numerical solutions can be obtained by solving Equation 11.7 using e.g. a Newton-Rhapson approach, or by solving the differential equation 11.4. In the latter case, one has to be careful around r_s, for which the equation as written reduces to $0 = 0$.

takes a vastly larger supply of gas, such as provided directly by a stellar companion in a binary system, the stellar envelope of a collapsing system or the disc surrounding an AGN, in order for accretion to be visible at cosmic distances. In these cases we have accretion *discs*, where the resulting gas flow geometry is no longer spherically symmetric and the analysis of the problem becomes accordingly more complicated. The energy release in these cases also tends to become directed, taking the form of astrophysical *jets*.

PROBLEM 11.3
Sonic points

Compute values of the sonic points for the entries of Table 11.1, both for a typical ISM ($\rho = 1m_p$ cm^{-3}) and an equally cold but far denser environment with $n = 10^{10}$ cm^{-3}. Take the 'typical' temperature to be $T = 10$ K (in reality, the range of 'typical' temperatures is quite large). Compare your results against the radius of the object. Assume $\hat{\gamma} = 4/3$.

PROBLEM 11.4
Bondi accretion profile at small radius

Take the limiting case of Equation 11.8 at small radius (i.e. $r/r_s \ll 1$) to obtain an expression for $\rho(r)$ in this limit. Assuming this limit to be applicable, find the value for the density at radius $r_s/10$ using typical ISM value ($\rho = 1m_p$ cm^{-3}). Assume $\hat{\gamma} = 4/3$. How does this compare to the value for ρ at $r = r_s$ and $r \uparrow \infty$?

PROBLEM 11.5
Isothermal Bondi accretion

For this problem you will need the following constants. The solar mass $M_. = 1.98892 \times 10^{33}$ g, Boltzmann's constant $k_B = 1.380658 \times 10^{-16}$ erg K^{-1}, and the hydrogen mass $m_H = 1.6733 \times 10^{-24}$ g.

i) For an ideal gas, why does a polytropic exponent value $\hat{\gamma} = 1$ in $p = C\rho^{\hat{\gamma}}$ correspond to an *isothermal* polytropic gas?

ii) Using the definition of the sound speed, show that it is constant for an isothermal gas.

iii) Recall that Euler's equation is given by

$$\frac{\partial \mathbf{v}}{\partial t} + \mathbf{v} \cdot \nabla \mathbf{v} + \frac{1}{\rho} \nabla p = \frac{\mathbf{f}}{\rho}.$$

By using gravity as the accreting force and by assuming spherical symmetry, show that the equation for Bondi accretion of an isothermal gas can be written as

$$\frac{1}{2} v^2 + c_s^2 \ln \left[\frac{\rho}{\rho(\infty)} \right] - \frac{GM}{r} = 0.$$

iv) The sonic point is given by

$$r_s = \frac{GM}{2c_s^2},$$

also for an isothermal gas. Using this and your previous results, show that

$$\dot{M} = \pi G^2 M^2 \frac{\rho(\infty)}{c_s^3} \exp[3/2].$$

v) Calculate the accretion rate of a neutron star with mass $M = M_\odot$, radius $R = 10^6$ cm, accreting from an interstellar medium with density $\rho = m_H$ and temperature of 200 K. Assume $\hat{\gamma} = 1$.

11.3 THE EDDINGTON LUMINOSITY

One of the reasons that we see high accretion luminosity being produced in the presence of discs rather than in response to spherically symmetric infalling matter is that there exists a natural cap on the radiation that spherical accretion can produce: at some point the accretion flow will be stalled by the pressure from the radiation that it generates. The luminosity at which this balance point occurs is called the *Eddington Luminosity*, and can be derived as follows.

The gravitational force on a single ionized hydrogen atom is given by

$$F = \frac{-GM(m_p + m_e)}{r^2} \approx \frac{-GMm_p}{r^2}, \qquad (11.14)$$

where we can ignore the electron mass m_e relative to the proton mass m_p.

At the same time there is a force due to momentum transfer from outgoing photons. Assuming spherical symmetry both for the accretion flow and the resulting source luminosity, the fraction of momentum absorbed by one hydrogen atom on the sky will be proportional to $\sigma_T/4\pi r^2$ at distance r, where $\sigma_T = 6.7 \times 10^{-25}$ cm^2 the Thomson cross section of the electron. This cross section is a factor m_p/m_e smaller for protons, so here we can ignore the protons relative to the electrons, rather than the other way around as for gravity. The factor $1/4\pi r^2$ translates from luminosity to flux, energy per second per unit area. Since the magnitude of the momentum of a photon is given by E/c, the radiation pressure exerts a force given by the time derivative of the momentum that is intercepted

$$F = \frac{\dot{E}}{c} \frac{\sigma_T}{4\pi r^2} = \frac{L\sigma_T}{4\pi r^2 c}. \qquad (11.15)$$

The two forces can be equated to find the luminosity at which they are in balance, i.e.

$$L_{Edd} = \frac{4\pi GMm_p c}{\sigma_T} \approx 1.3 \times 10^{38} \left(\frac{M}{M_\odot}\right) \text{ erg s}^{-1}, \qquad (11.16)$$

where $M_\odot = 1.98892 \times 10^{33}$ g the mass of the sun.

PROBLEM 11.6
Bondi accretion & Eddington luminosity

For this Problem, you will need the Thomson cross section $\sigma_T = 6.7 \times 10^{-25}$ cm^2, proton mass $m_p = 1.6726231 \times 10^{-24}$ g, light speed $c = 2.99792458 \times 10^{10}$ cm s^{-1}, gravitational constant $G = 6.67259 \times 10^{-8}$ dyne cm^2 g^{-2}, solar mass $M_\odot = 1.98892 \times 10^{33}$ g. We will compare accretion energy release and Eddington luminosity for a few sources.

i) Consider a white dwarf star with radius $R_* = 10^9$ cm, and mass comparable to the sun. The star exists in a close binary system and accretes from its binary companion at a steady rate of 10^{16} g s^{-1}. By how many orders of magnitude can it increase this baseline accretion rate before it hits the Eddington limit?

ii) Answer the same question as previous, now for the case of a neutron star of radius 10 km, again of mass comparable to the sun.

iii) Finally, consider an active galactic nucleus (AGN) emitting energy at a luminosity of 10^{47} erg s^{-1} (such luminosities are indeed observed for these sources). Let us say this is done through Bondi accretion onto its central black hole, with efficiency $\eta = 0.1$. What is the corresponding mass accretion rate at this luminosity?

iv) What is the mass of the black hole of this AGN in order for it to radiate exactly at its Eddington luminosity?

11.4   ACCRETION DISCS

Spherical symmetry assumes the accretion flow to proceed exactly along radial lines. In practice, it does not take much for a fluid element being pulled towards an accretion source to miss its target. While this might seem obvious, given the distance scales involved, one could consider for example plasma approaching a solar-mass black hole to get a feeling for the actual numbers. Assume for example that the infalling plasma has reached the sonic radius $r_s = GM/2c_s(r_s)^2$, coming from an 'infinity' where $c_s(\infty) = 10^7$ cm s^{-1}, while heading for a target with (Schwarzschild) radius $R = 2GM/c^2$. It will only take an angle $\theta \approx (2c_s(r_s)/c)^2 \approx 10^{-5}$ degrees relative to a direct approach to avoid the black hole event horizon (R and r_s are 3×10^5 cm and 3×10^{11} cm, respectively, in this example).

But if infalling matter (initially) misses its target, the fact that gravity is a conservative forces puts a limit on how close the matter can eventually get to the accreting object without some mechanism acting to dissipate its angular momentum. The kinetic and potential energy of a test particle of mass m and velocity $\mathbf{v}$ will have to add up to a value below zero for an orbit that is not unbound (a sum larger than zero) or describing a grazing encounter (a sum equal to zero). We have

$$E_g + E_k = -\frac{GMm}{r} + \frac{1}{2}mv^2 < 0, \tag{11.17}$$

at a distance r to an accreting object of mass M. The gravitational force on the test particle obeys

$$\mathbf{F} = m\mathbf{a} = -\nabla E_g = -\frac{GMm}{r^2}\hat{\mathbf{e}}_r = -\frac{GM}{r^3}\mathbf{r}, \tag{11.18}$$

where $r = x\hat{\mathbf{e}}_x + y\hat{\mathbf{e}}_y + z\hat{\mathbf{e}}_z$ the position vector introduced back in Section 1.1. The time-change in angular momentum $\mathscr{L} \equiv m\mathbf{r} \times \mathbf{v}$ can easily shown to be zero:

$$\frac{d\mathscr{L}}{dt} = \frac{d}{dt}(m\mathbf{r} \times \mathbf{v}) = m\mathbf{v} \times \mathbf{v} + m\mathbf{r} \times \mathbf{a} = 0. \tag{11.19}$$

In the energy equation, the squared velocity v^2 can be split into a contribution from a velocity component $v_\parallel$ along r (i.e. towards the accreting object) and from a component $v_\perp$ perpendicular to r:

$$-\frac{GMm}{r} + \frac{1}{2}mv_\perp^2 + \frac{1}{2}mv_\parallel^2 < 0. \tag{11.20}$$

Because the angular momentum is given by $|\mathscr{L}| = mrv_\perp$, thanks to the cross-product being zero for parallel vectors, we can rewrite the previous expression as

$$-\frac{GMm}{r} + \frac{\mathscr{L}^2}{2mr^2} + \frac{1}{2}mv_\parallel^2 < 0. \tag{11.21}$$

Only the leftmost term in this expression is negative, but the middle term on the LHS depends on radius more strongly. Therefore, if we keep decreasing the radius of the test particle, at some point it becomes impossible to maintain a sum smaller than zero. At the energy minimum, $v_\parallel^2 = 0$ (because any non-zero contribution will be positive). An extremum for r from the remaining terms in the LHS can be found by solving for r in the expression

$$\frac{d}{dr}\left(-\frac{GMm}{r} + \frac{\mathscr{L}^2}{2mr^2}\right) = 0. \tag{11.22}$$

This leads to $\mathscr{L}^2 = GMm^2 r$, or

$$v_\perp^2 = \frac{GM}{r}, \tag{11.23}$$

which can be recognized as describing a circular Keplerian orbit. The angular momenta of the infalling test particles (or fluid elements, when considered in bulk) are likely to have at least somewhat of a preferred orientation, so the natural result for gas flowing towards an accreting source is the formation of a circular accretion disc. In this context, the radius of the Keplerian orbit described above is known as the *circularization radius*.

11.4.1   THIN DISCS

If we assume the accretion disc to be *thin*, we can use the momentum equation from gas dynamics to draw some conclusions about the nature of the gas flow in such a disc. Let us assume that in a *thin* disc, the width H of the disc will be far smaller than a local radius r for all points of interest on the disc, i.e. $H \ll r$ (the approximation will break down close to the centre of the disc), as illustrated in Figure 11.3. The momentum equation

$$\frac{\partial \rho \mathbf{v}}{\partial t} + \nabla \cdot (\rho \mathbf{vv}) + \nabla p = \mathbf{f}, \tag{11.24}$$

has a z-component obeying

$$\frac{\partial \rho v_z}{\partial t} + \nabla \cdot (\rho \mathbf{v}v_z) + \frac{\partial p}{\partial z} = f_z. \tag{11.25}$$

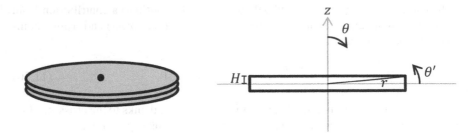

Figure 11.3 A thin accretion disc of height H. Left, sketch of disc. Right, definitions of angles and length scales.

If we assume zero velocity in the z-direction and the external force f to be gravity, we obtain (using Equation 11.18)

$$\frac{\partial p}{\partial z} = -\frac{GM\rho}{r^3}z = -\frac{GM\rho}{r^3}r\sin\theta',\qquad(11.26)$$

with angle θ' as illustrated in Figure 11.3. Using this expression to form an estimate of the disc properties, we take it to imply

$$\frac{\Delta p}{\Delta z} \approx -\frac{GM\rho}{r^2}\sin\theta',\qquad(11.27)$$

with $\Delta p \approx -p$, $\Delta z \approx H$ and $\sin\theta' \approx H/r$. Using that for circular Keplerian orbits we have $v_\perp^2 = GM/r$, we can further rewrite this expression as follows:

$$\frac{p}{H} \approx \frac{GMH\rho}{r^3} \Rightarrow \frac{p}{\rho} \approx \frac{H^2v_\perp^2}{r^2} \Rightarrow c_s^2 \sim \frac{H^2v_\perp^2}{r^2} \Rightarrow \frac{H}{r} \sim \frac{c_s}{v} \sim \frac{1}{\mathscr{M}},\qquad(11.28)$$

estimating the sound speed $c_s^2 \sim p/\rho$ (i.e. the factor $\hat{\gamma}$ is beside the point here, given that it is of order unity). What this result tells us is that the gas spins highly *supersonically* with $\mathscr{M} \gg 1$ if the assumption H/r is to hold. This is all internally consistent: while the pressure p sets the scale height H of the disc, this height is very small and the mass being accreted effectively behaves as test particles in Keplerian orbits in the plane, whose dynamics are governed by classical mechanics rather than fluid dynamics (i.e. the influence of pressure becomes negligible, and the disc will not expand in height due to pressure forces).

11.4.2 ACCRETION DISCS AND VISCOSITY

So far we have not actually provided a prescription for how gas eventually *does* reach the surface of the accreting object. In order to move inward, masses in Keplerian orbits will need to shed some of their velocity. This requires some measure of

viscosity to exist in the flow, as produced for example by *shear* forces resulting from
a stratification of velocity with radius. The shear force density is given by

$$f = \eta r \frac{\partial \omega}{\partial r}, \qquad (11.29)$$

where ω the angular frequency and η the *shear* in g cm^{-1} s^{-1}. The kinematic viscosity v is defined as $v = \eta/\rho$. Recall from Section 9.5.1 that the *Reynold's number*
of the fluid is given by Re $\approx \frac{L^2}{vT} \approx VL/v$, where L a characteristic length scale, T
a characteristic time scale and V a characteristic velocity scale. For accretion discs,
Re > 1, which indicates that the fluid is *not* viscous, seemingly at odds with the viscosity requirement. However, in practice Re $\gg 1$ for accretion discs, and at this point
the flow becomes turbulent. This turbulence then provides its own form of viscosity
(as a means to distribute large scale velocities down to smaller scales, effectively
slowing down the bulk flow). It can be calculated that a significant fraction of the
energy of the accreting material is already released as it travels inwards through the
accretion disc.

A commonly used analytical approach to capture this effect, is to assume that the
turbulent viscosity follows

$$v = \alpha c_s H, \qquad (11.30)$$

where the parameter α is a tunable parameter that can be set to mimic the far more
complex physics underlying turbulent accretion flows. Using this approach, analytical solutions to the structure of thin discs can be found. For *thick* discs, where the
requirement $H/r \ll 1$ is dropped, no such solutions are available, but numerical simulations can be used.

To explore the profile of thin discs (without solving fully for the case of α-
viscosity), let us set up a steady-state thin disc and integrate out the vertical density
profile in order to concentrate on the radial profile of the disc in the xy-plane. For
that, we introduce an integrated density

$$\Sigma \equiv \int \rho dz. \qquad (11.31)$$

In cylindrical coordinates, the vertically integrated continuity equation then becomes
(see also Equation C.28):

$$\frac{\partial \Sigma}{\partial t} + \frac{1}{\varsigma} \frac{\partial}{\partial \varsigma} (\varsigma \Sigma v^\varsigma) = 0, \qquad (11.32)$$

if we assume the disc is symmetric around the z-axis. The partial derivative with
respect to z in the continuity equation is integrated out, leaving values at the boundaries of the integration domain that we can take to be zero anyway if we extend our
integration domain beyond the height of the thin disc. In a steady state, we assume a
constant mass accretion rate $\dot{M}$, same as for Bondi accretion. Because the mass now
all needs to come in through the disc, we have

$$\dot{M} = -2\pi \varsigma \Sigma v^\varsigma. \qquad (11.33)$$

The thin-disc approximation allows us to use ς rather than r here as part of the approximation. With the fluid elements in the disc spinning in close to Keplerian orbits, we have the velocity in the ϕ-direction given by $v_\perp$ as defined earlier, $v^\phi = v_\perp = \sqrt{GM/\varsigma}$. The corresponding angular momentum density $l^\phi = \varsigma v^\phi = \varsigma^2 \omega$.

In steady state, we not only require a constant mass transfer $\dot{M}$, but also a constant amount of anguler momentum transfer $\mathscr{L}$ (now referring to the total angular momentum of all accreted gas, rather than that of a test particle or a fluid element). Of course, in the case of pure Keplerian orbits in the absence of friction, $\dot{M} = \mathscr{L} = 0$. For our axisymmetric disc in cylindrical coordinates, conservation of angular momentum takes the form (see Problem 11.7):

$$\frac{\partial \rho \varsigma v^\phi}{\partial t} + \frac{1}{\varsigma}\frac{\partial}{\partial \varsigma}\left(\varsigma \rho v^\varsigma \varsigma v^\phi - \eta \varsigma^3 \frac{\partial (v^\phi/\varsigma)}{\partial \varsigma}\right)$$

$$+ \frac{\partial}{\partial z}\left(\rho v^z \varsigma v^\phi - \eta \varsigma \frac{\partial v^\phi}{\partial z}\right) = \varsigma f^\phi. \tag{11.34}$$

Because gravity acts in the radial direction, we can take $f^\phi = 0$. We can replace the coefficient of shear viscosity η by the coefficient per unit mass v to make the dependency on ρ explicit, using $\eta = \rho v$. Integrating once again over z, we have

$$\frac{\partial \Sigma \varsigma v^\phi}{\partial t} + \frac{1}{\varsigma}\frac{\partial}{\partial \varsigma}\left(\varsigma^2 \Sigma v^\varsigma v^\phi - \Sigma v \varsigma^3 \frac{\partial (v^\phi/\varsigma)}{\partial \varsigma}\right) = 0. \tag{11.35}$$

We know that v^ϕ depends on ς according to $\propto \varsigma^{-1/2}$ (Equation 11.23), and thus we can compute that

$$\frac{\partial}{\partial \varsigma}\left(\frac{v^\phi}{\varsigma}\right) = -\frac{3}{2}\frac{v^\phi}{\varsigma^2}. \tag{11.36}$$

Using this in the previous equation and asserting steady state, implies

$$\frac{1}{\varsigma}\frac{\partial}{\partial \varsigma}\left(\varsigma^2 \Sigma v^\varsigma v^\phi + \frac{3}{2}\Sigma v \varsigma v^\phi\right) = 0. \tag{11.37}$$

This tells us that the term in brackets, which we can identify as minus the angular momentum flux once we multiply by 2π to account for all directions, has to be a constant in steady state:

$$\mathscr{L} = -2\pi \varsigma \Sigma \varsigma v^\phi v^\varsigma - 3\pi \varsigma \Sigma v v^\phi = \dot{M}\varsigma v^\phi - 3\pi \varsigma \Sigma v v^\phi. \tag{11.38}$$

This can be rewritten in the form

$$\Sigma v = \frac{\dot{M}}{3\pi}\left(1 - \frac{\mathscr{L}}{\dot{M}\varsigma v^\phi}\right). \tag{11.39}$$

In other words, if v can be approximated as constant, we have an expression for the density profile as a function of global properties $\dot{M}$ and $\mathscr{L}$. Other expressions for

v (such as the α-prescription) would require further information about the pressure profile.

PROBLEM 11.7
Deriving angular momentum density conservation for an accretion disc

Derive Equation 11.34 using the mathematical tools provided in Appendix C.

Further useful information can be learned from considering the energy conservation law in steady state. For a fast-spinning axisymmetric thin accretion disc, we can take the energy term to be dominated by kinetic energy and the stress tensor to be dominated by the shear viscosity between different annuli. In cylindrical coordinates, we then have an expression

$$\frac{1}{\varsigma}\frac{\partial}{\partial\varsigma}\left(\varsigma\frac{1}{2}\Sigma v^\phi v^\phi v^\varsigma - \varsigma\Sigma v\frac{\partial\left(v^\phi/\varsigma\right)}{\partial\varsigma}\varsigma v^\phi\right) = \Sigma\dot{q} + v^\varsigma\Sigma g^\varsigma, \tag{11.40}$$

for the vertically integrated disc profile. The first term on the RHS is a heat dissipation term (presumably radiation). The second term is work done by gravity, with g^ς the gravitational acceleration in the ς-direction. The shear term can be simplified because we know how v^ϕ depends on ς. We can rewrite the kinetic energy term using $\dot{M}$ (which does not change with radius) to eliminate v^ς and Σ:

$$\frac{1}{\varsigma}\frac{\partial}{\partial\varsigma}\left(-\frac{\dot{M}}{4\pi}v^\phi v^\phi + \Sigma v\frac{3}{2}v^\phi v^\phi\right) = \Sigma\dot{q} + v^\varsigma\Sigma g^\varsigma. \tag{11.41}$$

We can use Equation 11.39 to eliminate Σv from the shear term in order to be left with terms for which we know how they depend on cylindrical radius ς. Doing this and applying the partial derivative takes us in a few steps to

$$-\frac{1}{\varsigma}\frac{\dot{M}}{4\pi}\frac{v^\phi v^\phi}{\varsigma} + \frac{1}{\varsigma}\frac{3}{2}\frac{\mathscr{L}}{2\pi}\frac{v^\phi}{\varsigma^2} = \Sigma\dot{q} - \frac{\dot{M}}{2\pi\varsigma}\frac{GM}{\varsigma^2}. \tag{11.42}$$

We also used the opportunity to recast the gravity term using the definition of $\dot{M}$ and $g^\varsigma = GM/\varsigma^2$. Finally, we can use $v^\phi = \sqrt{GM/\varsigma}$ (Equation 11.23) to eliminate v^ϕ. What is left can then straightforwardly be reorganized in the form

$$\Sigma\dot{q} = -\frac{3}{4\pi}\frac{GM}{\varsigma^3}\dot{M}\left(1 - \frac{\mathscr{L}}{\dot{M}\sqrt{GM\varsigma}}\right). \tag{11.43}$$

What is particularly appealing here is that we have arrived at an expression for the dissipation term that is independent of the details of the dissipation mechanism (owing to Equation 11.39). If the dissipation is indeed equated to radiation, the disc will have a luminosity

$$L = -2\pi\int\varsigma\Sigma\dot{q}d\varsigma, \tag{11.44}$$

with each side emitting half of this amount. If the emission generated by the disc is thermal emission with effective temperature T, then the total local energy loss per unit time is given by the Stefan-Boltzmann law. When $\Sigma \dot{q} \propto T^4$, Equation 11.43 therefore gives us the temperature profile of the disc: $T \propto \varsigma^{-3/4}$ (with corrections where the second term in brackets in Equation 11.43 can ot be neglected).

12 Concepts in Plasma Physics

12.1 INTRODUCTION

Up until now we have steered clear of one of the three states of matter that can be described using fluid dynamics: the plasma state. A plasma is a fluid that consists of charged particles. Plasmas play a large role in astrophysics, because in many astrophysical systems the electrons have been decoupled from the protons. The fact that plasmas consist of charged particles, rather than neutral ones (like e.g. interstellar clouds that are not necessarily ionized) means that, even if there is no net charge, plasmas are susceptible to internal electromagnetic effects and the action of external large scale magnetic fields. We therefore need a framework that combines both fluid dynamics and electrodynamics. When we limit ourselves to the fluid scale, this framework is called *magnetohydrodynamics* (or 'MHD' for short). More generally, the collective behaviour of large groups of charges is the subject of plasma physics, and for example plasma kinetic theory delves down to the smaller scales in a manner similar to what we explored in Chapter 2: Boltzmann's transport equation is replaced by an analogous equation covering electromagnetic forces, known as the *Fokker-Planck* equation. This equation lies beyond the scope of this book, and we will not present it here.

12.2 INCORPORATING MAXWELL'S EQUATIONS

In plasmas, charged particles are typically free to move around. We therefore do not gain much by distinguishing between 'free' and 'bound' charge and current densities, and it is reasonable to use the vacuum version of Maxwell's equations. In *Gaussian cgs* units, these are

$$\nabla \cdot \mathbf{E} = 4\pi q, \tag{12.1}$$

$$\nabla \cdot \mathbf{B} = 0, \tag{12.2}$$

$$\nabla \times \mathbf{E} = -\frac{1}{c}\frac{\partial \mathbf{B}}{\partial t}, \tag{12.3}$$

$$\nabla \times \mathbf{B} = \frac{4\pi}{c}\mathbf{j} + \frac{1}{c}\frac{\partial \mathbf{E}}{\partial t}. \tag{12.4}$$

The symbols have the following meaning: $\mathbf{E}$ is the electric field, $\mathbf{B}$ the magnetic field, q the electric charge density, $\mathbf{j}$ the electric current density. Note how in this choice of units the *permittivity of free space* ε_0 and the *permeability of free space* μ_0 do not occur, having been absorbed in the definitions of $\mathbf{B}$ and $\mathbf{E}$. By contrast, in SI units,

DOI: 10.1201/9781003095088-12

we would have had

$$\nabla \cdot \mathbf{E} = \frac{q}{\varepsilon_0} \text{ (SI units)}, \tag{12.5}$$

$$\nabla \cdot \mathbf{B} = 0 \text{ (SI units)}, \tag{12.6}$$

$$\nabla \times \mathbf{E} = -\frac{\partial \mathbf{B}}{\partial t} \text{ (SI units)}, \tag{12.7}$$

$$\nabla \times \mathbf{B} = \mu_0 \mathbf{j} + \mu_0 \varepsilon_0 \frac{\partial \mathbf{E}}{\partial t} \text{ (SI units)}. \tag{12.8}$$

Which units are more natural or easy to work with is subjective. Equations in Gaussian cgs units are less cluttered with additional symbols while also being the standard in astrophysics, SI units are more standard elsewhere. Permittivity and permeability of free space are related according to $1/(\varepsilon_0 \mu_0) \equiv c^2$, reflecting that, no matter which units are used, when wave equations are derived from Maxwell's equations, electromagnetic waves will be found to travel at the speed of light (in the classical picture, they *are* light). Gaussian cgs notation further has the advantage that it treats spatial derivatives on an apparently equal footing to derivatives with respect to ct rather than t, as one would expect in a relativistic framework. Maxwell's equations can indeed be shown to be fully consistent with special relativity (regardless of which unit system is used, of course).

In addition to Maxwell's equations, an important equation in electrodynamics is the *Lorentz force law* for Lorentz force density $\mathbf{f}_L$:

$$\mathbf{f}_L = q\mathbf{E} + q\frac{\mathbf{v}}{c} \times \mathbf{B}. \tag{12.9}$$

The *electric conductivity* σ_E is a measure of how easy it is for a force acting on charges to induce a current:

$$\mathbf{j} = \sigma_E \frac{\mathbf{f}}{q}. \tag{12.10}$$

The larger the electric conductivity of the conducting material, the higher the current density that results from a given force per unit charge $\mathbf{f}/q$. In cgs units, charge density q has the units $[g^{\frac{1}{2}}][cm^{-\frac{3}{2}}][s^{-1}]$, while the electric conductivity σ_E has units $[s^{-1}]$. If the force density is provided by the Lorentz force, we have

$$\mathbf{j} = \sigma_e \left(\mathbf{E} + \frac{\mathbf{v}}{c} \times \mathbf{B} \right). \tag{12.11}$$

This expression was first formulated as *Ohm's law* in a frame where velocity $\mathbf{v}$ is zero:

$$\mathbf{j} = \sigma_e \mathbf{E}. \tag{12.12}$$

PROBLEM 12.1
Charge flow

Show how Equation 12.11 and Maxwell's equations can be used to derive a continuity equation for electric charge flow of the form

$$\frac{\partial q}{\partial t} + \nabla \cdot \mathbf{j} = 0.$$

12.2.1 THE MAGNETIC FIELD EQUATION

For the largest scale of interest in a plasma (the macroscopic scale of fluid behaviour), we can look for a way to merge the physics of Maxwell's equations with that of the hydrodynamical equations. Let us take another look at Equation 12.4. As it turns out, the changes in the electric field in a plasma tend to be very small in practice, given that the electric field as a whole stays close to zero due to (near) charge neutrality. This means that normally Equation 12.4 can be approximated as

$$\nabla \times \mathbf{B} \approx \frac{4\pi}{c}\mathbf{j}. \tag{12.13}$$

Equation 12.13 is known as *Ampère's law* and predates Maxwell's unification of the equations of electrodynamics. As a matter of fact, the approximation discards Maxwell's solution to the problem that Ampère's law only works for electrostatics: without Maxwell's added term for the evolution of the electric field, the equation

$$\nabla \cdot \left(\frac{4\pi}{c}\mathbf{j}\right) = \nabla \cdot (\nabla \times \mathbf{B}) = 0, \tag{12.14}$$

can be combined with a continuum equation for charge flow (see Problem 12.1) to imply that:

$$\frac{\partial q}{\partial t} + \nabla \cdot \mathbf{j} = 0 \Rightarrow \frac{\partial q}{\partial t} = 0. \tag{12.15}$$

In plasmas that are net-neutral ($q = 0$), this is not a problem: it is what we assumed in the first place. Typically, neutrality is a very reasonable assumption, although a simple scale argument does indicate that we should remain careful in the case of plasma flow moving with relativistic velocity. Equation 12.3 suggests that $E/L \sim B/(cT)$, with L a length scale and T a time scale. Similarly, Equation 12.4 suggests $B/L \sim E/(cT)$ if, for the sake of argument, the time-change of the electric field is assumed to be dominant over the contribution from the current term on the RHS. Combining the two suggests that this scenario is consistent with $L^2/T^2 \sim c^2$. If L/T is considered to reflect a velocity scale of the plasma, then the implication is that for relativistic fluid flow, Ampère's law is not necessarily sufficient. Near-light speed changes in the fluid state now potentially occur on the same time scales that it takes for charge-neutrality to (re-)establish itself.

That caveat aside, the approximation from Equation 12.13 is useful in that it will allow us to eliminate **E** from our equations in favour of **B**. This is achieved by first combining Equation 12.11 law and Equation 12.4, which are now both expressions for **j**:

$$(\nabla \times \mathbf{B})\frac{c}{4\pi} = \sigma_e\left(\mathbf{E} + \frac{\mathbf{v}}{c} \times \mathbf{B}\right),$$

$$\Rightarrow \quad \mathbf{E} = \frac{c}{4\pi\sigma_e}\nabla \times \mathbf{B} - \frac{\mathbf{v}}{c} \times \mathbf{B},$$

$$\Rightarrow \quad \mathbf{E} = \frac{\eta_e}{c}\nabla \times \mathbf{B} - \frac{\mathbf{v}}{c} \times \mathbf{B}. \tag{12.16}$$

In the last step we defined the *electric resistivity* $\eta_e \equiv c^2/(4\pi\sigma_e)$ (i.e. the higher the conductivity, the lower the resistivity). Taking the curl of both sides of the result above yields

$$\nabla \times \mathbf{E} = \nabla \times \left(\frac{\eta_e}{c} \nabla \times \mathbf{B} \right) - \nabla \times \left(\frac{\mathbf{v}}{c} \times \mathbf{B} \right). \tag{12.17}$$

A comparison with Equation 12.3 reveals that

$$\frac{\partial \mathbf{B}}{\partial t} = -\nabla \times (\eta_e \nabla \times \mathbf{B}) + \nabla \times (\mathbf{v} \times \mathbf{B}). \tag{12.18}$$

This result tells us that in plasmas where $\partial \mathbf{E}/\partial(ct) \approx 0$ we can limit ourselves to solving an additional equation for the magnetic field alone, in order to know all about the electromagnetic fields. The expression can be simplified a little further by using that the magnetic field has to remain divergence-free (i.e. $\nabla \cdot \mathbf{B} = 0$). First, recall the vector identity

$$\nabla \times (\nabla \times \mathbf{B}) = -\nabla^2 \mathbf{B} + \nabla (\nabla \cdot \mathbf{B}). \tag{12.19}$$

PROBLEM 12.2
Cross-product vector identity

Demonstrate that the vector identity from Equation 12.19 holds in Cartesian coordinates (it of course holds for arbitrary coordinate systems).

Second, assume the electric resistivity to be approximately constant at the length scales of interest. Applying this then yields

$$\frac{\partial \mathbf{B}}{\partial t} = \eta_e \nabla^2 \mathbf{B} + \nabla \times (\mathbf{v} \times \mathbf{B}). \tag{12.20}$$

Two terms dictate how the magnetic field changes over time. The first term on the RHS is a magnetic diffusion term. The second term describes convection of the field by the fluid. If η_e is sufficiently small (i.e. low electric resistivity), the diffusion term can be ignored. In this case, we are considering *ideal magnetohydrodynamics*, just like ignoring viscosity in fluids allowed for a description in terms of ideal fluid dynamics.

12.2.2 THE EQUATIONS OF MAGNETOHYDRODYNAMICS

Above, we saw that the magnetohydrodynamics assumption that the electric field can be ignored is typically justified, which made it sufficient to formulate an expression to describe the evolution of $\mathbf{B}$ only. However, to express the full set of equations of magnetohydrodynamics, we still need to know how the magnetic field influences the fluid state and vice versa (the latter via the motion of charged particles).

First of all, if we have finite rather than infinite conductivity, some energy must be lost during the transportation of charges. This energy has to go somewhere. It ends up dissipated into the fluid and therefore must show up on our energy balance sheet as a heating term:

$$\frac{\partial \mathscr{E}}{\partial t} + \nabla \cdot \left([\mathscr{E} + p] \mathbf{v} \right) - \mathbf{v} \cdot \mathbf{f} = \frac{|\mathbf{j}|^2}{\sigma_e} = \frac{4\pi \eta_e}{c^2} |\mathbf{j}|^2, \qquad (12.21)$$

The RHS heating term is akin to the power in the well-known *Joule Heating law* $P = RI^2$ (where P the power, I the current and R the resistance). Because magnetic fields do no work, we do not need to worry about any work term due to this field in the energy equation.

But even if there is no work done, we still have the Lorentz force, so we can anticipate it playing a role in the momentum and velocity fluid equations, which contain force terms. In a plasma with net neutral charge $q = 0$, but there might still be net currents $\mathbf{j} \neq 0$, e.g. when protons move in one direction while electrons move in the other direction. The Lorentz force density is

$$\mathbf{f}_L = \frac{1}{c} \mathbf{j} \times \mathbf{B} = \frac{1}{4\pi} \left(\nabla \times \mathbf{B} \right) \times \mathbf{B} = \frac{1}{4\pi} \left(\mathbf{B} \cdot \nabla \right) \mathbf{B} - \frac{1}{8\pi} \nabla \left(B^2 \right). \qquad (12.22)$$

The second step is an application of Equation 12.4. The last step makes use of the vector identity

$$\nabla \left(\mathbf{X} \cdot \mathbf{Y} \right) = \mathbf{X} \times \left(\nabla \times \mathbf{Y} \right) + \mathbf{Y} \times \left(\nabla \times \mathbf{X} \right) + \left(\mathbf{X} \cdot \nabla \right) \mathbf{Y} + \left(\mathbf{Y} \cdot \nabla \right) \mathbf{X}. \qquad (12.23)$$

This force term can now be included in the momentum equations. The term $(4\pi)^{-1} (\mathbf{B} \cdot \nabla) \mathbf{B}$ is related to magnetic tension caused by the curvature of magnetic field lines and disappears if the magnetic field lines are straight. The second term is akin to a magnetic pressure effect. The full set of *magnetohydrodynamical equations* is now:

$$\frac{\partial \rho}{\partial t} + \nabla \cdot (\rho \mathbf{v}) = 0, \qquad (12.24)$$

$$\frac{\partial \rho \mathbf{v}}{\partial t} + \nabla \cdot (\rho \mathbf{v} \mathbf{v}) + \nabla \left(p + \frac{B^2}{8\pi} \right) = \mathbf{f} + \frac{(\mathbf{B} \cdot \nabla) \mathbf{B}}{4\pi}, \qquad (12.25)$$

$$\frac{\partial \mathscr{E}}{\partial t} + \nabla \cdot ([\mathscr{E} + p] \mathbf{v}) = \mathbf{v} \cdot \mathbf{f} + \frac{4\pi \eta_e}{c^2} |\mathbf{j}|^2, \qquad (12.26)$$

$$\frac{\partial \mathbf{B}}{\partial t} = -\nabla \times (\eta_e \nabla \times \mathbf{B}) + \nabla \times (\mathbf{v} \times \mathbf{B}). \qquad (12.27)$$

Simulating an astrophysical plasma in the continuum limit amounts to solving these equations, or their (special) relativistic generalization.

12.3 THE NATURE OF PLASMAS

The electromagnetic force is vastly stronger than gravity. Between two electrons of charge $-e$, we have a repulsive electromagnetic force

$$F_L = \frac{e^2}{r^2}, \qquad (12.28)$$

(note again the use of cgs units). Comparing this to the gravitational attraction between the two electrons, we see that

$$\frac{|\mathbf{F}_L|}{|\mathbf{F}_g|} = \left(\frac{e^2}{r^2}\right)\left(\frac{m_e^2 G}{r^2}\right)^{-1} = \frac{e^2}{m_e^2 G} \sim 4 \times 10^{42}. \tag{12.29}$$

Gravity is only relevant when there is effectively charge neutrality in the plasma up to extremely high order of precision. To dwell on this a bit more, consider a case where we have both protons e and electrons p with a very small difference in number between the two: $N_e = (1+\delta)N_p$ (the sun for example has $N_e = 10^{57}$ electrons and a practically equal number of ions N_+, mostly hydrogen). For the forces to balance completely, we would need

$$\frac{|\mathbf{F}_L|}{|\mathbf{F}_g|} = 1 \Rightarrow \frac{e^2(N_e - N_p)}{r^2} = \frac{m_e m_p G}{r^2} \Rightarrow \delta \sim 10^{-40}. \tag{12.30}$$

In practice, the sun has $\delta \sim 10^{-36}$, which provides the surface pressure to counteract gravity.

We can analyze the effect of a charge perturbation in the plasma using the same methods that we applied when investigating sound waves. We assume a base state at rest of equal electron and ion number density $n_{e,0}$ and $n_{+,0}$ respectively. We have $n_e = n_{e,0} + n_{e,1}$, with $n_{e,1} \ll n_{e,0}$. The perturbed state has a velocity $\mathbf{v} = \mathbf{v}_{e,1}$. Under these assumptions, the continuity equation, in terms of number density rather than mass density, can be approximated as

$$\frac{\partial}{\partial t}(n_{e,0} + n_{e,1}) + \nabla \cdot (n_{e,0}\mathbf{v}_{e,1} + n_{e,1}\mathbf{v}_{e,1}) = 0 \Rightarrow \frac{\partial n_{e,1}}{\partial t} + n_{e,0}\nabla \cdot \mathbf{v}_{e,1} = 0. \tag{12.31}$$

Dividing by the base state number density and taking another partial derivative with respect to time, we obtain

$$\frac{1}{n_{e,0}}\frac{\partial^2 n_{e,1}}{\partial t^2} + \nabla \cdot \frac{\partial \mathbf{v}_{e,1}}{\partial t} = 0. \tag{12.32}$$

We can use the Lorentz force to provide the acceleration that leads to a time change in the velocity, i.e.

$$\mathbf{F}_L = m_e \frac{\partial \mathbf{v}_e}{\partial t} = -e\mathbf{E}, \tag{12.33}$$

where e again the magnitude of the electron charge, leading to

$$\frac{1}{n_{e,0}}\frac{\partial^2 n_{e,1}}{\partial t^2} - \frac{e}{m_e}\nabla \cdot \mathbf{E} = 0. \tag{12.34}$$

Furthermore, we have a Maxwell equation $\nabla \cdot \mathbf{E} = -4\pi n_{e,1}e$ (the other charges cancel). This we can use to obtain

$$\frac{1}{n_{e,0}}\frac{\partial^2 n_{e,1}}{\partial t^2} + \frac{4\pi e^2}{m_e}n_{e,1} = 0. \tag{12.35}$$

This is indeed an expression allowing for oscillating solutions of the type $n_{e,1} \propto \cos(\omega_p t)$. It is not exactly like a sound wave because the oscillation does not get to go anywhere: what will happen is that local charge imbalances get compensated rapidly by freely moving charges at the level of individual particles. Since the charge imbalance is real, an oscillation behaviour is triggered. Here

$$\omega_p = \sqrt{\frac{4\pi n_{e,0} e^2}{m_e}}, \qquad (12.36)$$

is the *plasma angular frequency*. The corresponding plasma frequency is

$$\nu_p = \frac{\omega_p}{2\pi} \sim 9 \times 10^3 \sqrt{n_{e,0}} \text{ Hz}. \qquad (12.37)$$

For example, the ionosphere of the Earth (60 -1000 km) is a dilute plasma surrounding the Earth, with electron number density $n_e \sim 10^6$ cm^{-3}. The plasma frequency, in this case, will be $\nu_p \sim 10^7$ Hz or about 10 MHz. Radiation below this frequency will have difficulty passing through the ionosphere because the plasma response time is too fast relative to the intrinsic timescale of the perturbing radiation. This means that radio waves below 10 MHz are effectively absorbed by the atmosphere. Nowadays we have large scale observatories that get close to probing this limit (e.g. the Low Frequency Array, or LOFAR, which covers 10 MHz to 240 MHz; the Square Kilometre Array, or SKA, will cover from 50 MHz to 30 GHZ).

With a frequency comes a length scale and the typical plasma response length scale $\lambda \approx \frac{u_e}{\omega_p}$, where u_e the typical velocity of an electron. If we equate this typical velocity to the electron temperature T_e using $m_e u_e^2 \sim k_B T_e$, we obtain

$$\lambda_D = \sqrt{\frac{k_B T_e}{4\pi e^2 n_{e,0}}}, \qquad (12.38)$$

which is known as the *Debye* length. This is a measure of the penetration depth of electromagnetic waves at frequencies below the plasma frequency.

12.4   FIELD FREEZING

We can define the *magnetic flux* Φ through a surface according to

$$\Phi_m \equiv \int_S \mathbf{B} \cdot d\mathbf{S}. \qquad (12.39)$$

If the surface is a closed surface (e.g. enclosing a fluid parcel), we have

$$\frac{d\Phi_m}{dt} = \frac{d}{dt} \oint \mathbf{B} \cdot d\mathbf{S} = \frac{d}{dt} \int_V (\nabla \cdot \mathbf{B}) \, dV = \frac{d}{dt} 0 = 0, \qquad (12.40)$$

by virtue of the divergence theorem. However, in the case of very large electric conductivity σ_e (or small η_e), we can make a stronger statement, which is that even the

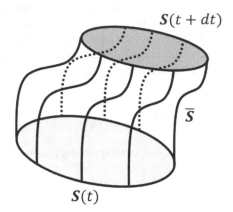

Figure 12.1 The volume traced out by an evolving surface element over time. $\bar{S}$ denotes the surface created by the movement of the contours of surface S over time.

number of magnetic field lines threading any *sub-surface* of a fluid parcel is conserved.

For an arbitrary evolving surface $S(t)$, we have

$$\frac{d\Phi_m}{dt} \equiv \lim_{dt \to 0} \left(\int_{S(t+dt)} \mathbf{B}(t+dt) \cdot d\mathbf{S} - \int_{S(t)} \mathbf{B}(t) \cdot d\mathbf{S} \right) \frac{1}{dt}, \tag{12.41}$$

keeping in mind that the surface area itself is a function of time. As a practical point, if we want to evaluate this expression, it would be helpful if we had an expression for $S(t)$ alone, rather than one including both $S(t)$ and $S(t+dt)$. This would simplify the interpretation of the magnetic flux in terms of Maxwell's equations (which are written in Eulerian form, not Lagrangian). To eliminate $S(t+dt)$, we consider the volume traced by $S(t)$ over time. This is shown in Figure 12.1, where we define a side area $\bar{S}$ as well. The magnetic flux through the *closed* surface defining the full boundary of this volume will be zero by definition:

$$-\int_{S(t)} \mathbf{B}(t+dt) \cdot d\mathbf{S} + \int_{\bar{S}} \mathbf{B}(t+dt) \cdot d\bar{\mathbf{S}} + \int_{S(t+dt)} \mathbf{B}(t+dt) \cdot d\mathbf{S} = 0. \tag{12.42}$$

Note how the minus sign is used to ensure the outward orientation of the surface element. The expression holds for the magnetic field at any time t', and we just picked $t' = t + dt$ in order to be able to write:

$$\int_{S(t+dt)} \mathbf{B}(t+dt) \cdot d\mathbf{S} = \int_{S(t)} \mathbf{B}(t+dt) \cdot d\mathbf{S} - \int_{\bar{S}} \mathbf{B}(t+dt) \cdot d\bar{\mathbf{S}}. \tag{12.43}$$

This we can substitute in our expression for the flux time derivative and obtain

$$\frac{d\Phi_m}{dt} = \lim_{dt \to 0} \left(\int_{S(t)} (\mathbf{B}(t+dt) - \mathbf{B}(t)) \cdot d\mathbf{S} - \int_{\bar{S}} \mathbf{B}(t+dt) \cdot d\bar{\mathbf{S}} \right) \frac{1}{dt}. \tag{12.44}$$

The first term on the RHS contains

$$\lim_{dt \to 0} \frac{\mathbf{B}(t+dt) - \mathbf{B}(t)}{dt}\bigg|_{\text{fixed location}} \equiv \frac{\partial \mathbf{B}}{\partial t}. \tag{12.45}$$

Here *fixed location* reflects that the time derivative is taken on $S(t)$ throughout, which means that using the partial derivative with respect to time is appropriate. The second term on the RHS can be rewritten using that the orientation of the surface element $\bar{\mathbf{S}}$ has to be perpendicular both to the line element along the edge $d\mathbf{r}$ and the flow velocity $\mathbf{v}$, i.e.

$$d\bar{\mathbf{S}} = d\mathbf{r} \times \mathbf{v}dt. \tag{12.46}$$

Using this, we find

$$\lim_{dt \to 0} \left(\int_{\bar{S}} \mathbf{B}(t+dt) \cdot d\mathbf{r} \times \mathbf{v}dt \right) \frac{1}{dt} = \oint \mathbf{B} \cdot d\mathbf{r} \times \mathbf{v},$$

$$= \oint (\mathbf{v} \times \mathbf{B}) \cdot d\mathbf{r},$$

$$= \int_{S(t)} \nabla \times (\mathbf{v} \times \mathbf{B}) \cdot d\mathbf{S}, \tag{12.47}$$

by using a standard vector identity and applying Stokes' theorem. Collecting terms, we have

$$\frac{d\Phi_m}{dt} = \int_{S(t)} \left(\frac{\partial \mathbf{B}}{\partial t} - \nabla \times (\mathbf{v} \times \mathbf{B}) \right) \cdot d\mathbf{S}. \tag{12.48}$$

For very large electric conductivity, $\eta_e \approx 0$, so according to Equation 12.27 the integrand of this expression is zero and this result implies

$$\frac{d\Phi_m}{dt} = 0, \tag{12.49}$$

regardless of the size or geometry of area S. In terms of fluid parcels, this phenomenon is known as *field freezing*. Every fluid element (surface) retains its magnetic field as it moves, such that the fluid and field move together as a unit. The magnetic field can be visualized as a collection of highly elastic bands that either drag plasma particles or are dragged by them, depending on which term dominates in the coupling between fluid and field in the momentum equation. In astrophysics, this situation is not uncommon, and examples can be found in the solar wind and magnetosphere, the interaction between Jupiter and its moon Io, regular stellar collapse and neutron star collapse, to name but a few.

12.5 MAGNETOHYDRODYNAMIC WAVES

The presence of magnetic fields adds a new dimension to the behaviour of sound waves in plasmas. The see what types of waves occur in MHD, our starting point is the usual one, going back to Section 5.1 and following very similar steps. We consider the usual homogeneous and motionless base state, now including a magnetic field and specified by $\rho = \rho_0$, $p = p_0$, $\mathbf{B} = \mathbf{B}_0$, $\mathbf{v} = 0$. To this state we add a small perturbation such that $\rho = \rho_0 + \rho_1$, $p = p_0 + p_1$, $\mathbf{B} = \mathbf{B}_0 + \mathbf{B}_1$, $\mathbf{v} = \mathbf{v}_1$. To linear order in the perturbations, mass conservation and the velocity equation become

$$\frac{\partial \rho_1}{\partial t} + \rho_0 \nabla \cdot \mathbf{v}_1 = 0, \tag{12.50}$$

$$\frac{\partial \mathbf{v}_1}{\partial t} = -\frac{1}{\rho_0} c_s^2 \nabla \rho_1 + \frac{1}{4\pi\rho_0} \left(\nabla \times \mathbf{B}_1 \right) \times \mathbf{B}_0. \tag{12.51}$$

New here is the inclusion of an expression for the Lorentz force density from Equation 12.22. We also already used the definition of the sound speed $c_s^2 = dp/d\rho|_0$ to eliminate p from the original velocity equation. The two expressions can be combined to obtain an expression for $\mathbf{v}_1$ by taking the gradient of the first, the partial derivative with respect to time from the second and combining the results:

$$\frac{\partial^2 \mathbf{v}_1}{\partial t^2} - c_s^2 \nabla \left(\nabla \cdot \mathbf{v}_1 \right) = \frac{1}{4\pi\rho_0} \left(\nabla \times \frac{\partial \mathbf{B}_1}{\partial t} \right) \times \mathbf{B}_0. \tag{12.52}$$

At this point we can apply our assumption that the perturbations are all of the form $constant \times \exp\left[i \left(\omega - \mathbf{k} \cdot \mathbf{r} \right) \right]$, which allows for the substitutions $\partial/\partial t \to i\omega$ and $\nabla \to -i\mathbf{k}$. We will further write the magnetic field as $\mathbf{B} = B_0 \left(\hat{\mathbf{n}} + \mathbf{b}_1 \right)$, where B_0 sets the overall scale of the magnetic field, $\hat{\mathbf{n}}$ the direction of the unperturbed field and (small) $\mathbf{b}_1$ carries information both about the direction and relative size of the perturbation. As a result, we have

$$-\omega^2 \mathbf{v}_1 + c_s^2 + c_s^2 \mathbf{k} \left(\mathbf{k} \cdot \mathbf{v}_1 \right) = \frac{B_0^2}{4\pi\rho_0} \left(k \times \omega \mathbf{b}_1 \right) \times \hat{\mathbf{n}}. \tag{12.53}$$

If we assume infinite conductivity ($\eta_e = 0$) for simplicity, Equation 12.20 suggests a means to eliminate $\mathbf{b}_1$ from the expression above:

$$\frac{\partial \mathbf{B}}{\partial t} = \nabla \times \left(\mathbf{v} \times \mathbf{B} \right),$$

$$\Rightarrow \frac{\partial \mathbf{B}_1}{\partial t} = \nabla \times \left(\mathbf{v}_1 \times \mathbf{B}_0 \right),$$

$$\Rightarrow \omega \mathbf{b}_1 = -\mathbf{k} \times \left(\mathbf{v}_1 \times \hat{\mathbf{n}} \right). \tag{12.54}$$

If we apply this, we find

$$\omega^2 \mathbf{v}_1 - c_s^2 \mathbf{k} \left(\mathbf{k} \cdot \mathbf{v}_1 \right) + \frac{B_0^2}{4\pi\rho_0} \left(\mathbf{k} \times \left[\mathbf{k} \times \left(\hat{\mathbf{n}} \times \mathbf{v}_1 \right) \right] \right) \times \hat{\mathbf{n}} = 0. \tag{12.55}$$

This equation can be simplified further in order to provide us with insight in MHD waves. The long series of repeated cross products is the most cumbersome aspect of this expression, and can systematically be reduced by applying vector identities B.11

and B.12 from appendix B. First addressing the cross products within the straight brackets and then the remaining cross products, it can be shown that

$$(\mathbf{k} \times [\mathbf{k} \times (\hat{\mathbf{n}} \times \mathbf{v}_1)]) \times \hat{\mathbf{n}} = [\hat{\mathbf{n}} (\mathbf{k} \cdot \mathbf{v}_1) - \mathbf{v}_1 (\mathbf{k} \cdot \hat{\mathbf{n}})] \mathbf{k} \cdot \hat{\mathbf{n}}$$
$$- \mathbf{k} [(\mathbf{k} \cdot \mathbf{v}_1) - \hat{\mathbf{n}} \cdot \mathbf{v}_1 (\mathbf{k} \cdot \hat{\mathbf{n}})]. \tag{12.56}$$

Anticipating the nature of MHD waves, we define the *Alfvén* speed c_A as follows

$$c_A^2 \equiv \frac{B_0^2}{4\pi\rho_0}. \tag{12.57}$$

Applied to Equation 12.55, we then get

$$\left[\omega^2 - c_A^2 (\mathbf{k} \cdot \hat{\mathbf{n}})^2\right] \mathbf{v}_1 + \mathbf{k} \left[c_A^2 (\mathbf{k} \cdot \hat{\mathbf{n}}) (\mathbf{v}_1 \cdot \hat{\mathbf{n}}) - (c_s^2 + c_A^2) (\mathbf{k} \cdot \mathbf{v}_1)\right]$$
$$+ \hat{\mathbf{n}} c_A^2 (\mathbf{k} \cdot \hat{\mathbf{n}}) (\mathbf{k} \cdot \mathbf{v}_1) = 0. \tag{12.58}$$

This expression still looks quite cumbersome, so in order to make progress with understanding its implications we need to start looking at particular orientations for the relevant vectors $\mathbf{k}$, $\hat{\mathbf{n}}$ and $\mathbf{v}_1$. Let us first consider the case where the wave vector $\mathbf{k}$ is perpendicular to the orientation $\hat{\mathbf{n}}$ of the unperturbed field. Because, in this case, $\mathbf{k} \cdot \hat{\mathbf{n}} = 0$, the equation becomes a lot more manageable:

$$\omega^2 \mathbf{v}_1 + \mathbf{k} \left[- (c_A^2 + c_s^2) (\mathbf{k} \cdot \mathbf{v}_1)\right] = 0. \tag{12.59}$$

If we further assume that $\mathbf{k}$ is aligned with $\mathbf{v}_1$, which indicates the direction of the displacements of the fluid elements, we find a wave equation for a *fast MHD wave*:

$$\omega^2 - k^2 (c_A^2 + c_s^2) = 0. \tag{12.60}$$

This particular case describes a longitudinal wave very similar to a normal sound wave except for its propagation speed. The reason it propagates faster than a sound wave in an unmagnetized fluid is because of the added pressure due to the magnetic field. Recalling the definition of the Alfvén speed (Equation 12.57) and looking back at MHD Equation 12.25, we see that the term $\nabla B^2/(8\pi\rho)$ is to c_A^2 what $\nabla p/\rho$ is to c_s^2. In a polytropic gas, $c_s^2 = \hat{\gamma} p_0/\rho_0$, with the magnetic pressure p_B providing an analogous $c_A^2 = 2p_{B,0}/\rho_0$ if $p_B \equiv B^2/(8\pi)$. Considering the analogy further, the factor 2 implies that $B^2 \propto \rho^2 \Rightarrow B \propto \rho$ in the case of a one-dimensional longitudinal wavefront in a direction perpendicular to the background magnetic field: during passage of the wave, the field lines get compressed in a manner proportional to the fluid density.

What about the more general case, where $\mathbf{k}$ and $\hat{\mathbf{n}}$ are not necessarily perpendicular? We can make Equation 12.58 somewhat easier to read by at least making some concrete choices about how to align our coordinate system relative to $\mathbf{k}$ and $\hat{\mathbf{n}}$. We consider a wave vector $\mathbf{k}$ in the x-direction and background field orientation $\hat{\mathbf{n}}$ in the xy-plane at an angle ϕ to the x-axis, i.e. $\hat{\mathbf{n}} = \cos\phi\hat{\mathbf{e}}_x + \sin\phi\hat{\mathbf{e}}_y$. In vector notation,

Equation 12.58 then takes the form

$$\begin{bmatrix} \left(\omega^2 - k^2 c_s^2 - k^2 c_A^2 \sin^2\phi\right) v_1^x + k^2 c_A^2 \cos\phi \sin\phi v_1^y \\ k^2 c_A^2 \cos\phi \sin\phi v_1^x + \left(\omega^2 - k^2 c_A^2 \cos^2\phi\right) v_1^y \\ \left(\omega^2 - k^2 c_A^2 \cos^2\phi\right) v_1^z \end{bmatrix} = \begin{bmatrix} 0 \\ 0 \\ 0 \end{bmatrix} \qquad (12.61)$$

If for example $v_1^z \neq 0$, we have according to the z-entry of this vector equation that

$$\omega^2 - k^2 c_A^2 \cos^2\phi = 0. \qquad (12.62)$$

Plugging this into the remaining two entries shows that $v_1^x = v_1^y = 0$ is required for consistency. what we have here is a *transverse* wave where the z-direction displacements of fluid elements (and therefore the 'vibrations' of the magnetic field lines) are perpendicular to the x-direction of the wave. Such waves are called *Alfvén* waves, moving with phase and group velocity $c_A \cos\phi$. Here, the $\cos\phi$ term accounts for the relative angle between the wave front and the baseline magnetic field orientation. If $\phi = 0$, the Alfvén wave travels along the field lines. If $\phi = \pi/2$, no wave front occurs with this velocity. Instead, it is the aforementioned fast MHD waves that allow for wave solutions where $\mathbf{k}$ and $\hat{\mathbf{n}}$ are perpendicular to each other, but since these are longitudinal they require $v_1^z = 0$. The three scenarios discussed thus far are illustrated in Figure 12.2.

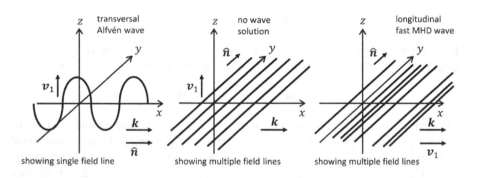

Figure 12.2 Three example orientations for the wave vector $\mathbf{k}$, perturbation displacement vector $\mathbf{v}_1$ and unperturbed magnetic field line direction $\hat{\mathbf{n}}$.

Finally, looking at the case where $v_1^z = 0$, Equation 12.61 can be reduced to a two-dimensional problem:

$$\begin{bmatrix} \omega^2 + k^2 c_A^2 \cos^2\phi - k^2\left(c_A^2 + c_s^2\right) & k^2 c_A^2 \sin\phi\cos\phi \\ k^2 c_A^2 \sin\phi\cos\phi & \omega^2 - k^2 c_A^2 \cos^2\phi \end{bmatrix} \begin{bmatrix} v_1^x \\ v_1^y \end{bmatrix} = \begin{bmatrix} 0 \\ 0 \end{bmatrix}. \qquad (12.63)$$

Either the velocity perturbation vector is zero (which means no wave at all), or the determinant of the matrix is zero. This latter condition leads to a dispersion relation of the form

$$\omega^4 - k^2\left(c_A^2 + c_s^2\right)\omega^2 + k^2 v_A^2 c_s^2 \cos^2\phi = 0. \qquad (12.64)$$

The solution of this equation is

$$\frac{\omega^2}{k^2} = \frac{c_A^2 + c_s^2 \pm \sqrt{\left(c_A^2 + c_s^2\right)^2 - 4c_A^2 c_s^2 \cos^2 \phi}}{2}. \tag{12.65}$$

The solutions where the square root gets added describe the range of wave speeds for *fast* MHD waves. Our previous example of a fast wave can be recovered by setting $\phi = \pi/2$ (field orientation and wave propagation are perpendicular). Solutions where the square root is subtracted describe the range of wave speeds of *slow* MHD waves.

Looking, as a final example, at the case where $v_1^z = 0$, $\phi = 0$, the dispersion relation reduces to

$$\frac{\omega^2}{k^2} = \frac{c_A^2 + c_s^2 \pm \left(c_A^2 - c_s^2\right)}{2}. \tag{12.66}$$

The two solutions, in this case, correspond to waves that either have wave speed c_s or c_A. In the former case, v_1^y needs to be zero for consistency according to Equation 12.63 (unless c_s happens to be exactly equal to c_A). We therefore have a normal sound wave propagating in the direction of the orientation of the magnetic field lines. In the latter case, it is v_1^x that needs to be zero in general, and we again have an Alfvén wave, now with the displacements lying in the xy-plane rather than in the z-direction.

12.6 TWO EXAMPLE SIMILARITY PARAMETERS IN PLASMA PHYSICS

The inclusion of plasma effects in fluid dynamics opens up additional options for similarity parameters. One such parameter is the magnetization σ (not to be confused with the viscosity tensor σ):

$$\sigma \equiv \frac{B^2/8\pi}{\rho c^2} = \frac{\text{'magnetic energy'}}{\text{'rest mass energy'}}. \tag{12.67}$$

In the case of high magnetization $\sigma \gg 1$, the fluid is driven by MHD effects, whereas for $\sigma \ll 1$ a pure fluid dynamics approach might well be sufficient.

Another important similarity parameter is the magnetic Reynold's number Re_M, defined using

$$\text{Re}_M = \frac{VL}{\eta_e c/4\pi}, \tag{12.68}$$

for a given velocity scale V and length scale L. Recalling the discussion from Section 12.2.1, a large magnetic Reynold's number tells us that the resistivity of the plasma can be neglected and the plasma can be described using the ideal MHD equations.

13 Computational Fluid Dynamics

It often does not take much for a fluid dynamics problem to become unsolvable by analytical means and to require some form of approximation. With the advent of powerful computers, computational fluid dynamics (CFD) has become an important tool in the toolkit of the (astro-)physicists or engineer. As an area of research CFD actually predates computers, if it is taken to describe a systematic algorithmic approach to solving Euler's equation through discretization and approximation (even if by hand, rather than on an actual computer). By now CFD is a vast field of research of great interest not only to (astro-)physicists and engineers, but also to mathematicians and computer scientists. The aim of this chapter is to provide an overview of some techniques currently in use within astrophysics that the reader can use as a starting point for further literature study. We build up towards a viable computer code for solving Euler's equations that contain all the ingredients for getting started and numerically solving simple fluid dynamics problems. We focus on the *finite volume method*, which is used by a range of widely available hydrodynamics packages currently deployed in astrophysics research[1].

13.1 EULER'S EQUATIONS IN TERMS OF A STATE VECTOR EQUATION

We begin by casting Euler's equation in a generic form using a *state vector* notation. Not to be confused with three-dimensional space vectors or four-dimensional space-time vectors, they represent a grouping together of the conserved quantities of fluid dynamics that will help us focus on the problem of how to discretize and solve a set of coupled partial differential equations (PDEs) without getting too distracted by the exact details of the physics they represent. First, recall Euler's equations:

$$\frac{\partial \rho}{\partial t} + \nabla \cdot (\rho \mathbf{v}) = 0, \qquad (13.1)$$

$$\frac{\partial \rho \mathbf{v}}{\partial t} + \nabla \cdot (\rho \mathbf{v} \mathbf{v} + p \mathbb{1}) = \mathbf{f} \qquad (13.2)$$

$$\frac{\partial \mathscr{E}}{\partial t} + \nabla \cdot ([\mathscr{E} + p] \mathbf{v}) = \rho \dot{q} + \mathbf{v} \cdot \mathbf{f}, \qquad (13.3)$$

where all terms have their usual meaning as introduced in Chapter 2. These conservation laws are a set of coupled PDEs that describe the time evolutions of the trio

---

[1] The topic of finite volume CFD is discussed in depth in a number of well-known textbooks, including [25, 53]. High-profile examples of codes using finite volume methods in astrophysics are FLASH [15], PLUTO [34] and AREPO [50].

DOI: 10.1201/9781003095088-13

of conserved quantities mass density, momentum density and energy density. These quantities change over time either due to an inbalance between incoming and outgoing flux (represented by the divergences), through sink or source terms (shown here are external forces, heating and cooling, and work done on the fluid), or through the impact of internal degrees of freedom within the fluid (the pressure terms).

Considering the case of one spatial dimension first, we can recast the set of conservation laws in terms of the evolution of a state vector Q and a state flux vector F (leaving the bold font type for actual vectors) as follows:

$$\frac{\partial}{\partial t}Q(x,t) + \frac{\partial}{\partial x}F(Q) = 0, \tag{13.4}$$

where,

$$Q \equiv \begin{bmatrix} \rho \\ \rho v \\ \mathscr{E} \end{bmatrix} \equiv \begin{bmatrix} Q_1 \\ Q_2 \\ Q_3 \end{bmatrix}, \tag{13.5}$$

$$F(Q) \equiv \begin{bmatrix} \rho v \\ \rho v^2 + p \\ v(\mathscr{E} + p) \end{bmatrix} \equiv \begin{bmatrix} Q_2 \\ \frac{Q_2^2}{Q_1} + p(Q) \\ Q_2\frac{(Q_3 + p(Q))}{Q_1} \end{bmatrix}. \tag{13.6}$$

We have assumed no sink and source terms, setting heating and external forces to zero for the time being. The equation above still requires the presence of an additional equation of state for p, as usual. Throughout this chapter we will use the EOS in terms of adiabatic exponent $\hat{\gamma}_{ad}$ first encountered for an ideal gas of point particles (see Section 2.4.3):

$$p = (\hat{\gamma}_{ad} - 1)\rho\varepsilon. \tag{13.7}$$

For ideal monatomic non-relativistic gases, the adiabatic exponent obeys $\hat{\gamma}_{ad} = 5/3$. A numerical implementation of fluid dynamics means that we need to formulate a time update approach to solving Equation 13.4, making us of the EOS whenever needed.

The state vector equation can be generalized to multiple dimensions. In this case, we have

$$\frac{\partial}{\partial t}Q(\mathbf{x},t) + \nabla \cdot F(Q) = 0, \tag{13.8}$$

where,

$$Q \equiv \begin{bmatrix} \rho \\ \rho\mathbf{v} \\ \mathscr{E} \end{bmatrix} \equiv \begin{bmatrix} Q_1 \\ \mathbf{Q}_2 \\ Q_3 \end{bmatrix}, \tag{13.9}$$

$$F(Q) \equiv \begin{bmatrix} \rho\mathbf{v} \\ \rho\mathbf{v}\mathbf{v} + p\mathbb{1} \\ \mathbf{v}(\mathscr{E} + p\mathbb{1}) \end{bmatrix} \equiv \begin{bmatrix} \mathbf{Q}_2 \\ \frac{\mathbf{Q}_2\mathbf{Q}_2}{Q_1} + p(Q)\mathbb{1} \\ \mathbf{Q}_2\frac{(Q_3 + p(Q))}{Q_1} \end{bmatrix} \tag{13.10}$$

Be careful with the divergence here. It acts on the actual vectors embedded within $F(Q)$, i.e., $\mathbf{Q}_2$, and not on the state flux vector (which is five-dimensional, not three-dimensional).

13.2   RUDIMENTARY FINITE DIFFERENCE SCHEMES

In CFD, we are trying to find a numerical solution to the PDE that models the evolution of the state vector (assuming one spatial dimension for now):

$$\frac{\partial}{\partial t}Q(x,t) = -\frac{\partial}{\partial x}F(Q). \tag{13.11}$$

This means that we need to have numerical schemes for derivatives with respect to space and time. The definition of the derivative of a function F at a point x suggests a first approach:

$$\frac{dF(x)}{dx} \equiv \lim_{h \to 0} \frac{F(x+h) - F(x)}{h}. \tag{13.12}$$

This is known as the limiting case of a *forward difference* scheme, because in order to obtain the derivative at x we include a step forward to $x+h$. Mathematically, this is no different from *backward difference* and *centred difference* schemes, given respectively by

$$\frac{dF}{dx} \equiv \lim_{h \to 0} \frac{F(x) - F(x-h)}{h}, \tag{13.13}$$

$$\frac{dF}{dx} \equiv \lim_{h \to 0} \frac{F(x+h) - F(x-h)}{2h}, \tag{13.14}$$

at least in the absence of singular points in the domain $(x-h, x+h)$. Numerically, one might expect the algorithms to implement the different schemes to lead to the same end result, but perhaps surprisingly we will see that this is not the case.

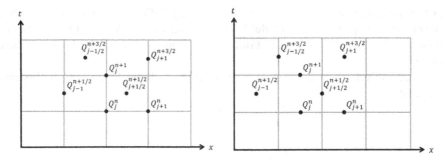

Figure 13.1   Possible discretizations of space and time. Space indices get a subscript, time indices a superscript. On the left, vertices are labelled with integer entries to describe a finite difference method. On the right, the spatial centres of cells are labelled with integer numbers, which is the more common convention for a finite volume method.

For now we will just proceed with a naive discretization of the PDE. Consider a grid as depicted in Figure 13.1. Values are typically defined either at cell corners (e.g. Q_j^n in the figure), the edges of cells (e.g. in the midpoint between Q_j^n and Q_{j+1}^n on the figure) or at cell centres[2] (e.g. $Q_{j+1/2}^{n+1/2}$). The grid shown in the figure is a *regular* grid, which helps to keep the math simple but is not a requirement for discretization (as a matter of fact, irregular grids can sometimes be vastly superior in applications). If the discretization is done such that different choices for the function locations are used together (e.g. cell corners *and* cell centres), the grid is a *staggered* grid. For now, we follow the grid as defined on the left of Figure 13.1, with integer j values probing specific locations in the fluid at given times n. The simplest discretization of the spatial derivative takes the form

$$\frac{\partial F(x_j)}{\partial x} \rightarrow \frac{F_{j+1}^n - F_j^n}{\Delta x}, \tag{13.15}$$

in a forward difference implementation on a grid with spacing Δx. In this expression, neither n nor j have yet been fixed to a particular value, and for example

$$\frac{\partial F(x_{j-1})}{\partial x} \rightarrow \frac{F_j^n - F_{j-1}^n}{\Delta x}, \tag{13.16}$$

etc. are all equivalent expressions to describe the same discretization. This type of representation represents a *finite difference* approach, where the grid points represent the *local* state of the fluid at that point. Later, we will move on to a *finite volume* approach, where the grid points are taken to represent the state across a given cell instead of just a single point within that cell (at which point the grid as defined on the right of Figure 13.1 becomes the more obvious discretization).

The aim of a numerical solver for the PDEs is to systematically advance the timestep from known initial conditions until we reach the time of interest. This may also be a series of times, in case it is the dynamical evolution itself that we care about. Replacing the RHS of Equation 13.11 by a single symbol Y for the moment, we are looking for a numerical approach to

$$\frac{\partial Q(x,t)}{\partial t} = Y(Q,x,t). \tag{13.17}$$

As with the spatial differencing schemes, we have multiple options for taking the derivative and advancing a single timestep Δt:

$$Q_j^{n+1} = Q_j^n + Y^n \Delta t, \qquad \text{(forward differencing)}, \tag{13.18}$$

$$Q_j^{n+1} = Q_j^n + Y^{n+1} \Delta t, \qquad \text{(backward differencing)}, \tag{13.19}$$

$$Q_j^{n+1} = Q_j^n + \frac{Y^{n+1} + Y^n}{2} \Delta t, \qquad \text{(centred differencing)}. \tag{13.20}$$

Because Y^n is a function of Q^n, only one of the three options above (forward time differencing) amounts to an explicit expression for Q^{n+1} in terms of values purely at time n. The other two methods are *implicit* methods, which suggests that somehow

---

[2]It is also possible of course to shift all indices by $1/2$, such that the cell centres get the integer index labels.

one already needs to know Q^{n+1} in order to compute Q^{n+1}, or that at minimum some further rewriting of the expression is necessary in order to obtain a recipe for generating Q^{n+1}. We will focus on explicit methods for now, only remarking briefly on implicit methods later on.

Note that we are trying to solve a somewhat different problem for the timestep discretization than for the spatial discretization. In the case of implementing the timestep, we assume that the complete spatial distribution of the fluid is known at a given time, and we want to obtain an unknown state at the next time. For computing the spatial derivative, we assume instead that we already know the current state at both sides of the Δx divide.

In view of the above, a first implementation of an *explicit* scheme with both forward time differencing and forward spatial differences would include the following update recipe for the state vector at a given position:

$$Q_j^{n+1} = Q_j^n - \frac{\Delta t}{\Delta x} \left(F_{j+1}^n - F_j^n \right).\tag{13.21}$$

However, if we apply this to a simple shocktube problem, the results will turn out to be an unmitigated disaster, as the snapshots from Figure 13.2 demonstrate!

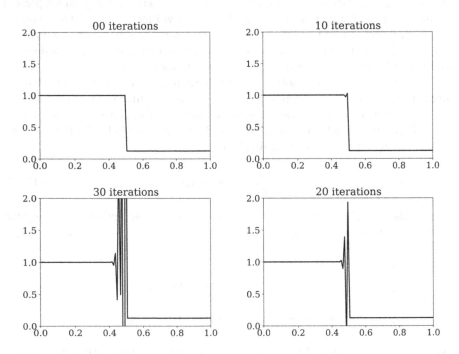

Figure 13.2 Applying the naive forward differencing scheme in time and space to a basic shocktube. The density snapshots quickly reveal a catastrophic instability. The different times are ordered clockwise.

13.3 STABILITY, ACCURACY AND DIFFUSION

A good numerical solver needs to meet a number of criteria. It should be as *accurate* as we can make it without becoming prohibitively computationally expensive (i.e. take forever to run), with the smallest possible impact of Δx and Δt being finite rather than infinitesimally small. It should have as little artificial *diffusion* as possible, meaning that in the absence of actual diffusion sharp fluid profiles such as shock fronts do not degrade over time under the cumulative effect of small but unphysical terms being introduced in the numerical form of the PDE. And, as demonstrated above, it should be *stable*, and given that this is the biggest problem by far of our naive approach, we will discuss that first.

13.3.1 STABILITY

Analysing the stability of a numerical solver is conceptually related to analysing e.g. whether a given fluid configuration is stable in theoretical fluid dynamics (and the methods deployed in Chapter 10 are very much relevant here). In a *Von Neumann* stability analysis, which we here demonstrate in a limited form applicable to the constant velocity advection problem, we insert a trial oscillation solution into our updating scheme and see under which conditions its amplitude grows. Consider the limited case of the density evolution at constant flow velocity v:

$$\rho_j^{n+1} = \rho_j^n - v \frac{\Delta t}{\Delta x} \left(\rho_{j+1}^n - \rho_j^n \right). \tag{13.22}$$

We can insert a trial solution $\rho_j^n = A^n \exp[ij\theta]$, with θ a phase angle and $i^2 \equiv -1$ as usual rather than a grid index. The phase angle θ relates to the perturbation wave number k and the cell size Δx according to $\theta = k\Delta x$. This leads to

$$A^{n+1} = A^n \left(1 - ce^{i\theta} + c \right), \qquad c \equiv v \frac{\Delta t}{\Delta x}. \tag{13.23}$$

This solution requires that

$$\|A^{n+1}\| \leq \|A^n \left(1 - ce^{i\theta} + c \right)\|, \tag{13.24}$$

if the amplitude of the solution is not to grow each time step. In turn, this implies

$$\left(1 - ce^{i\theta} + c \right)^\dagger \left(1 - ce^{i\theta} + c \right) \leq 1. \tag{13.25}$$

Because this can be rewritten in the form

$$1 + 2c \left(1 - \cos\theta \right) + 2c^2 \left(1 - \cos\theta \right) \leq 1, \tag{13.26}$$

which does not hold in general for θ, we conclude that the solver is *unconditionally unstable*. This unfortunate property is shared with the centred scheme

$$\rho_j^{n+1} = \rho_j^n - \frac{c}{2} \left(\rho_{j+1}^n - \rho_{j-1}^n \right), \tag{13.27}$$

where substituting the same trial solution shows a condition

$$1 + c^2 \sin^2 \theta \leq 1, \tag{13.28}$$

that is also violated in general. On the other hand, the scheme

$$\rho_j^{n+1} = \rho_j^n - c \left(\rho_j^n - \rho_{j-1}^n \right), \tag{13.29}$$

leads to a condition

$$1 + 2c (c - 1)(1 - \cos \theta) \leq 1, \tag{13.30}$$

which is only violated generally when $c > 1$ (or $c < 0$, but let us assume for the moment that $v \geq 0$). This scheme is therefore *conditionally stable*. From the definition of c, the condition here is that $v \leq \Delta x / \Delta t$. The interpretation is that over a single time step Δt the fluid locally has to flow no further than between two nearest neighbours on the grid if an update algorithm using only nearest neighbours is to remain stable.

13.3.2   ACCURACY

It is not sufficient for a scheme to be (conditionally) stable. We would also like it to be as *accurate* as possible, which we will quantify here.

Another way of looking at the definition of the derivative is to view it as part of a Taylor series expansion. Consider the spatial derivative in one dimension (i.e. the RHS of Equation 13.11, limited to one dimension),

$$Q_{j+1}^n = Q_j^n + \frac{\partial Q}{\partial x}|_j \Delta x + \frac{1}{2} \frac{\partial^2 Q}{\partial x^2}|_j (\Delta x)^2 + O \left((\Delta x)^3 \right). \tag{13.31}$$

A minor rewrite leads to an expression for the derivative:

$$\frac{\partial Q}{\partial x}|_j = \frac{Q_{j+1}^n - Q_j^n}{\Delta x} - \frac{1}{2} \frac{\partial^2 Q}{\partial x^2}|_j \Delta x + O \left((\Delta x)^2 \right). \tag{13.32}$$

What this equation tells us is that the forward difference quotient scheme is only accurate up to first order in Δx (i.e. the dominant error term is of order Δx). Likewise, we find for the backward difference quotient that starting from a Taylor series with a negative difference $-\Delta x$ means going from

$$Q_{j-1}^n = Q_j^n - \frac{\partial Q}{\partial x}|_j \Delta x + \frac{1}{2} \frac{\partial^2 Q}{\partial x^2}|_j (\Delta x)^2 - O \left((\Delta x)^3 \right). \tag{13.33}$$

to

$$\frac{\partial Q}{\partial x}|_j = \frac{Q_j^n - Q_{j-1}^n}{\Delta x} + \frac{1}{2} \frac{\partial^2 Q}{\partial x^2}|_j \Delta x - O \left((\Delta x)^2 \right). \tag{13.34}$$

This is again accurate to first order (the minus sign in the rightmost term is not meaningful and could have been absorbed in the O-symbol).

If instead we sum expressions 13.32 and 13.34 for the forward and backward difference quotient, we find that

$$\frac{\partial Q}{\partial x}|_j = \frac{Q_{j+1}^n - Q_{j-1}^n}{2\Delta x} + O \left((\Delta x)^2 \right). \tag{13.35}$$

The centred difference scheme is formally accurate to *second* order. It would have been superior to the other two, had it not been for the stability issue pointed out in the previous section. The art now is to find a numerically feasible scheme that is both accurate to higher order *and* stable.

13.3.3 DIFFUSION AND DISPERSION

Closely related to the issue of accuracy (actually, manifestations of numerical limitations on accuracy) are spurious diffusion and dispersion due to numerical schemes. To illustrate these effects, we again consider the basic advection equation (fluid flow at constant velocity, if you will):

$$\frac{\partial \rho}{\partial t} = -v \frac{\partial \rho}{\partial x}. \tag{13.36}$$

The general solution has the form $\rho (vt - x)$, which describes an arbitrary profile propagating unchanged through the grid with velocity v. But if we introduce a first-order accurate numerical scheme for the spatial derivative this equation becomes

$$\frac{\partial \rho}{\partial t} = -v \frac{\partial \rho}{\partial x} - v \frac{\Delta x}{2} \frac{\partial^2 \rho}{\partial x^2}, \tag{13.37}$$

when taking the forward difference scheme from Equation 13.32 as an example (we've dropped terms of higher order in Δx and Δt, assuming that the time step scheme is at least second-order accurate). The problem is that this is a qualitatively different kind of equation, and the second derivative of ρ in this expression introduces a *diffusion* term which will diffuse the features of the fluid profile over the grid. This is a form of *numerical viscosity*, and over time its impact can grow to completely wash out small scale details of the initial conditions. When the leading order noise term is a third spatial derivative rather than a second derivative, there still remains the effect of *numerical dispersion* where wave speeds across the grid become frequency-dependent (unlike the frequency-independent speed of sound) and the fluid profile also deteriorates relative to its exact solution as a result.

Since these effects scale with powers of Δx, the bottom line of this section is that they should provide further impetus to focus on high-order accurate solvers, typically combined with a brute-force approach that keeps Δx and Δt as small as possible (leading to increases in computing duration and memory requirements as an inevitable result). Below we dwell a bit further on the two differential equations above, but this basic PDE mathematics excursion is not essential to understanding the points already made about diffusion.

It is generally true that a function $q(x,t)$ can equally be considered a function of any arbitrary two parameters $\xi(x,t)$, $\eta(x,t)$, as long as ξ and η are not trivially related (e.g. $\xi \equiv 2\eta$). If we pick $\xi \equiv vt - x$ and $\eta \equiv t$, we obtain

$$\frac{\partial}{\partial t} = \frac{\partial \xi}{\partial t} \frac{\partial}{\partial \xi} + \frac{\partial \eta}{\partial t} \frac{\partial}{\partial \eta} = v \frac{\partial}{\partial \xi} + \frac{\partial}{\partial \eta}, \tag{13.38}$$

$$\frac{\partial}{\partial x} = \frac{\partial \xi}{\partial x} \frac{\partial}{\partial \xi} + \frac{\partial \eta}{\partial x} \frac{\partial}{\partial \eta} = -\frac{\partial}{\partial \xi}. \tag{13.39}$$

Substituting this in the basic advection equation, we obtain

$$v\frac{\partial \rho(\xi,\eta)}{\partial \eta} = 0. \tag{13.40}$$

This states that ρ is only a function of ξ, not a function of η (NB, this is not the same as saying ρ is time-independent, since there remains a time dependency through ξ). This confirms our earlier statement that the most general solution obeys $\rho = \rho(\xi = vt - x)$. We could have quickly concluded the same thing arguing from physics: a frame transformation to a fixed-velocity frame that eliminates v renders the RHS side of the advection equation equal to zero, meaning that q is only non-constant by being a function of this frame transformation $x' = x - vt$.

Another approach to finding how x and t are potentially coupled in a solution to the advection equation would be by separating the variables and trying the solution $\rho(x,t) = A(x)B(t)$:

$$A\frac{\partial B}{\partial t} = -vB\frac{\partial A}{\partial x} \Rightarrow \frac{1}{B}\frac{\partial B}{\partial t} = -\frac{v}{A}\frac{\partial A}{\partial x}. \tag{13.41}$$

Since in the last step the LHS is a function only of t while the RHS is a function only of x and x and t are independent coordinates, both sides should be constant. This means that we have

$$B = e^{Ct}, \tag{13.42}$$

with C a constant not yet specified. Similarly, we get

$$A = e^{-\frac{C}{v}x}. \tag{13.43}$$

with the *same* but still unspecified constant C. We might as well write C in units of v and imaginary i (neither of which does takes away anything from its arbitrariness, and we can always render C imaginary to compensate again for i), and write

$$\rho(x,t) = e^{Ci(vt-x)}. \tag{13.44}$$

This is *a* solution, and it will remain a solution when multiplied with a constant pre-factor or when a different C value is chosen. More generally, we therefore have

$$\rho(x,t) = \sum_{\lambda=0}^{\infty} \alpha_\lambda e^{C_\lambda i(vt-x)}, \tag{13.45}$$

which is leading us towards a Fourier expansion of an arbitrary function of $(vt - x)$, with α_λ and C_λ coefficients that will end up being determined by the initial state of the fluid.

If we attack the equation containing what we have identified as a diffusion term first with the transformation of variables $x,t \to \xi,\eta$, we will need

$$\frac{\partial^2}{\partial x^2} = -\frac{\partial}{\partial \xi}\left(-\frac{\partial}{\partial \xi}\right) = \frac{\partial^2}{\partial \xi^2}. \tag{13.46}$$

The resulting expression for ρ is

$$\frac{\partial \rho}{\partial \eta} + \frac{v\Delta x}{2}\frac{\partial^2 \rho}{\partial \xi^2} = 0. \tag{13.47}$$

This expression is identical in form to the famous *heat equation*, which is *the* go-to example of a diffusive process and justifies our labeling of the extra terms as a diffusion term. A separation of variables approach to this equation follows the same lines as for the advection equation, and leads to a general result

$$\rho(x,t) = \sum_{\lambda=0}^{\infty} \alpha_\lambda e^{\frac{\Delta xv}{2}C_\lambda^2 t + iC_\lambda(vt-x)}. \tag{13.48}$$

Note that we have an inevitable ninety degree phase shift in the complex plane between the t and $vt - x$ terms. This is what is responsible for the diffusive behaviour of this solution.

13.3.4   THE LAX-FRIEDRICHS AND LAX-WENDROFF SCHEMES

Before moving on fully to finite volume methods, we first introduce some finite difference schemes that we will subsequently rederive from a finite volume approach. For now, we limit ourselves to the advection equation with constant v instead of the full set of fluid equations. Recall the scheme

$$\rho_j^{n+1} = \rho_j^n - c\left(\rho_j^n - \rho_{j-1}^n\right), \qquad c \equiv \frac{v\Delta t}{\Delta x}. \tag{13.49}$$

This was stable as long as $0 \le c \le 1$. Since the exact solution obeys $\rho(x,t) = \rho(x - v\Delta t, t - \Delta t)$, the requirement that $c \ge 0$ for stability should come as no surprise. The exact solution tells us that the solution propagates with velocity v, and if $v < 0$ we would have a solution that propagates through the grid coming from the *right*, while the grid updates draw on points ρ_{j-1} to the *left* of ρ_j. The numerical scheme is 'facing in the wrong direction' when looking to gather information for its update. Conversely, the scheme

$$\rho_j^{n+1} = \rho_j^n - c\left(\rho_{j+1}^n - \rho_j^n\right), \qquad c \equiv \frac{v\Delta t}{\Delta x}, \tag{13.50}$$

is only stable as long as $-1 \le c \le 0$. Schemes that properly account for the direction from which information ('wind') flows are called *upwind* schemes. A guaranteed upwind scheme that allows for both signs of c, is given by

$$\rho_j^{n+1} = \rho_j^n - c^+\left(\rho_j^n - \rho_{j-1}^n\right) - c^-\left(\rho_{j+1}^n - \rho_j^n\right), \tag{13.51}$$

if we define

$$c^\pm = \frac{1}{2}\left(c \pm |c|\right), \tag{13.52}$$

as a mechanism to set one of the terms to zero depending on the sign of c.

A well-known scheme that does not depend on a discrete choice to ensure an upwind method, is the Lax-Friedrichs scheme. We obtain it by starting from the unconditionally unstable scheme

$$\rho_j^{n+1} = \rho_j^n - \frac{c}{2}\left(\rho_{j+1}^n - \rho_{j-1}^n\right),$$ (13.53)

which draws in equal measure from the points to the left and right of the point to be updated. If we replace the time derivative term ρ_j^n by an average over its surrounding points, we obtain the Lax-Friedrichs scheme

$$\rho_j^{n+1} = \frac{1}{2}\left(\rho_{j-1}^n + \rho_{j+1}^n\right) - \frac{1}{2}c\left(\rho_{j+1}^n - \rho_{j-1}^n\right)$$
$$= \frac{1}{2}(1+c)\rho_{j-1}^n + \frac{1}{2}(1-c)\rho_{u+1}^n.$$ (13.54)

Trying out a few values for c, we can see that this scheme no longer always equally draws on the left and right neighbouring points:

$$c = \frac{1}{2}:\ \rho_j^{n+1} = \frac{3}{2}\rho_{j-1}^n + \frac{1}{4}\rho_{j+1}^n,$$ (13.55)

$$c = -\frac{1}{2}:\ \rho_j^{n+1} = \frac{1}{4}\rho_{j-1}^n + \frac{3}{2}\rho_{j+1}^n,$$ (13.56)

$$c = 0:\ \rho_j^{n+1} = \frac{1}{2}\rho_{j-1}^n + \frac{1}{2}\rho_{j+1}^n.$$ (13.57)

This scheme is conditionally stable as long as $|c| \le 1$.

If instead of replacing the time derivative term with an average, we choose to average between the forwards and backwards spatial derivative schemes with weights β_1, β_2, we obtain

$$\rho_j^{n+1} = \rho_j^n - \beta_1 c\left(\rho_j^n - \rho_{j-1}^n\right) - \beta_2 c\left(\rho_{j+1}^n - \rho_j^n\right).$$ (13.58)

For the particular choice $\beta_1 = \frac{1}{2}(1+c)$ and $\beta_2 = \frac{1}{2}(1-c)$ we obtain the Lax-Wendroff scheme:

$$\rho_j^{n+1} = \rho_j^n - \frac{1}{2}(1+c)c\left(\rho_j^n - \rho_{j-1}^n\right) - \frac{1}{2}(1-c)c\left(\rho_{j+1}^n - \rho_j^n\right)$$
$$= \frac{1}{2}c(1+c)\rho_{j-1}^n + \left(1-c^2\right)\rho_j^n - \frac{1}{2}c(1-c)\rho_{j+1}^n.$$ (13.59)

Interestingly, this scheme is both conditionally stable[3] and *second*-order accurate in space and time.

All of the schemes presented here are variations on a theme, where the time-updated value for ρ_j depends on some weighed sum over a series of zones around

---

[3]Keep in mind however that this stability is with respect to the advection equation, not the full set of Euler's equations. The Lax-Wendroff scheme applied to Euler's equations would not be able to pass all the shock-tube tests from this chapter.

position j at the previous time step, i.e.,

$$\rho_j^{n+1} = \sum_{k=j-j_L}^{j+j_R} B_k \rho_k^n, \tag{13.60}$$

with weights B_j and j_L and j_R indicating some range (1 in the examples so far). If all the terms $B_j \geq 0$, then the scheme is considered *monotone* (and will be at most first-order accurate)[4]. If some terms are negative, the scheme can be accurate to higher order at the cost of introducing oscillations in the solution. As long as the stability criterion is met, these oscillations will remain minor.

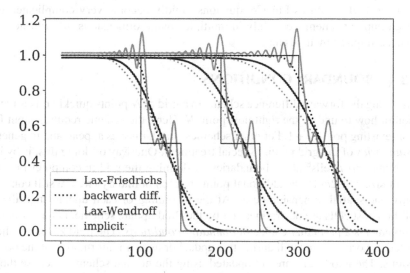

Figure 13.3 Comparison of solutions to the advection equation with $v = 100$. Snapshots are shown at $t = 1, 2, 3$. For the explicit methods $\Delta t = 0.002$, for the implicit method $\Delta t = 0.1$.

The different schemes are compared[5] in Figure 13.3. Note how only the Lax-Wendroff scheme (which is non monotone) shows oscillations. Because an implicit scheme is still straightforward to implement in the case of the advection equation, this is shown as well. The implicit scheme follows a simple backwards differencing scheme for the spatial derivative (and therefore relies on $v > 0$):

$$\rho_j^{n+1} = \rho_j^n - c\left(\rho_j^{n+1} - \rho_{j-1}^{n+1}\right), \qquad c \equiv \frac{v\Delta t}{\Delta x} \tag{13.61}$$

This can be rewritten as

$$\rho_j^{n+1} = \frac{\rho_j^n + c\rho_{j-1}^{n+1}}{1+c}. \tag{13.62}$$

---

[4]The question whether a given scheme is monotone or not is obviously a useful and important one in computational fluid dynamics. As for most topics in this chapter, the interested reader is referred to e.g. [53] for further details.

[5]This particular setup is used to demonstrate numerical diffusion in Bodenheimer2007.

Note how each entry at the new time depends on *all* entries to its left evaluated also at the new time. By fixing the leftmost point to its boundary condition, we are able to avoid an infinite regression. Because the new point depends on everything to its left, the method is unconditionally stable (still assuming $v > 0$ when using a backward differencing spatial scheme). We therefore show a time step $\Delta t = 0.1$ in the figure, which would be ten times over the maximum allowed time step for the explicit methods set by the requirement that $|c| < 1$. We could even have used a time step $\Delta t = 1$, but the price that one pays is increased diffusion. Using a time step identical to the one used for the implicit schemes would yield a result almost indistinguishable from the explicit backward differencing scheme. Because multiple coupled PDEs such as Euler's equations rapidly become very complicated to cast in an implicit scheme, certainly in multiple spatial dimensions, we will not further discuss implicit methods here.

13.4   BOUNDARY CONDITIONS

Applying the forward difference scheme to a grid of N points quickly raises the question of how to update the rightmost point N. Here, the scheme requires input from a non-existing point $N + 1$. For other schemes, similar issues appear, and in general the *boundaries* of the grid require special treatment. One way of doing this, is by implementing *ghost* cells[6] at the boundaries by defining the grid in computer memory to be of size $N + 2m$ for m additional points at each boundary (we are still considering only one spatial dimension here). At the beginning of each update step, the states of the ghost cells are determined not by the fluid update scheme, but externally: this implements a particular choice of *boundary conditions*. The fluid zones are then updated as normally and will during this update draw from information at the boundary points. The ghost cells are not updated using the normal scheme, because their values will once again be reset using external conditions at the beginning of the next iteration. Generally, we can identify the following types of boundary conditions:

- **Outflow.** In outflow or open boundary conditions, the fluid state of the outermost actual grid cells is duplicated onto the ghost zones. As a result, the impact of the boundaries is not felt during the update step and the behaviour of the fluid at the boundary remains unchanged. Fluid flows either into or out of the grid depending on the sign of the velocity relative to the direction of the boundary.
- **Periodic.** When the left ghost cells are filled with duplicates of the rightmost actual grid cell, and the right ghost cells filled with duplicates of the leftmost actual grid cell, a periodic boundary is established. Whatever moves out of the grid on the right, appears again on the left and vice versa.
- **Reflecting.** In reflecting boundary conditions the ghost cells contain duplicates of their neighbouring cells within the grid, except with the sign of

---

[6]I am using the terms *cells* and *points* interchangeably for the moment. In terms of the *finite difference* approach it might make more sense to talk of points. In terms of a numerical implementation in a C array, we are talking about the cells or entries of the array. Once we move to a *finite volume* approaches, the term *cell* becomes unambiguously preferable.

velocity flipped. As a consequence of flipping the sign of the velocity, fluid pushing onto the boundaries of the grid receives an equal push back and cannot leave the grid. The boundary is closed and the fluid is trapped in a 'box'.

- **User-defined.** User-defined boundary conditions can be anything. If the ghost cells are for example continuously set to contain an energetic fluid with a certain velocity, this can be used to model a physical process occurring at the edge of the grid (e.g. the launching of a stellar wind, or a blast wave).

13.5   FINITE VOLUME METHODS

13.5.1   FINITE DIFFERENCE METHODS VERSUS FINITE VOLUME METHODS

So far, we have limited ourselves to *finite difference methods*. Their weak spot is that they only probe the fluid at a number of individual positions. For conservation laws, this is not a natural approach. Instead, an approach which is extremely popular in numerical hydrodynamics is the *finite volume method*. Here, we take the value at a given spatial index i to represent not just that single location in space, but the entire integrated content of a cell centred at i (see also Figure 13.4). This approach gets us far closer to numerically conserving the total amount of each conserved quantity (mass, momentum, energy) on the grid, because all the flux leaving one zone is directly entered into the neighbouring zone and vice versa. In practice, the cell volumes are often divided out again from the numerical scheme if they are of constant size, but this does not impact the point that finite volume methods, in general, are more naturally suited to conservation laws. For simple schemes, the differences between finite difference and finite volume approaches is sometimes merely philosophical. In such cases, the same numerical prescription will follow regardless of whether one

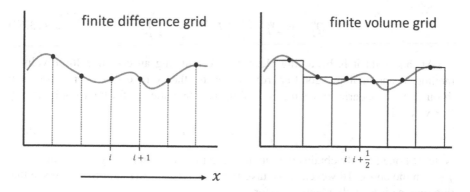

Figure 13.4   A comparison between finite difference methods and finite volume methods, where grid points respectively capture the fluid state at a specific position and across a given volume element.

considers the indices to label cell-averages or single points in space. We will however also discuss more complex schemes where a finite volume method really comes
into its own.

13.5.2   BASIC FINITE VOLUME METHODS

To move beyond finite difference methods, we need to consider the solution to

$$\frac{\partial Q}{\partial t} + \frac{\partial F}{\partial x} = 0, \tag{13.63}$$

for an entire spacetime volume element. Integrating the expression from x_0 to x_1 in
space, we have

$$\frac{\partial}{\partial t} \int_{x_0}^{x_1} Q(x,t)dx + F(x_1,t) - F(x_0,t) = 0. \tag{13.64}$$

Integrating over time as well, this in turn becomes

$$\int_{x_0}^{x_1} Q(x,t_1)dx - \int_{x_0}^{x_1} Q(x,t_0)dx + \int_{t_0}^{t_1} F(x_1,t)dt - \int_{t_0}^{t_1} F(x_0,t)dt = 0. \tag{13.65}$$

This we re-arrange and divide by $\Delta x \equiv x_1 - x_0$:

$$\frac{1}{\Delta x} \int_{x_0}^{x_1} Q(x,t_1)dx = \frac{1}{\Delta x} \int_{x_0}^{x_1} Q(x,t_0)dx$$
$$- \frac{\Delta t}{\Delta x} \left(\frac{1}{\Delta t} \int_{t_0}^{t_1} F(x_1,t)dt - \frac{1}{\Delta t} \int_{t_0}^{t_1} F(x_0,t)dt \right). \tag{13.66}$$

We take $x_0 = x_{j-1/2}$, $x_1 = x_{j+1/2}$, t_0 the time at time coordinate n and t_1 the time
at $n+1$. If we view the expression above in terms of a spatial average of the state
vector at cell j, and a time average of the state flux vector, the expression states

$$Q_j^{n+1} = Q_j^n - \frac{\Delta t}{\Delta x} \left(F_{j+\frac{1}{2}}^{avg} - F_{j-\frac{1}{2}}^{avg} \right). \tag{13.67}$$

At this point it is becoming clear that formulating an effective finite-volume
method means finding a good approximation for the average flux over a time step
through the boundaries of each cell. A simplistic first attempt for the right boundary
flux would be

$$F_{j+\frac{1}{2}}^{avg} \equiv \frac{1}{2} \left(F_j^n + F_{j+1}^n \right). \tag{13.68}$$

Note that one directly obtains the flux through the left boundary by substituting $j \to$
$j-1$ in the above. However, if we take the flux to be $F = vq$ and fix v, we see that
this scheme is actually identical in form to

$$Q_j^{n+1} = Q_j^n - \frac{v\Delta t}{\Delta x} \frac{1}{2} \left(Q_{j+1}^n - Q_{j-1}^n \right), \tag{13.69}$$

i.e. the centred scheme from Equation 13.27, which was shown to be unconditionally unstable. Taking a finite volume method approach is therefore not a guarantee for stability.

A more productive scheme is given by

$$F_{j+\frac{1}{2}}^{avg} \equiv \frac{1}{2}\left(F_j^n + F_{j+1}^n\right) + \frac{1}{2}\frac{\Delta x}{\Delta t}\left(Q_j^n - Q_{j+1}^n\right), \qquad (13.70)$$

which includes an additional term to salvage the unconditionally unstable scheme. As a matter of fact, when substituting the advection equation flux qv in this expression, the resulting scheme can be shown to be equivalent to the Lax-Friedrichs scheme that we encountered before and which is conditionally stable under the usual time step condition.

If we can obtain a finite volume version of the Lax-Friedrichs scheme, we might also be able to get a version of the Lax-Wendroff scheme for coupled equations. Indeed, this is obtained when taking

$$F_{j+\frac{1}{2}}^{avg} \equiv F\left(Q_{j+\frac{1}{2}}^{n+\frac{1}{2}}\right), \qquad (13.71)$$

meaning the flux as determined from an intermediate state q defined as

$$Q_{j+\frac{1}{2}}^{n+\frac{1}{2}} \equiv \frac{1}{2}\left(Q_j^n + Q_{j+1}^n\right) + \frac{1}{2}\frac{\Delta t}{\Delta x}\left(F_j^n - F_{j+1}^n\right). \qquad (13.72)$$

This can be shown to reduce to Equation 13.59 in the case of $F = vq$.

13.5.3 TIME STEP CRITERION AND A WORKING CODE

At this point nearly all the pieces are in place to quickly write a functional hydro-dynamics code for solving Euler's equations in one spatial dimension. This amounts to implementing the flowchart from Figure 13.5. The only thing that we have not touched upon is determining the time step size for Euler's equations. The advection equation had fixed v, and all the conditionally stable methods we have implemented so far required that $v\Delta t/\Delta x \leq 1$ which in turn limits Δt once the other quantities are known. Taking this requirement as inspiration, we can introduce as condition that $\Delta t < \Delta x/a$, where a the maximum signal speed on the grid. This type of condition on the maximum size of the time step is known as the *Courant-Friedrichs-Lewy* (CFL) condition. For Euler's equations, the maximum possible wave speed in the grid frame is given by the largest value of $a = |v| + |c_s|$. For a polytropic gas, the sound speed c_s^2 obeys $c_s^2 = \hat{\gamma}p/\rho$. In practice, you will want to build in a royal extra margin (e.g. scale Δt by $1/3$) to account for shocks, when using a simplistic estimate for a based purely on the sound speed. If the signal speed a is estimated in a more sophisticated manner, this margin can be made tighter. There will be some trade-off between the size of the margin and the extra time and/or computer memory needed for a more detailed estimate of the signal speeds.

Some code demonstrations are shown in Figure 13.6. These are examples of a standard test, the *shock tube* first encountered in Section 7.2, where left and right

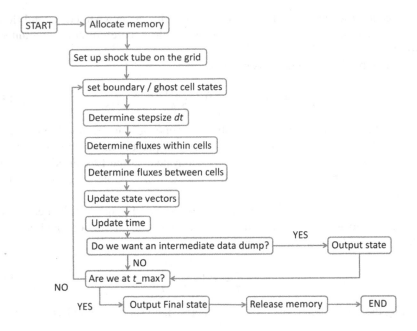

Figure 13.5   A flowchart for solving a standard shock tube test.

states are initialized that will lead to a solution showing the key features of a shocks, rarefaction waves and contact discontinuities. Note how again it is the Lax-Wendroff scheme that is better able to approximate the exact solution by being second-order accurate (at the cost of oscillations).

13.5.4   THE HLL METHOD AND GODUNOV APPROACH

Figure 13.6 shows a common way to show the performance of a hydrodynamics code, by testing it against a shock tube problem of which the solution is known analytically. Scenarios where a discontinuous jump between two fluid states is set up at $t = 0$, and the resulting pattern of shocks/waves and a contact discontinuity are solved for, are called Riemann problems, the solution of which is known analytically even for relativistic fluid flow. Once time starts, the contact discontinuity separating the initial states will be bracketed by two wave fronts. These can be either rarefaction waves, shock fronts, or a combination of the two. In Figure 13.6 we see a rarefaction wave that has moved to the left. Behind the front, there is a smoothly changing fluid profile (in contrast to a shock, there are no abrupt first-order discontinuities in the profile for a wave). The contact discontinuity has drifted rightwards to $x \approx 0.55$ or so. The rightmost drop at 0.7 is a shock front.

The fact that this profile (and any other Riemann result) can be computed analytically proves extremely helpful in practice for the purpose of computing fluxes between cells, even when this solution is only approximated. Once we start

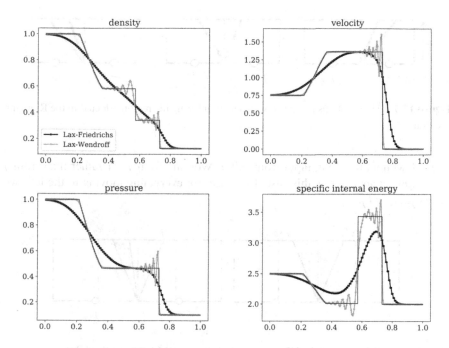

Figure 13.6 Standard shock tube test for Lax-Friedrichs and Lax-Wendroff methods. The initial conditions are that the left state ($x < 0.3$ at $t = 0$) has $\rho = 1$, $p = 1$ and $v = 0.75$, whereas the right state has $\rho = 0.125$, $p = 0.1$ and $v = 0$. Both solvers use 200 zones (not counting ghost zones) between $x = 0 \ldots 1$. The snapshots are taken at $t = 0.2$, the adiabatic exponent $\hat{\gamma}_{ad} = 1.4$.

interpreting the states Q_j as describing the state throughout a computational zone (i.e. the finite-volume approach), we realise that between every two zones we basically have a Riemann problem at the beginning of each time step. This Riemann problem will dictate the form that the flux across the cell boundary will take. Hydrodynamics solvers that use a finite volume approach and solve for the fluxes between cells starting from the assumption that we have a Riemann problem across each cell boundary, are called Godunov methods, after the first person to successfully formulate this approach.

There are three main options for the trajectories of the two outermost wave fronts, depending on the relative jump in fluid conditions across the cell boundary. In Figure 13.6 we have a front propagating to the left and one to the right. Alternatively, we could have both fronts propagating to the right or both to the left (e.g. if the initial states start with a large initial velocity and the fluid is rapidly flowing through the boundary already at $t = 0$, regardless of any additional shock fronts that subsequently emerge), as shown in Figure 13.7. If e.g. the rightmost case in the figure (both waves moving to the right) is realized between cells $j - 1$ and j, then the flux $F_{j-1/2}$ is simply given by F_{j-1}. The flux F_{j-1} can be determined directly from the state vector

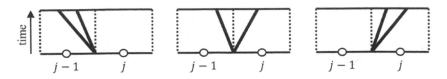

Figure 13.7 Three options for the wave fronts bracketing the region affected in the Riemann problem.

at $j-1$ without requiring input from cell j. Working from $j-1$ rather than from j is the *upwind* approach in this case. Likewise, for everything moving to the left, we have $F_{j-1/2}=F_j$.

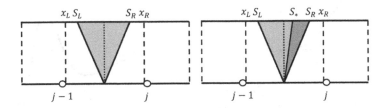

Figure 13.8 Schematic of the HLL approach (left) and HLLC approach (right).

The middle case in Figure 13.7 is more complicated, and requires some accounting for the Riemann problem to obtain a value for $F_{j-1/2}$. We proceed to derive the *HLL* scheme (after Harten, Lax & Van Leer), which is based on an approximation that assumes the state in between the wave fronts to be homogeneous. We start once more from the conservation laws in integral form. We centre the grid origin at the grid boundary between cells $j-1$ and j, and we consider an interval between x_L and x_R that comfortable encloses the distances that the waves have covered (see the left image in Figure 13.8), moving with outer wave speeds S_L and S_R respectively. Later we will expand this algorithm to explicitly account for the motion of the contact discontinuity as well (the HLLC method, with the CD assigned a wave speed S_* as illustrated in the right image in Figure 13.8). Between times t_0 and t_1 (a time interval Δt going from n to $n+1$), we have

$$\int_{x_L}^{x_R} Q(x,t_1)\,dx = \int_{x_L}^{x_R} Q(x,t_0)\,dx - \int_{t_0}^{t_1} F(x_R,t)\,dt + \int_{t_0}^{t_1} F(x_L,t)\,dt$$

$$= x_R Q_R^n - x_L Q_L^n - \Delta t F_R + \Delta t F_L. \tag{13.73}$$

Since the position x_R lies within cell j and has not yet been reached by any of the waves triggered by the Riemann problem, we have $Q_R = Q_j$ and $F_R = F_j$. Likewise, we have $Q_L = Q_{j-1}$ and $F_L = F_{j-1}$. On the other hand, we could also rewrite the

LHS above by splitting up the integration domain:

$$\int_{x_L}^{x_R} Q(x,t_1)dx = \int_{x_L}^{\Delta t S_L} Q(x,t_1)dx + \int_{\Delta t S_L}^{\Delta t S_R} Q(x,t_1)dx + \int_{\Delta t S_R}^{x_R} Q(x,t_1)dx$$

$$= (\Delta t S_1 - x_L)Q_L^n + \int_{\Delta t S_L}^{\Delta t S_R} Q(x,t_1)dx + (x_R - \Delta t S_R)Q_R^n \quad (13.74)$$

Equating these two expressions and solving for the remaining unevaluated integral, we obtain

$$\int_{\Delta t S_L}^{\Delta t S_R} Q(x,t_1)dx = \Delta t S_R Q_R^n - \Delta t S_L Q_L^n + \Delta t F_L^n - \Delta t F_R^n. \quad (13.75)$$

If we divide left and right by the domain interval $\Delta t(S_R - S_L)$, we can interpret the resulting LHS as describing an average state Q^{hll}:

$$Q^{hll} = \frac{S_R Q_R^n - S_L Q_L^n + F_L^n - F_R^n}{S_R - S_L}. \quad (13.76)$$

Remember that we are after an expression for $F_* \equiv F_{j-1/2}$ for the case when the boundary lies between the two waves S_L and S_R. This flux will presumably turn out to be some function of Q^{hll}. To bring this flux term into the picture, let us consider the interval between 0 (the boundary at the origin of our ad-hoc coordinate system) and $\Delta t S_R$ and again cast the conservation laws in integral form:

$$\int_0^{\Delta t S_R} Q(x,t_1)dx = \int_0^{\Delta t S_R} Q(x,t_0)dx - \int_{t_0}^{t_1} F(\Delta t S_R, t)dt + \int_{t_0}^{t_1} F(0,t)dt, \quad (13.77)$$

which can be trivially reorganized to get an expression for the flux through the origin over time:

$$\int_{t_0}^{t_1} F(0,t)dt = \int_{t_0}^{t_1} F(\Delta t S_R, t)dt - \int_0^{\Delta t S_R} Q(x,t_0)dx + \int_0^{\Delta t S_R} Q(x,t_1)dx. \quad (13.78)$$

We have simple expressions for all three terms on the RHS. The first is a flux term evaluated outside of the reach of the wave S_R and is therefore equal to $\Delta t F_R$, the constant flux multiplied by the time interval. The second RHS term also has a constant and known integrand $Q(x,t_0)$ on account of being evaluated at t_0 before the waves start moving. The integrand in the third RHS term is given by state that we just approximated by the constant Q^{hll}. This leads to

$$\int_{t_0}^{t_1} F(0,t)dt = \Delta t F_R - \Delta t S_R Q_R^n + \Delta t S_R Q^{hll}, \quad (13.79)$$

where we use the superscript n to indicate temporal iteration point n coinciding with t_0. If we divide left and right by integration domain Δt, we again have an average quantity on the LHS, this time F^{hll}:

$$F^{hll} = F^R - S_R Q_R^n + S_R Q^{hll}$$

$$= \frac{S_R F_L - S_L F_R + S_R S_L (Q_R^n - Q_L^n)}{S_R - S_L}. \quad (13.80)$$

Now all we need are some estimates for the wave speeds S_L and S_R, and we have a recipe for $F_{j-1/2}$ for all cases of interest as listed in Fig 13.7 (which can be implemented with a decision tree based on the signs of S_L and S_R, i.e., $S_L, S_R < 0$ means both wave fronts left-moving, etc.). Luckily, we can be quite sloppy in how we estimate them before this becomes a source of inaccuracy on par with the starting assumption of a flat fluid state between the two waves.

13.5.5 WAVE SPEED ESTIMATES

What we haven't covered so far is how to actually compute the wave speeds S_R and S_L. These can be estimated directly from the solution to the Riemann problem, or indirectly using for example the pressure in the region between by the waves. It would take us beyond the scope of the book to delve too deeply into motivating a particular approach or into fully deriving an estimate once a few guiding principles are established. Instead, we take the easy route, and merely present an estimate based on the pressure p_* between the waves that can be slotted into a numerical implementation of the methods presented in this chapter. For pressure p_* we have

$$p_* = max\left(0, \frac{1}{2}(p_L + p_R) - \frac{1}{2}(v_R - v_L)\frac{1}{2}(\rho_L + \rho_R)\frac{1}{2}(c_{s,L} + c_{s,R})\right). \qquad (13.81)$$

If $p_* \leq p_L$, we take for S_L:

$$S_L = v_L - c_{s,L}, \qquad (13.82)$$

i.e., a sound wave moving to the left. If $p_* > p_L$, we use instead

$$S_L = v_L - c_{s,L}\sqrt{1 + \frac{\hat{\gamma}_{ad} + 1}{2\hat{\gamma}_{ad}}(p_*/p_L - 1)}, \qquad (13.83)$$

more appropriate to a shock wave. If $p_* \leq p_R$, we take for S_R:

$$S_R = v_R + c_{s,R}, \qquad (13.84)$$

again a sound wave. If $p_* > p_R$, we use again instead

$$S_R = v_R + c_{s,R}\sqrt{1 + \frac{\hat{\gamma}_{ad} + 1}{2\hat{\gamma}_{ad}}(p_*/p_R - 1)}. \qquad (13.85)$$

13.5.6 THE HLLC METHOD

The HLL method offers a massive improvement over the Lax-Friedrich and Lax-Wendroff schemes, as illustrated in Figure 13.9. The HLL method however still involves two severe approximations. The first is that we assume the states between S_L and S_R to be constant both to the left and right of the contact discontinuity (CD) that must lie somewhere within this range, the second is that the constant states to the left and right of this CD are equal to each other. It is quite feasible to relax the latter

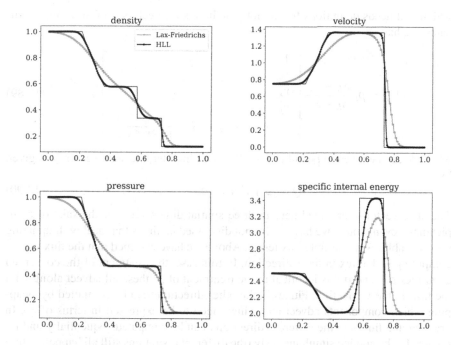

Figure 13.9 Standard shock tube test for Lax-Friedrichs and HLL methods. The initial conditions are that the left state ($x < 0.3$ at $t = 0$) has $\rho = 1$, $p = 1$ and $v = 0.75$, whereas the right state has $\rho = 0.125$, $p = 0.1$ and $v = 0$. Both solvers use 200 zones (not counting ghost zones) between $x = 0 \ldots 1$. The snapshots are taken at $t = 0.2$, the adiabatic exponent $\hat{\gamma}_{ad} = 1.4$.

constraint and explicitly account for the density jump across the CD. An example of such an approach is the *HLLC* scheme (the 'C' refers to the contact discontinuity). Its impact on the performance on the shock tube shown in Figure 13.9 will be negligible, but in other cases the difference between HLL and HLLC can be quite stark.

An implementation of the HLLC scheme involves computing an estimate for the velocity S_* of the contact discontinuity, and expanding the decision tree for $F_{j-1/2}^{hllc}$ to include *two* intermediate flux states vectors $F_{*,L}$ and $F_{*,R}$ instead of a single F_*. Having already discussed the conceptual background for the HLL scheme, we limit ourselves here to a brief summary. Assume that we have computed S_L and S_R as described above. For the CD velocity, we have

$$S_* = \frac{p_R - p_L + \rho_L v_L(S_L - v_L) - \rho_R v_R(S_R - v_R)}{\rho_L(S_L - v_L) - \rho_R(S_R - v_R)}. \tag{13.86}$$

For the new fluxes, we use

$$F_{*,L} = F_L + S_L(Q_{*,L} - Q_L), \tag{13.87}$$

$$F_{*,R} = F_R + S_R(Q_{*,R} - Q_R), \tag{13.88}$$

which still needs definitions for the intermediate state vectors. In Cartesian coordinates, we have:

$$Q_{*,K} = \rho_K \frac{S_K - v_K}{S_K - S_*} \begin{bmatrix} 1 \\ S_* \\ v_K^y \\ v_K^z \\ \frac{\mathscr{E}_K}{\rho_K} + (S_* - v_K)\left(S_* + \frac{p_K}{\rho_K(S_K - v_K)}\right) \end{bmatrix}, \tag{13.89}$$

where K can be taken to be either L or R. As a reminder, the state vector Q is given by

$$Q = [\rho, \rho v^x, \rho v^y, \rho v^z, \mathscr{E}]^T. \tag{13.90}$$

The state vector is provided here in three spatial dimensions for the sake of completeness, even though we have so far not discussed methods that deal with updating state variables with multiple flux terms. Above we have assumed that the flux we are computing is the flux in the x-direction. In this case, the treatment of the conserved quantities ρv^y and ρv^z is identical to the treatment of ρ: these all advect along with the flow in the x-direction. Fluxes in the other directions can be computed by swapping which components advect and which need to be expressed in terms of S_*. In practice, the fluxes in the different directions can be computed sequentially and do not need to be applied simultaneously (the different operations still all happen within the same time span Δt, though).

In Figure 13.10, we show a comparison between HLL and HLLC for a shock problem where the differences between the two stand out clearly. Note that, as expected, the difference between the two manifests itself in the density profile (and thus in specific internal energy, which includes a division by density).

13.6 HIGHER ORDER IN TIME AND SPACE

The minimum size of Δx (and as a result, Δt) will be limited by practical considerations about e.g. the amount of computer memory available or constraints on the maximum running time of the program. To nevertheless still improve accuracy, Godunov schemes can be made accurate to *higher order* in space and time, and there exists a rich literature on methods to achieve this. Often such schemes describe an approach to simultaneously increase the order of accuracy in Δx and Δt, and it is generally advisable to have the order of accuracy in both to be comparable if not equal: there is little point in spending a lot of effort going to higher order in one of the two beyond the point where the other has already become the bottleneck in what accuracy the program will achieve. Nevertheless, it is perfectly feasible to split the approach for achieving higher order in Δx from that for achieving higher order in Δt. Because this is a popular approach in computational hydrodynamics and because it allows us to describe the two separately, we will discuss separate higher-order space and time methods below.

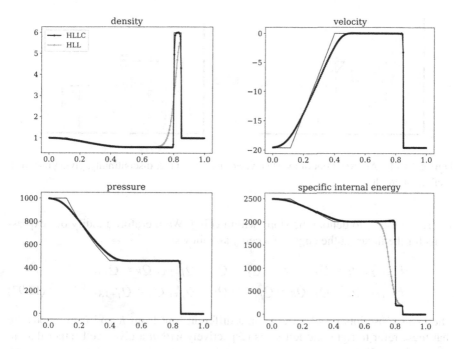

Figure 13.10 Standard shock tube test for HLL and HLLC methods. The initial conditions are that the left state ($x < 0.8$ at $t = 0$) has $\rho = 1$, $p = 1000.0$ and $v = -19.59745$, whereas the right state has $\rho = 1.$, $p = 0.01$ and $v = -19.59745$. Both solvers use 200 zones (not counting ghost zones) between $x = 0 \ldots 1$. The snapshots are taken at $t = 0.012$, the adiabatic exponent $\hat{\gamma}_{ad} = 1.4$.

13.6.1 HIGHER ORDER IN SPACE

Take Q_j and x_j to denote respectively the cell-centred state vectors and coordinates (to lowest order, piecewise flat, Q_j is indeed equal to the cell-averaged value so we do not confuse matters by this notational choice). However, in the Godunov approach we are specifically trying to solve for fluxes through cell boundaries, and the state vector is no longer flat within each cell once we go to higher order. In the piecewise linear method (PLM) we generalize the state vector notation to allow for fluctuations within each cell as follows:

$$Q_j(x) = Q_j + (x - x_j) \left(\frac{\partial Q}{\partial x} \right)_j. \tag{13.91}$$

In particular, near the right and left edges respectively of a cell j, this implies

$$Q_{j,\mathrm{r}} = Q_j + \frac{\Delta x}{2} \left(\frac{\partial Q}{\partial x} \right)_j = Q_j + \frac{\Delta x}{2} s_j,$$

$$Q_{j,\mathrm{l}} = Q_j - \frac{\Delta x}{2} \left(\frac{\partial Q}{\partial x} \right)_j = Q_j - \frac{\Delta x}{2} s_j, \tag{13.92}$$

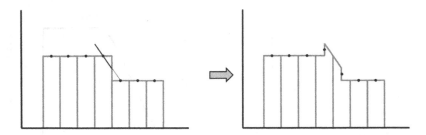

Figure 13.11   An overshoot of the state vector near a contact discontinuity, giving rise to an oscillating profile.

where we use s_j to denote the slope within cell j. We therefore modify our expressions for the fluxes at the edges of a cell j as follows:

$$F_{j-1/2} : Q_L = Q_{j-1}, \ Q_R = Q_j \rightarrow Q_L = Q_{j-1,\mathfrak{r}}, \ Q_R = Q_{j,\mathfrak{l}},$$
$$F_{j+1/2} : Q_L = Q_j, \ Q_R = Q_{j+1} \rightarrow Q_L = Q_{j,\mathfrak{r}}, \ Q_R = Q_{j+1,\mathfrak{l}}. \qquad (13.93)$$

The extra subscripts $\mathfrak{r}$ and $\mathfrak{l}$ are set in a different font from R and L to emphasize that these refer to right and left sides respectively *within* a given cell rather than to surrounding cells.

We still need to define s_j in numerical terms. As possible choices the usual suspects make their appearance again (upwind, downwind and centred, where upwind and downwind are relative to the flow direction):

$$s_j = \frac{Q_j - Q_{j-1}}{\Delta x}, \qquad s_j = \frac{Q_{j+1} - Q_j}{\Delta x}, \qquad s_j = \frac{Q_{j+1} - Q_{j-1}}{2\Delta x}. \qquad (13.94)$$

There is the usual issue here about upwind versus downwind, in particular, when we are not solving an advection equation with a fixed velocity of known direction, but the full set of Euler's equations. Additionally, there is the potential for the slope to lead to an overshoot in a state vector value at the boundary of a cell in the presence of a discontinuity. This effect is illustrated in Figure 13.11 for the Lax-Wendroff scheme applied to the advection equation[7]. In this advection problem the slope in zone j is determined from zones $j + 1$ and j, which is a downwind approach for a right-moving velocity. Once the profile in the left image of Figure 13.11 is advected for some time using the Lax-Wendroff scheme, it gives rise to the oscillation depicted in the right figure. Similar issues occur near discontinuities for other flux-computing methods and slope reconstruction schemes.

Ideally, we want a method that knows when not to overcomplicate matters and reduces to a simpler (i.e. lower-order) scheme in the presence of shocks. To implement this, we need a *slope limiter*, an algorithm for computing s_i that includes the possibility of a smaller or even nonexistent slope where prudent. On this topic too a

---

[7]The figure is similar to an illustration from [25], where further discussion of this issue can be found.

rich literature exists. We show a variety of the *minmod* algorithm:

$$a \equiv \frac{Q_j - Q_{j-1}}{\Delta x}, \qquad b \equiv \frac{Q_{j+1} - Q_j}{\Delta x}, \qquad c \equiv \frac{Q_{j+1} - Q_{j-1}}{2\Delta x},$$

$$s_j \equiv \frac{1}{4}\text{sign}(a)\,(\text{sign}(a) + \text{sign}(b))\,(\text{sign}(a) + \text{sign}(c))\min(|a|,|b|,|c|). \quad (13.95)$$

The last expression can be rephrased in a way that brings out more clearly what it does:

$$\text{minmod}(a,b,c) = \begin{cases} \min(a,b,c), & \text{if } a,b,c > 0, \\ \max(a,b,c), & \text{if } a,b,c < 0, \\ 0, & \text{otherwise.} \end{cases} \quad (13.96)$$

The algorithm sets s_j to zero when state Q_j represents an extremum (i.e. it is larger or smaller than both its neighbours). Otherwise, it picks out the slope with the smallest magnitude. In the formulation above, this will be either a or b, since c as defined above is equivalent to the average $(a + b)/2$. Therefore c is never picked and this formulation is a bit silly for appearing to include c as an option. However, when the prescriptions for the amplitudes of a and b are tweaked, there actually may be a case where c ends up the smallest of the three. An example of the latter is the *monotonized central-difference limiter*, which replaces $a \to 2a$ and $b \to 2b$ in the above.

When considering the full set of Euler's equations, there remain a few loose ends to address at this point. We have not one quantity that we want to find a higher order spatial accuracy expression for, but as many as there are components to the state vector of conserved variables (lets say three, assuming one spatial dimension for now). Yet we also have derived quantities such as velocity and pressure. Do we compute these slopes for all five separately? Or do we pick three and derive the results for the others from them (e.g. $v_j(x) \equiv [\rho v]_j(x)/\rho_j(x)$)? If so, which three? (Or do we do something even more complicated, and use some linear combination that exactly captures the wave families emerging from the Riemann problem between two zones, i.e., the *characteristics* of the wave pattern?)

In practice the answer here is often that we compute the set ρ, v and p using an expression for the slope within the cell, and then derive the remaining ones, $\mathscr{E}$ and ρv, from them. This is not so much borne out of rigorous mathematical analysis, but out of accumulated experience in the field. It appears a fairly safe choice when it comes to avoiding an occasional appearance of an unphysical (i.e. negative) value of the pressure term to crash the program.

The other thing one might wonder is how the slope-limiter approach generalizes to multiple components. Here too, the answer is more the result of practice than of rigorous mathematical analysis. For uniform and Cartesian grids, it turns out to be fine just to treat the variables as independent for the purpose of computing their piecewise linear form, and to give each variable its own slope limiter check[8].

---

[8]This does not necessarily hold for curvilinear coordinates. When experimenting with those, it might be worthwhile to give [32] a read.

13.6.2 HIGHER ORDER IN TIME

Separately implementing a higher-order time step is an exercise in applying a Runge-Kutta algorithm. At lowest order we have forward Euler:

$$Q^{n+1} = Q^n + \Delta t Y(t^n, Q^n), \tag{13.97}$$

where we drop the subscript j to simplify the notation. Going to second order in a Runge-Kutta scheme suggests instead:

$$K_1 = \Delta t Y(t^n, Q^n),$$
$$K_2 = \Delta t Y(t^n + \Delta t, Q^n + K_1),$$
$$Q^{n+1} = Q^n + \frac{1}{2}(K_1 + K_2). \tag{13.98}$$

If you are familiar with Runge-Kutta, the generalization to third order (and even higher) is straightforward. The one thing to note here is that you will need room to store your tentative time step(s), and you will likely have to allocate memory for an intermediate state Q' and/or intermediate fluxes f', depending on exactly how you implement the scheme. A comparison between HLLC solvers at lowest order and second order is shown in Figure 13.12.

13.7 ALTERNATIVES AND EXTENSIONS TO THE FINITE VOLUME APPROACH

In this chapter we have discussed finite difference and finite volume methods on a fixed mesh. Various alternative approaches to solving Euler's equations exist. While an in-depth treatment lies beyond the scope of this book (which is after all not a treatise on numerical methods), we mention Lagrangian approaches to finite volume methods as well as conceptually distinct *Smoothed Particle Hydrodynamics* (SPH) and *finite element methods*[9].

13.7.1 ADAPTING THE FIXED MESH

A fixed mesh approach to CFD reflects a Eulerian framing of the problem, and the discussion in Section 13.5.2 might have brought to mind our initial derivation of Euler's equations in Section 2.1. Typically, flow patterns of interest within a fluid dynamics problem are not expected to be equally spread and of equal scale throughout the computational domain. Instead, when simulating an astrophysical scenario, be it accretion flow, jet propagation, blast waves, stellar pulsations, galaxy evolution, cosmology or anything else, different parts of the domain will at times be more relevant than others (with studies of turbulence perhaps sometimes offering an exception). A first generalization of the methods described previously in this chapter

---

[9]More detail on SPH methods can be found in e.g. [3, 45]. Finite element methods are treated and contrasted to finite volume methods in e.g. [8].

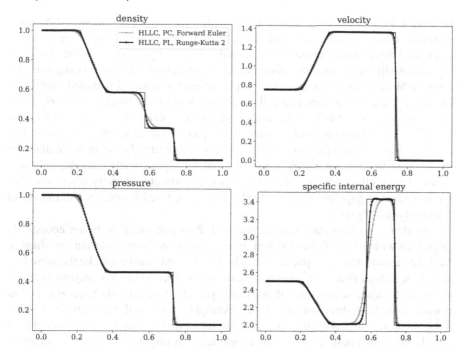

Figure 13.12 Standard shock tube test for HLLC methods at lowest order versus 2nd order Runge-Kutta time step and Piecewise Linear Method spatial reconstruction. The initial conditions are that the left state ($x < 0.3$ at $t = 0$) has $\rho = 1$, $p = 1$ and $v = 0.75$, whereas the right state has $\rho = 0.125$, $p = 0.1$ and $v = 0$. Both solvers use 200 zones (not counting ghost zones) between $x = 0 \ldots 1$. The snapshots are taken at $t = 0.2$, the adiabatic exponent $\hat{\gamma}_{ad} = 1.4$.

would therefore be to allow for grid cells of various size across the grid. It is in principle straightforward to replace Δx by Δx_j in the derivations, although one has to be careful to identify all cases where implicit use was made of the assumption that Δx does not vary between cells.

This way, many cells can for example be devoted to resolve a regions of interest within an astrophysical jet, while leaving the unshocked interstellar medium at lower resolution, thus saving computational time and memory. But this only works up to a point. For if the jet is not stationary, but e.g. pushing an expanding shock front outwards, then the region of interest changes over time. Within a Eulerian approach, such problems are best addressed by using a well-known approach called *adaptive mesh refinement* (AMR). In AMR solvers, time is devoted during each time step (or every few time steps) to redraw parts of the grid of increased or decreased interest. Adjacent cells are merged where the flow features do not change sharply (as can be assessed e.g. by looking at the spatial derivatives of conserved variables or other markers of the flow), while cells are split where extra resolution is needed. The gain in computational power can be quite substantial, with for example six levels of refinement over the base resolution along a direction leading already to an increase of resolution by a factor of $2^6 = 64$. A grid with a baselines resolution of 100 zones would now have an *effective* resolution of 6,400 zones even if only using a small fraction of this effective resolution at one time. The computational overhead of dealing with AMR, however, also rapidly becomes substantial. We can no longer rely on a single block in memory, or at least on a straightforwardly distributed number of blocks, and use the location of a cell in memory within that block to uniquely specify the position of the cell on the actual grid and its relations to its neighbours. That is, cell $j - 1$ will not necessarily precede cell j in computer memory if it is formed at a later stage by splitting an originally larger cell (certainly not in more than one dimension). Instead, we will now need to store this dynamically changing information along with the grid, copying cell data across computer memory when new cells are generated or gaps from removed cells need to be filled in order not to run out of allocated memory space.

Another complementary way to deal with flow patterns that do not necessarily align well with a fixed Cartesian grid, is to move away from Cartesian coordinates. Certainly in astrophysics, spherical or cylindrical coordinates might be the more logical choice. A jet that is axi-symmetric around the z-axis does not require three co-ordinates to describe and even if not fully symmetric might only require a modest resolution in the ϕ-direction. As another example, a spherical blast wave that shows a degree of self-similarity, like the cases discussed in Sections 7.5 and 8.6 might not only be described best using a radial coordinate r, it might even benefit from an increasing grid cell size with radius, e.g., a fixed radial size element $\Delta \ln r$ rather than Δr.

In such cases it is worthwhile to work out the extent to which it is possible to isolate the complications that appear in Euler's equations once the divergence terms are written in other than Cartesian coordinates, such as spherical or cylindrical co-ordinates (see Appendix C). While details are again beyond the scope of the book, the single example of spherical symmetry should hopefully suffice to get the basic concept across even if the situation can get more complex if more dimensions or less straightforward variables are used. Consider mass conservation:

$$\frac{\partial \rho}{\partial t} + \frac{1}{r^2} \frac{\partial}{\partial r} \left(r^2 \rho v \right) = 0 \Rightarrow \frac{\partial \rho}{\partial t} + \frac{\partial}{\partial r} \left(\rho v \right) = -\frac{2\rho v}{r}. \tag{13.99}$$

The latter expression is actually identical to mass conservation in one dimension in Cartesian coordinates and can be numerically solved as such, except for a term on the RHS that we call a *geometrical source term*. Source terms can be taken care of when solving the fluid dynamics equations in the same sequential manner that was suggested earlier for dealing with multi-dimensional fluxes. Each iteration, the fluid state is first updated based on the flux terms, then the source term gets added or subtracted. The time then gets updated only once, after both operations are performed.

13.7.2 ARBITRARY LAGRANGIAN-EULERIAN METHODS

Although a computationally powerful approach, using AMR to dynamically generate and merge cells across a fixed mesh in order to keep up with the pattern of the flow is like paying the price for the stubbornness of sticking with a Eulerian view where a Lagrangian view might be more natural. If instead of keeping cell positions fixed, the simulation allows the cells themselves to advect and resize with the flow, the simulation would naturally keep devoting its computational resources to the area of interest at least for subsonic flow patterns. It is for this reason that numerical stellar structure modelling often uses the mass coordinates from Section 3.3.

In more than one dimension, however, simply deforming the grid by moving the vertices of the grid structure comes with its own problems. Cells that neighbour one another at one point during the simulation, might not be the most natural neighbouring pair after some time has elapsed in the simulation (think for example of differential rotation in an accretion disc, or shear flow along the edge of a jet). Also, a given fluid cell mimicking a parcel might end up compressed by too much, and as a result end up a bottleneck when the CFL condition is used to determine the next time step. All this suggests that a pure Lagrangian approach is not necessarily ideal either, and that some regridding is likely needed in both approaches.

One way to address this is by using an *Arbitrary Langrangian-Eulerian* (ALE) method, which on the one hand relaxes the Eulerian constraint that cell boundaries remain fixed in space while on the other hand dropping the Langrangian assumption that cell boundaries advect with the flow. Instead, during each iteration the grid cell boundaries are displaced by whatever amount is practically useful to solve the problem at hand. It is probably best to think of this as a redrawing of the cell boundaries, rather than as displacing existing boundaries. Boundaries, after all, are not physical objects and can therefore be displaced by any arbitrary amount, 'moving' even faster than light speed if necessary. The change in cell content due to the motion of the boundaries acts like a flux term in the equations, so if boundaries are displaced with the local flow velocity (e.g. the average of the fluid velocities of the adjacent cells), this method would exactly preserve the mass content of cells and be fully Lagrangian. ALE methods are often implemented using irregular spatial meshes that strike a balance between cell shapes, number of boundaries between cells and the sizes of cells. A well-known example of such a tessellation is the *Voronoi tesselation*, where a user-defined total of seed points define separate cells to be defined by all points closer to a given seed point than to any of the others.

13.7.3 SMOOTHED PARTICLE HYDRODYNAMICS

SPH methods also follow a Lagrangian principle, taking their inspiration from numerical methods that are used in N-body methods where a limited number N of elements and their interactions are followed over time (e.g. a planetary system). A fluid state can be sampled by designating 'particles' that represent the local fluid state. Because hydrodynamics is a continuum theory, the particles can not be treated as discrete points, but can for example be assumed to designate the central points of smooth kernels describing some sort of mass distribution. A Gaussian kernel for particle i centred on location $\mathbf{r}_j$ could read

$$W(\mathbf{r}|\mathbf{r}_j, h) = \frac{1}{h^3 \pi^{3/2}} \exp\left[-\frac{|\mathbf{r} - \mathbf{r}_j|^2}{h^2}\right], \tag{13.100}$$

where h a smoothing length scale. The corresponding mass obeys $\rho_j(\mathbf{r}) = mW(\mathbf{r}|\mathbf{r}_j, h)$ and the local fluid mass is given by summing over particles, $\rho(\mathbf{r}) = \sum_j \rho_j$. This summation should formally include all particles in the system, but can in fact be truncated after some distance between $\mathbf{r}$ and $\mathbf{r}_j$. Other fluid variables, such as pressure, can then be either derived directly from the density (in the barotropic case, where $p = p(\rho)$) or by keeping track of particle thermal energy along with particle position and velocity during the simulation.

SPH methods work well in multiple dimensions and are best suited to problems without inherent symmetry that suggest an obvious choice of grid geometry. Furthermore, SPH codes work well when the astrophysical process of interest plays out against an otherwise near-empty background like the intergalactic medium or an interstellar medium that is not expected to have a noticeable impact on the system during the simulation. For these reasons, SPH methods have been traditionally been deployed in collision problems, such as the merging of two galaxies or the collision between two stars. The disadvantages of SPH methods are that due to their smooth nature, SPH particles are less suited to treat problems involving sharp shocks or where a fixed boundary is important. Setting up an SPH simulation can be more challenging as well. In a grid-based method, once the grid is chosen, its initialization is dictated completely by the required initial fluid state, while the freedom in setting up the initial distribution of mass kernels in an SPH method is more immediately apparent, increasing the risk of a suboptimal choice and an initial configuration that impacts the result of the simulation for longer than intended.

13.7.4 FINITE ELEMENT METHODS

Finite element methods represent yet another approach to solving fluid dynamical equations or other continuum problems. In a finite element method, the computational domain is subdivided in a series of non-overlapping cells just like in a finite volume method. However, now the cells also get assigned *nodes*, starting with nodes on shared boundaries between cells but at some point including notes within cells as well if the number of nodes per cell is increased. These nodes then serve to anchor approximate solutions to the local fluid state within a cell, cast as a weighted sum of judiciously chosen basis functions that is split in a time-dependent factor and a position-dependent factor. The more terms / basis functions to this sum (with each additional node introducing another free variable requiring an extra sum term) the higher the accuracy of the approximation to the fluid state. Finite element methods then implement an update algorithm to the weight factors, based on the hydrodynamics equations, with shared nodes between cells accounting for the impact of flux.

Although arguably more conceptually abstract than finite volume methods, the advantage of finite element methods is that they sidestep the need in finite volume methods to involve ever larger numbers of neighbouring cells when increasing the spatial accuracy to higher order. While not an issue in one dimension, the latter can be challenging to implement on irregular geometries in higher dimensions in a finite volume method.

13.8 COMPUTATIONAL HYDRODYNAMICS AND SPECIAL RELATIVITY

It is in principle straightforward to discretize the special relativistic equations of fluid dynamics presented in Chapter 8 and implement these in a numerical solver. The Lorentz factor adds additional couplings between the conservation laws, but the state and flux vectors can be defined as before. In three dimensions, we can define state vector Q according to

$$
Q = \begin{bmatrix} D \\ \tau \\ S^x \\ S^y \\ S^z \end{bmatrix} = \begin{bmatrix} \gamma\rho \\ w\gamma^2 - p - D \\ w\gamma^2\beta^x \\ w\gamma^2\beta^y \\ w\gamma^2\beta^z \end{bmatrix} = \begin{bmatrix} \gamma\rho \\ \gamma^2(\rho+p+e)-p-\gamma\rho \\ \gamma^2(\rho+p+e)\beta^x \\ \gamma^2(\rho+p+e)\beta^y \\ \gamma^2(\rho+p+e)\beta^z\gamma \end{bmatrix}, \quad (13.101)
$$

where we work in units of space and time where $c \equiv 1$. The lab frame density is denoted D (repeating a convention first introduced in Section 8.6). We use τ to indicate the top-left T^{00} term of the energy-momentum tensor, i.e. the term generalizing energy density, albeit without rest mass. The τ notation is often used in text books and papers on the subject (also, do not confuse this τ with proper time, also denoted by τ). The generalized momentum entries T^{0i} of the energy-momentum tensor are denoted S^x, S^y and S^z, again following common convention.

Omitting the rest mass from the energy equation is not an issue. Assuming no matter gets destroyed or created, the evolution of the rest mass is still accounted for by the continuity equation. It also avoids numerical round-off errors when the rest mass greatly dominates the other energy terms (i.e. the fluid is actually not relativistically hot). The corresponding state flux vectors are given by

$$
F^x = \begin{bmatrix} D\beta^x \\ S^x - D\beta^x \\ S^x\beta^x + p \\ S^y\beta^x \\ S^z\beta^x \end{bmatrix}, \quad F^y = \begin{bmatrix} D\beta^y \\ S^y - D\beta^y \\ S^x\beta^y \\ S^y\beta^y + p \\ S^z\beta^y \end{bmatrix}, \quad F^z = \begin{bmatrix} D\beta^z \\ S^z - D\beta^z \\ S^x\beta^z \\ S^y\beta^z \\ S^z\beta^z + p \end{bmatrix}. \quad (13.102)
$$

As before, the set of conservation laws is then given by,

$$
\frac{\partial Q}{\partial t} + \frac{\partial}{\partial x}F^x + \frac{\partial}{\partial y}F^y + \frac{\partial}{\partial z}F^z = 0. \quad (13.103)
$$

Please be reminded that the state flux vectors are five-dimensional, and that, say, F^x is not 'an entry of F' but rather a distinct five-dimensional state flux vector itself. Note also how, as in the non-relativistic case, pressure features in the state flux vectors entries for momentum flux. The x-component of momentum density merely advects in the y- and z-directions, with pressure only featuring in the x-direction. Likewise for the y and z components.

The finite-volume methods from this chapter such as the HLLC method can be adapted to special relativity as well, with special relativistic estimates for the wave speeds and the intermediate states[10]. One final thing worth noting here is the equation of state. If the flow is expected to be relativistically hot throughout, a simple EOS with fixed $\hat{\gamma}_{ad} = 4/3$ would suffice, and the corresponding linear relation between pressure and energy density means that one can easily be written in terms of the other and vice versa. But if the EOS is more complicated, it is not generally a given that it can be inverted analytically, or that any existing analytical inversion is not computationally expensive in practice. In such cases, the EOS can be tabulated in memory and pressure can be obtained via interpolation. Alternatively, a numerical root finder algorithm is deployed, such as Newton-Rhapson, or a bisection method. For a numerical root finder, the previous value of the pressure in the cell at hand, is usually a convenient starting point.

Special relativistic hydrodynamics (SRHD) codes can be tested in the same manner as non-relativistic codes: using shock tubes. Like the non-relativistic Riemann problem, the solution to the relativistic Riemann problem is known exactly[11], so any simulated shock tube flow can be compared directly to the 'truth'. SRHD codes are used to study a variety of astrophysical problems, such as accretion flow and astrophysical jet flow from active galactic nuclei and gamma-ray bursts.

---

[10]For an example from the literature where this is demonstrated, see [33].

[11]See for example, [29].

A Concepts from Thermodynamics

A.1 THE FIRST AND SECOND LAW OF THERMODYNAMICS

The first law of thermodynamics is a statement of conservation of energy and defines the concept of *heat*. In terms of specific (i.e., per unit mass) quantities, we have

$$d\varepsilon = dq + dw, \tag{A.1}$$

where ε the specific internal energy of a system of particles, q the specific heat and w specific work. In this book, we typically limit ourselves to work done through changes in volume, so we'll exclude here other processes that do work such as chemical processes that impact the number densities of a species of particle contained in a multi-species system. We can then write a change in work as $dw = -pd\mathscr{V}$, where the minus signs reflects that it typically takes effort to compress the system into a smaller volume (when $d\mathscr{V} < 0$ the specific internal energy ε therefore increases as a consequence of compression).

Casting heat in terms of specific entropy s and temperature T, we can write $dq \leq T ds$ for a local change in specific heat. According to the second law, the total entropy of an isolated system can only stay the same or increase. *Reversible* processes are processes for which the total entropy of the isolated system does not irrevocably increase. It should be noted that for reversible processes it is still possible to locally have $ds \neq 0$, as long as we maintain $dq = T ds$ and the total entropy of the entire isolated system stays the same. For a reversible process, therefore, we can write the first law as

$$d\varepsilon = T ds - pd\mathscr{V}. \tag{A.2}$$

This expression includes definitions of both temperature and pressure, according to

$$d\varepsilon = \left(\frac{\partial \varepsilon}{\partial s}\right)_{\mathscr{V}} ds + \left(\frac{\partial \varepsilon}{\partial \mathscr{V}}\right)_{s} d\mathscr{V} \Rightarrow T = \left(\frac{\partial \varepsilon}{\partial s}\right)_{\mathscr{V}}, \quad p = -\left(\frac{\partial \varepsilon}{\partial \mathscr{V}}\right)_{s}. \tag{A.3}$$

The internal energy differential is an exact differential, which implies:

$$\left(\frac{\partial T}{\partial \mathscr{V}}\right)_{s} = -\left(\frac{\partial p}{\partial s}\right)_{\mathscr{V}}. \tag{A.4}$$

There are various other identities that can be derived from exact differentials, such as the following starting from ds (this one will come in useful in appendix A.5, which

DOI: 10.1201/9781003095088-A

adds a derivation in support of Section 4.4):

$$
\begin{aligned}
ds &= \frac{dq}{T} = \frac{1}{T}\left(\frac{\partial q}{\partial \mathcal{V}}\right)_T d\mathcal{V} + \frac{1}{T}\left(\frac{\partial q}{\partial T}\right)_\mathcal{V} dT \\
&= \frac{1}{T}\left[\left(\frac{\partial \varepsilon}{\partial \mathcal{V}}\right)_T + p\right] d\mathcal{V} + \frac{1}{T}\left(\frac{\partial \varepsilon}{\partial T}\right)_\mathcal{V} dT,
\end{aligned}
\tag{A.5}
$$

which can be used to show

$$
\frac{\partial^2 s}{\partial T \partial \mathcal{V}} = \frac{\partial^2 s}{\partial \mathcal{V} \partial T},
$$

$$
-\frac{1}{T^2}\left(\frac{\partial \varepsilon}{\partial \mathcal{V}}\right)_T + \frac{1}{T}\frac{\partial^2 \varepsilon}{\partial T \partial \mathcal{V}} - \frac{p}{T^2} + \frac{1}{T}\left(\frac{\partial p}{\partial T}\right)_\mathcal{V} = \frac{1}{T}\frac{\partial^2 \varepsilon}{\partial \mathcal{V} \partial T},
$$

$$
T\left(\frac{\partial p}{\partial T}\right)_\mathcal{V} - p = \left(\frac{\partial \varepsilon}{\partial \mathcal{V}}\right)_T.
\tag{A.6}
$$

Unlike $d\varepsilon$, ds and $d\mathcal{V}$, the rate of heat change dq and rate of work dw (both specific) are both *inexact* differentials.

Being an exact differential, Equation A.2 implies that in principle there exists a function $\varepsilon(s, \mathcal{V})$ providing a constraint such that a thermodynamic system in equilibrium can be characterized fully using two out of the set of three variables ε, s and $\mathcal{V}$ (this would be an equation of state). These three variables are specific versions of *extensive* functions of state internal energy E_{int}, total entropy S and total volume V (another extensive function of state in thermodynamics is total particle number N). The extensive functions are additive, growing linearly with the system size. Combining e.g. two systems of N particles each results in a system with $2N$ particles. In the context of fluid dynamics, rather than thermodynamics, the use of specific variables makes sense in that it scales out the size of a fluid parcel (dictated by parcel mass M) and allows us to concentrate on the intrinsic properties of the fluid.

Pressure p and temperature T on the other hand, are *intensive* variables (merging two systems with an identical pressure p will result in a system again with pressure p). From the definitions of p and T provided above, it can be seen which intensive functions and which extensive functions form conjugate pairs: p pairs with V, while T pairs with S. Note that the first law does not contain dT and dp terms, even though T and p are present. This confirms that p and T are not independent of their conjugate partners and that they can be obtained from relevant equations of state instead of being independently specified (one such example would be the ideal gas law, which we return to below).

There are various terms to describe processes that leave a particular state variable invariant, as summarized below:

- **adiabatic:** no change in heat, $dq = 0$
- **incompressible:** no change in density, $d\mathcal{V} = 0$.
- **isentropic:** no change in entropy, $ds = 0$.
- **isothermal:** no change in temperature, $dT = 0$.
- **isobaric:** no change in pressure, $dp = 0$.

A.2 LEGENDRE TRANSFORMS

Instead of characterizing a thermodynamic equilibrium using s and $\mathcal{V}$ in conjuction with an expression for energy like the first law, it is perfectly possible to replace s and $\mathcal{V}$ by intensive functions T and p. This can be demonstrated through *Legendre transformations* where we subtract the conjugate pair $(\partial \varepsilon / \partial s)_\mathcal{V}\, s = Ts$ and/or $(\partial \varepsilon / \partial \mathcal{V})_s\, \mathcal{V} = -p\mathcal{V}$ from the specific internal energy. This also reveals various alternative measures of the energy of a system. The specific Gibbs free energy g for example is defined according to

$$g \equiv \varepsilon - Ts + p\mathcal{V}, \tag{A.7}$$

leading to

$$dg = -sdT + \mathcal{V}dp. \tag{A.8}$$

If we only subtract $-p\mathcal{V}$ or Ts we get expressions for respectively specific enhalpy h and specific Helmholtz free energy $\hbar$:

$$h \equiv \varepsilon + p\mathcal{V} \rightarrow dh = Tds + \mathcal{V}dp, \tag{A.9}$$

$$\hbar \equiv \varepsilon - Ts \rightarrow d\hbar = -sdT - pd\mathcal{V}. \tag{A.10}$$

Being exact differentials, these imply the following:

$$\left(\frac{\partial s}{\partial p} \right)_T = -\left(\frac{\partial \mathcal{V}}{\partial T} \right)_p, \tag{A.11}$$

$$\left(\frac{\partial T}{\partial p} \right)_s = -\left(\frac{\partial \mathcal{V}}{\partial s} \right)_p, \tag{A.12}$$

$$\left(\frac{\partial s}{\partial \mathcal{V}} \right)_T = -\left(\frac{\partial p}{\partial T} \right)_\mathcal{V}, \tag{A.13}$$

A.3 HEAT CAPACITIES

We will on occasion make use of the specific heat capacities at constant pressure and specific volume:

$$c_p \equiv \left(\frac{\partial q}{\partial T} \right)_p = T \left(\frac{\partial s}{\partial T} \right)_p, \tag{A.14}$$

$$c_\mathcal{V} \equiv \left(\frac{\partial q}{\partial T} \right)_\mathcal{V} = T \left(\frac{\partial s}{\partial T} \right)_\mathcal{V}. \tag{A.15}$$

When substituting Tds for dq in a reversible process, keep in mind that the following notations are equivalent:

$$\left(\frac{\partial q}{\partial T} \right)_p = \left(\frac{dq}{dT} \right)_p, \tag{A.16}$$

since the switch from full derivative d to partial ∂ is notationally redundant when we already explicitly indicate that the other variable is to be kept constant. Assuming the total mass M to remain unchanged (i.e. following a fluid parcel), means that the assumption of 'constant volume' is interchangeable with assuming 'constant density' or 'constant specific volume', so $c_V = c_{\mathscr{V}} = c_\rho$.

Because of the relation between heat, energy and work implied by the first law, $dq = d\varepsilon + p\,d\mathscr{V}$, it follows that the heat capacity at constant specific volume $d\mathscr{V} = 0$ can also be written as

$$c_{\mathscr{V}} = \left(\frac{\partial \varepsilon}{\partial T}\right)_{\mathscr{V}}. \tag{A.17}$$

A.4   THE IDEAL GAS LAW AND PERFECT GASES

In ideal gases, the well-known ideal gas law equation of state applies:

$$pV = Nk_BT, \tag{A.18}$$

where $k_B \equiv 1.380658 \times 10^{-16}$ erg K^{-1} is Boltzmann's constant connecting temperature to energy. As with the first law of thermodynamics, the temperature T in the ideal gas law is the *thermodynamic* temperature. When cast in terms of specific variables, the ideal gas law takes the form

$$p\mathscr{V} = \frac{k_BT}{\mu m} \Rightarrow p = \rho\frac{k_BT}{\mu m}. \tag{A.19}$$

Here μ is the mean molecular weight of the gas in question and $m = 1.6605402 \times 10^{-24}$ g the atomic mass unit.

By definition, the particles that make up an ideal gas have uncorrelated motion. With the exception of implied elastic collisions that sustain the isotropy of the gas, there are no ranged interactions between the particles. Furthermore, the spatial extent of the particles is assumed to be negligible when compared to their relative distance. Liquids, therefore, do not obey the ideal gas law. Even if they are assumed to be of negligible size, the particles in an ideal gas are not necessarily point particles, and might still harbour degrees of freedom other than translational degrees. Note also that the particles are not required to obey a Maxwell-Boltzmann distribution (which features a *kinetic* temperature that would be equal to the thermodynamic temperature in the case of thermal equilibrium), as long as the other conditions above are met.

A gas that obeys the ideal gas law and also has constant heat capacities is called a *perfect* gas (although the terminology 'ideal' and 'perfect' is sometimes used interchangeably in the literature). Being a thermodynamic state variable, the specific internal energy can be expressed as function of two other thermodynamic variables, for example $\varepsilon(p,\rho)$. For a perfect gas, because $c_{\mathscr{V}}$ is constant we have

$$c_{\mathscr{V}} = \left(\frac{\partial \varepsilon}{\partial T}\right)_{\mathscr{V}} \Rightarrow \varepsilon = c_{\mathscr{V}}T + constant(\rho). \tag{A.20}$$

By $constant(\rho)$ we indicate here a constant of integration for an integration path along fixed density whose value may still depend on this density ρ. Starting from the

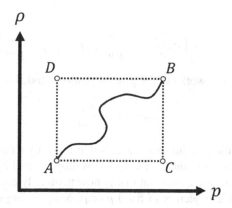

Figure A.1 An evolutionary path for specific internal energy $\varepsilon(p,\rho)$, starting from point A and moving to point B. The value of ε at point B will be independent of the path that is taken.

first law of thermodynamics and considering work done by a change in density, we have

$$\left(\frac{\partial \varepsilon}{\partial T}\right)_p = \left(\frac{\partial q}{\partial T}\right)_p - p\left(\frac{\partial v}{\partial T}\right)_p = c_p + \frac{p}{\rho^2}\left(\frac{\partial \rho}{\partial T}\right)_p. \tag{A.21}$$

In this expression we can substitute the ideal gas law Equation A.19 before and after taking the derivative of ρ and obtain

$$\left(\frac{\partial \varepsilon}{\partial T}\right)_p = c_p - \frac{p}{\rho^2}\frac{p\mu m}{k_B T^2} = c_p - \frac{k_B}{\mu m}. \tag{A.22}$$

This tells us that $(\partial \varepsilon/\partial T)_p$ too is a constant, such that

$$\left(\frac{\partial \varepsilon}{\partial T}\right)_p = c_p - \frac{k_B}{\mu m} \Rightarrow \varepsilon = \left(c_p - \frac{k_B}{\mu m}\right)T + constant(p). \tag{A.23}$$

Taken together, Equation A.20 and A.23 have an important implication for the nature of $\varepsilon(p,\rho)$. Consider Figure A.1, illustrating a change of ε going from a point A in state variable space to point B. Because $d\varepsilon$ is an exact differential, the value of ε at point B is actually independent of the path taken from A to B. The following expressions are therefore equivalent:

$$\varepsilon_B = \varepsilon_A + \int_A^B d\varepsilon = \varepsilon_A + \int_A^C d\varepsilon + \int_C^B d\varepsilon = \varepsilon_A + \int_A^D d\varepsilon + \int_D^B d\varepsilon. \tag{A.24}$$

We can compute the integrals along the two detours involving points C and D respectively, and the results should be the same. These routes follow either paths of constant p or constant ρ, so we can use Equations A.20 and A.23 directly. The additive constants drop out and we have that

$$\varepsilon_B = \varepsilon_A + \left(c_p - \frac{k_B}{\mu m}\right)T_B + \left(c_\gamma - c_p + \frac{k_B}{\mu m}\right)T_C - c_\gamma T_A, \tag{A.25}$$

should match

$$\varepsilon_B = \varepsilon_A + c_{\mathscr{V}} T_B + \left(c_p - c_{\mathscr{V}} - \frac{k_B}{\mu m} \right) T_D - \left(c_p - \frac{k_B}{\mu m} \right) T_A. \tag{A.26}$$

The only way for this to work out for general points A and B is if we require

$$c_p = c_{\mathscr{V}} + \frac{k_B}{\mu m}. \tag{A.27}$$

What's more, evaluating the integrals reveals that ε is only a function of temperature T (T_B in the above integrals, the T_C and T_D terms drop out and T_A represents an initial condition), up to an additive constant that is now genuinely constant and not merely so along some specific trajectory of fixed p or ρ. Suppressing the additive constant, we can write specific internal energy as

$$\varepsilon(T) = c_{\mathscr{V}} T. \tag{A.28}$$

This statement is also known as *Joule's second law*. It is consistent with for example the specific internal energy of an ideal gas of point particles, given by $\varepsilon = \frac{3}{2} \frac{k_B}{m} T$ (derived from kinetic theory in Section 2.4.3).

The fundamentals of thermodynamics briefly summarized in above have been in place for a very long time and there exists many text books on the subject that can be consulted for further reference[1].

A.5 THE ENERGY EQUATION IN STELLAR STRUCTURE MODELLING

In this section, we include a derivation of Equation 4.78 starting from Equation 4.77. If in Equation 4.77 we express the differential of specific internal energy in terms of $\mathscr{V}$ and T, we have

$$\frac{\partial L_r}{\partial m} = \dot{\varepsilon}_n - \dot{\varepsilon}_v - \left(\frac{\partial \varepsilon}{\partial \mathscr{V}} \right)_T \frac{d\mathscr{V}}{dt} - \left(\frac{\partial \varepsilon}{\partial T} \right)_{\mathscr{V}} \frac{dT}{dt} - p \frac{d\mathscr{V}}{dt}. \tag{A.29}$$

Using Equation A.6, we can rewrite this in the form

$$\frac{\partial L_r}{\partial m} = \dot{\varepsilon}_n - \dot{\varepsilon}_v - \left[T \left(\frac{\partial p}{\partial T} \right)_{\mathscr{V}} - p \right] \frac{d\mathscr{V}}{dt} - \left(\frac{\partial \varepsilon}{\partial T} \right)_{\mathscr{V}} \frac{dT}{dt} - p \frac{d\mathscr{V}}{dt}, \tag{A.30}$$

which we can simplify and where we can apply the definition of the specific heat capacity at constant volume to obtain

$$\frac{\partial L_r}{\partial m} = \dot{\varepsilon}_n - \dot{\varepsilon}_v - T \left(\frac{\partial p}{\partial T} \right)_{\mathscr{V}} \frac{d\mathscr{V}}{dt} - c_{\mathscr{V}} \frac{dT}{dt}. \tag{A.31}$$

Before we continue with the full expression, we first concentrate on the term $(\partial p / \partial T)_{\mathscr{V}}$. Thanks to the cyclic chain rule (which is derived as exercise 1.9), we

---

[1]This appendix has been compiled while drawing from [21, 36, 4, 20, 5] in particular.

have

$$\left(\frac{\partial p}{\partial T}\right)_{\mathscr{V}} = -\frac{\left(\frac{\partial \mathscr{V}}{\partial T}\right)_{p}}{\left(\frac{\partial \mathscr{V}}{\partial p}\right)_{T}}. \tag{A.32}$$

This can be cast in terms of the auxiliary quantity δ that also occurs in the end product of our derivation Equation 4.78 and a further auxiliary quantity α (staying with notation common to text books on the subject, e.g. [20]). Both serve to capture the features of an equation of state for density ρ:

$$d\ln\rho = \left(\frac{\partial \ln\rho}{\partial \ln p}\right)_{T} d\ln p + \left(\frac{\partial \ln\rho}{\partial \ln T}\right)_{p} d\ln T \equiv \alpha d\ln p - \delta d\ln T, \tag{A.33}$$

or, equivalently, specific volume $\mathscr{V}$:

$$d\ln\mathscr{V} = \left(\frac{\partial \ln\mathscr{V}}{\partial \ln p}\right)_{T} d\ln p + \left(\frac{\partial \ln\mathscr{V}}{\partial \ln T}\right)_{p} d\ln T \equiv -\alpha d\ln p + \delta d\ln T. \tag{A.34}$$

The resulting expression for $(\partial p/\partial T)_{\mathscr{V}}$ becomes

$$\left(\frac{\partial p}{\partial T}\right)_{\mathscr{V}} = \frac{p\delta}{T\alpha}. \tag{A.35}$$

Picking up where we left off at Equation A.31, we can make use of this result and the definitions of α and δ to move our derivation a few steps forward:

$$\begin{aligned}
\frac{\partial L_r}{\partial m} &= \dot{\varepsilon}_n - \dot{\varepsilon}_v - \frac{p\delta}{\alpha}\frac{d\mathscr{V}}{dt} - c_{\mathscr{V}}\frac{dT}{dt}, \\
&= \dot{\varepsilon}_n - \dot{\varepsilon}_v + \frac{p\delta}{\rho\alpha}\frac{d\ln\rho}{dt} - c_{\mathscr{V}}\frac{dT}{dt}, \\
&= \dot{\varepsilon}_n - \dot{\varepsilon}_v + \frac{p\delta}{\rho\alpha}\left(\frac{\alpha}{p}\frac{dp}{dt} - \frac{\delta}{T}\frac{dT}{dt}\right) - c_{\mathscr{V}}\frac{dT}{dt}, \\
&= \dot{\varepsilon}_n - \dot{\varepsilon}_v - \left(c_{\mathscr{V}} + \frac{p\delta^2}{\rho\alpha T}\right)\frac{dT}{dt} + \frac{\delta}{\rho}\frac{dp}{dt}.
\end{aligned} \tag{A.36}$$

Again, we have to leave the main derivation for a while in order to concentrate on a single term. Knowing our end goal (i.e. Equation 4.78), we can expect the term in brackets to simply reduce to c_p. To demonstrate this to be true, we look at the difference between the two specific heat capacities,

$$\begin{aligned}
c_p - c_{\mathscr{V}} &= \left(\frac{\partial \varepsilon}{\partial T}\right)_{p} + p\left(\frac{\partial \mathscr{V}}{\partial T}\right)_{p} - \left(\frac{\partial \varepsilon}{\partial T}\right)_{\mathscr{V}}, \\
&= \left(\frac{\partial \varepsilon}{\partial T}\right)_{\mathscr{V}}\left(\frac{\partial T}{\partial T}\right)_{p} + \left(\frac{\partial \varepsilon}{\partial \mathscr{V}}\right)_{T}\left(\frac{\partial \mathscr{V}}{\partial T}\right)_{p} + p\left(\frac{\partial \mathscr{V}}{\partial T}\right)_{p} - \left(\frac{\partial \varepsilon}{\partial T}\right)_{\mathscr{V}}, \\
&= \left(\frac{\partial \varepsilon}{\partial \mathscr{V}}\right)_{T}\left(\frac{\partial \mathscr{V}}{\partial T}\right)_{p} + p\left(\frac{\partial \mathscr{V}}{\partial T}\right)_{p}.
\end{aligned} \tag{A.37}$$

Making use of the identity from Equation A.6 once more, this further evaluates to

$$c_p - c_{\mathcal{V}} = T \left(\frac{\partial p}{\partial T} \right)_{\mathcal{V}} \left(\frac{\partial \mathcal{V}}{\partial T} \right)_p - p \left(\frac{\partial \mathcal{V}}{\partial T} \right)_p + p \left(\frac{\partial \mathcal{V}}{\partial T} \right)_p,$$

$$= T \left(\frac{\partial p}{\partial T} \right)_{\mathcal{V}} \frac{\delta}{\rho T}, \tag{A.38}$$

plugging in the definition of δ in the final step, alongside of some clean-up of cancelling terms. Referring to Equation A.35 we have the result that

$$c_p - c_{\mathcal{V}} = T \frac{p\delta}{T\alpha} \frac{\delta}{\rho T} = \frac{p\delta^2}{\alpha \rho T}. \tag{A.39}$$

Substituting this where we left off in our main derivation (Equation A.36), we can therefore indeed conclude the validity of Equation 4.78:

$$\frac{\partial L_r}{\partial m} = \dot{\varepsilon}_n - \dot{\varepsilon}_v - c_p \frac{dT}{dt} + \frac{\delta}{\rho} \frac{dP}{dt}. \tag{A.40}$$

B Vector Identities and Derivatives

Here we list various vector identities and derivatives that might come in useful.

B.1 VECTOR PRODUCT RULES

$$\nabla(fg) = f\nabla g + g\nabla f. \tag{B.1}$$

$$\nabla(\mathbf{v}\cdot\mathbf{w}) = \mathbf{v}\times(\nabla\times\mathbf{w}) + \mathbf{w}\times(\nabla\times\mathbf{v}) + (\mathbf{v}\cdot\nabla)\mathbf{w} + (\mathbf{w}\cdot\nabla)\mathbf{v}. \tag{B.2}$$

$$\nabla\cdot(f\mathbf{v}) = f(\nabla\cdot\mathbf{v}) + \mathbf{v}\cdot(\nabla f). \tag{B.3}$$

$$\nabla\cdot(\mathbf{v}\times\mathbf{w}) = \mathbf{w}\cdot(\nabla\times\mathbf{v}) - \mathbf{v}\cdot(\nabla\times\mathbf{w}). \tag{B.4}$$

$$\nabla\times(f\mathbf{v}) = f(\nabla\times\mathbf{v}) - \mathbf{v}\times(\nabla f). \tag{B.5}$$

$$\nabla\times(\mathbf{v}\times\mathbf{w}) = (\mathbf{w}\cdot\nabla)\mathbf{v} - (\mathbf{v}\cdot\nabla)\mathbf{w} + \mathbf{v}(\nabla\cdot\mathbf{w}) - \mathbf{w}(\nabla\cdot\mathbf{v}). \tag{B.6}$$

B.2 HIGHER-ORDER DERIVATIVES

$$\nabla\cdot(\nabla\times\mathbf{v}) = 0. \tag{B.7}$$

$$\nabla\times(\nabla f) = 0. \tag{B.8}$$

$$\nabla\times(\nabla\times\mathbf{v}) = \nabla(\nabla\cdot\mathbf{v}) - \nabla^2\mathbf{v}. \tag{B.9}$$

B.3 DOT AND CROSS-PRODUCT COMBINATIONS

$$\mathbf{u}\cdot(\mathbf{v}\times\mathbf{w}) = \mathbf{w}\cdot(\mathbf{u}\times\mathbf{v}) = \mathbf{v}\cdot(\mathbf{w}\times\mathbf{u}). \tag{B.10}$$

$$\mathbf{u}\times(\mathbf{v}\times\mathbf{w}) = \mathbf{v}(\mathbf{u}\cdot\mathbf{w}) - \mathbf{w}(\mathbf{u}\cdot\mathbf{v}). \tag{B.11}$$

$$(\mathbf{u}\times\mathbf{v})\times\mathbf{w} = \mathbf{v}(\mathbf{u}\cdot\mathbf{w}) - \mathbf{u}(\mathbf{v}\cdot\mathbf{w}). \tag{B.12}$$

B.4 SPHERICAL COORDINATES

Angle θ is the polar angle, starting along the z-axis at $\theta = 0$. Angle ϕ lies in the xy-plane, starting along the x-axis at $\phi = 0$.

DOI: 10.1201/9781003095088-B

$$\nabla f = \frac{\partial f}{\partial r}\hat{\mathbf{e}}_r + \frac{1}{r}\frac{\partial f}{\partial\theta}\hat{\mathbf{e}}_\theta + \frac{1}{r\sin\theta}\frac{\partial f}{\partial\phi}\hat{\mathbf{e}}_\phi. \tag{B.13}$$

$$\nabla^2 f = \frac{1}{r^2}\frac{\partial}{\partial r}\left(r^2\frac{\partial f}{\partial r}\right) + \frac{1}{r^2\sin\theta}\frac{\partial}{\partial\theta}\left(\sin\theta\frac{\partial f}{\partial\theta}\right) + \frac{1}{r^2\sin^2\theta}\frac{\partial^2 f}{\partial\phi^2}. \tag{B.14}$$

$$\nabla\cdot\mathbf{v} = \frac{1}{r^2}\frac{\partial}{\partial r}\left(r^2 v^r\right) + \frac{1}{r\sin\theta}\frac{\partial}{\partial\theta}\left(\sin\theta\, v^\theta\right) + \frac{1}{r\sin\theta}\frac{\partial v^\phi}{\partial\phi}. \tag{B.15}$$

$$\nabla\times\mathbf{v} = \frac{1}{r\sin\theta}\left[\frac{\partial}{\partial\theta}\left(\sin\theta\, v^\phi\right) - \frac{\partial v^\theta}{\partial\phi}\right]\hat{\mathbf{e}}_r$$
$$+ \frac{1}{r}\left[\frac{1}{\sin\theta}\frac{\partial v^r}{\partial\phi} - \frac{\partial}{\partial r}\left(rv^\phi\right)\right]\hat{\mathbf{e}}_\theta + \frac{1}{r}\left[\frac{\partial}{\partial r}\left(rv^\theta\right) - \frac{\partial v^r}{\partial\theta}\right]\hat{\mathbf{e}}_\phi. \tag{B.16}$$

B.5   CYLINDRICAL COORDINATES

the unit vector $\hat{\mathbf{e}}_\varsigma$ lies in the xy-plane, pointing away from the z-axis, the coordinate ς describes the distance to the z-axis. Angle ϕ is the same as in spherical coordinates.

$$\nabla f = \frac{\partial f}{\partial\varsigma}\hat{\mathbf{e}}_s + \frac{1}{\varsigma}\frac{\partial f}{\partial\phi}\hat{\mathbf{e}}_\phi + \frac{\partial f}{\partial z}\hat{\mathbf{e}}_z. \tag{B.17}$$

$$\nabla^2 f = \frac{1}{\varsigma}\frac{\partial}{\partial\varsigma}\left(\varsigma\frac{\partial f}{\partial\varsigma}\right) + \frac{1}{\varsigma^2}\frac{\partial^2 f}{\partial\phi^2} + \frac{\partial^2 t}{\partial z^2}. \tag{B.18}$$

$$\nabla\cdot\mathbf{v} = \frac{1}{\varsigma}\frac{\partial}{\partial\varsigma}\left(\varsigma v^\varsigma\right) + \frac{1}{\varsigma}\frac{\partial v^\phi}{\partial\phi} + \frac{\partial v^z}{\partial z}. \tag{B.19}$$

$$\nabla\times\mathbf{v} = \left[\frac{1}{\varsigma}\frac{\partial v^z}{\partial\phi} - \frac{\partial v^\phi}{\partial z}\right]\hat{\mathbf{e}}_\varsigma + \left[\frac{\partial v^\varsigma}{\partial z} - \frac{\partial v^z}{\partial\varsigma}\right]\hat{\mathbf{e}}_\phi + \frac{1}{\varsigma}\left[\frac{\partial}{\partial\varsigma}\left(\varsigma v^\phi\right) - \frac{\partial v^\varsigma}{\partial\phi}\right]\hat{\mathbf{e}}_z. \tag{B.20}$$

C Euler's Equations in Non-Cartesian Coordinate Systems

In this appendix we provide Euler's equations in cylindrical and spherical coordinates, along with some additional tools to derive these expressions. Readers who merely need the equations can skip past the derivation, which relies heavily on index notation and some additional ad-hoc notation and concepts[1].

As discussed in Section 1.1, the issue with non-Cartesian coordinates is that the basis vectors might well be position-dependent and do not even need to be mutually perpendicular. The issue is illustrated when taking the divergence of a vector $\mathbf{w}$:

$$\nabla \cdot \mathbf{w} = \nabla \cdot \left(w^i \mathbf{e}_i \right) = \mathbf{e}_i \cdot \nabla w^i + w^i \nabla \cdot \mathbf{e}_i \neq \frac{\partial}{\partial x^i} w^i, \qquad \text{(in general)}. \qquad \text{(C.1)}$$

Combinations of identical indices in subscript and superscript are summed over as per the Einstein summation convention. The last step assumes a coordinate system with coordinates x^i (that is, x^1, x^2, x^3) and denotes with w^i the components of the vector $\mathbf{w}$ when expressed in the natural basis corresponding to this coordinate system. Similarly, we have the following for the partial derivative of a vector $\mathbf{w}$ with respect to a coordinate x^i:

$$\frac{\partial}{\partial x^i} \mathbf{w} = \left(\frac{\partial}{\partial x^i} w^j \right) \mathbf{e}_j + w^j \frac{\partial}{\partial x^i} \mathbf{e}_j. \qquad \text{(C.2)}$$

In Cartesian coordinates, the last term on the RHS is zero. But, in general, the basis vectors can be changing functions of position themselves. We can express their partial derivatives in their own basis by introducing *connection coefficients* Γ^a_{bc}:

$$\frac{\partial}{\partial x^i} \mathbf{e}_j = \Gamma^k_{ji} \mathbf{e}_k. \qquad \text{(C.3)}$$

The connection coefficients are symmetrical in their covariant indices, $\Gamma^k_{ji} = \Gamma^k_{ij}$, as shown by the following (which has to hold for arbitrary basis vectors):

$$\Gamma^k_{ji} \mathbf{e}_k = \frac{\partial}{\partial x^i} \mathbf{e}_j = \frac{\partial^2 \mathbf{r}}{\partial x^i \partial x^j} = \frac{\partial^2 \mathbf{r}}{\partial x^j \partial x^i} = \Gamma^k_{ij} \mathbf{e}_k. \qquad \text{(C.4)}$$

Here we used the definition of the natural basis vectors in terms of the position vector $\mathbf{r}$ (thus supposing our coordinate system can at least be embedded within a Euclidean space where a position vector can be defined as in Section 1.1).

---

[1]The rather informal treatment here draws on [13] in particular, where a more careful discussion of the material can be found.

DOI: 10.1201/9781003095088-C

Looking for additional ways (other than via Equation C.3) to compute the connection coefficients for a given coordinate system, we can use the definition of the metric tensor g_{ij} to link it to the connection coefficient. We start from

$$\frac{\partial}{\partial x^k} g_{ij} = \frac{\partial}{\partial x^k} (\mathbf{e}_i \cdot \mathbf{e}_j) = \frac{\partial \mathbf{e}_i}{\partial x^k} \cdot \mathbf{e}_j + \mathbf{e}_i \cdot \frac{\partial \mathbf{e}_j}{\partial x^k}.$$
(C.5)

Using Equation C.3, this can be rewritten in the form

$$\frac{\partial}{\partial x^k} g_{ij} = \Gamma_{ik}^m g_{mj} + \Gamma_{jk}^m g_{im}.$$
(C.6)

We can combine this equation with versions of itself where the dummy indices have been relabelled and then contract with g^{ik} in order to obtain (see Problem C.1):

$$\Gamma_{ki}^l = \frac{1}{2} g^{lj} \left(\frac{\partial}{\partial x^k} g_{ij} + \frac{\partial}{\partial x^i} g_{jk} - \frac{\partial}{\partial x^j} g_{ki} \right).$$
(C.7)

Therefore, if we know the metric, we can generate the connection coefficients. If the metric does not depend on position, as in Cartesian coordinates, it directly follows that the connection coefficients are all zero.

PROBLEM C.1
The connection coefficient in terms of the metric tensor

Derive Equation C.7 by following the suggestion in the main text.

As reiterated at the start of this section, the gradients and divergences (such as those that occur in Euler's equations) will need to account for the potential position-dependence of the basis vectors. They therefore involve connection coefficients, and a derivative known as the *covariant derivative*, rather than mere partial derivatives. Before defining the covariant derivative, we mention an alternative notation for the partial derivative. In the literature, the partial derivative is often denoted with a comma, as in

$$\frac{\partial f}{\partial x^i} = f_{,i}, \qquad \frac{\partial w^i}{\partial x^j} = w^i{}_{,j}, \qquad \frac{\partial T^{ij}}{\partial x^k} = T^{ij}{}_{,k},$$
(C.8)

etc. This notation contrasts nicely with the notation for the covariant derivative introduced below, involving a semi-colon rather than a comma. For a scalar field, the covariant derivative and the partial derivative are still identical:

$$f_{;i} \equiv f_{,i}.$$
(C.9)

For the covariant derivative of a vector field $\mathbf{w}$ with coefficients w^i, we have

$$w^i{}_{;j} \equiv w^i{}_{,j} + \Gamma_{kj}^i w^k.$$
(C.10)

A tensor field with coefficients T^{ij} will have a covariant derivative defined according to

$$T^{ij}{}_{;k} \equiv T^{ij}{}_{,k} + \Gamma_{mk}^i T^{mj} + \Gamma_{mk}^j T^{im},$$
(C.11)

establishing a pattern that can be extended to an arbitrary number of contravariant indices. Gradients and divergences can now be cast in index notation according to

$$\nabla f \to f_{;i}, \qquad \nabla \cdot \mathbf{w} \to w^i_{;i}, \qquad \nabla \cdot \mathbf{T} \to T^{ij}_{;i}. \tag{C.12}$$

Although we will not make use of it, we mention for the sake of completeness (and without further justification) that for the covariant derivative of a covariant vector field w_i the definition involves a minus sign in front of the connection coefficient:

$$w_{i;j} \equiv w_{i,j} - \Gamma^k_{ij} w_k. \tag{C.13}$$

This generalizes to covariant tensors of different rank in a similar manner as the covariant derivative of contravariant tensor fields, albeit with minus signs instead. The connection coefficients Γ^a_{bc} are known as *Christoffel symbols of the first kind*. In spite of the indices, Γ^a_{bc} is itself *not* a tensor, even if it can be used to construct tensor quantities. It therefore does not transform as a tensor. The covariant derivative, by design, does (see Problem C.2 for an example).

PROBLEM C.2
The covariant derivative and the metric tensor

Show that $w_{i;j} = \left(g_{ik} w^k\right)_{;j} = g_{ik} \left(w^k\right)_{;j}$, while $\left(g_{ik} w^k\right)_{,j} \neq g_{ik} \left(w^k\right)_{,j}$ in general.

We can use the above to write down expressions for Euler's equations in general coordinates. We have to be careful in our notation here, though, because the physical interpretation we have attached to vector quantities tends to presuppose a given dimensionality. Angular velocity is not the same as linear velocity, for example. We will assume a set of generalized coordinates x^1, x^2, x^3 that may or may not have dimension of length (e.g. one or more of these might be angles instead). In Cartesian coordinates, of course $x^1 = x, x^2 = y, x^3 = z$.

If the generalized coordinates describe the properties of a fluid element or any other object for which we wish to define a generalized velocity, we take the generalized velocity u^i (*not* to be confused with *individual particle* velocity, which will not make an appearance in this appendix) to be given by $u^i = \dot{x}^i$. In Cartesian coordinates, the velocities will be the usual distances per unit time $u^1 = v^x, u^2 = v^y, u^3 = v^z$.

Finally, we will express force density as a more generalized $f^i = \rho \dot{u}^i$. In Cartesian coordinates, $f^1 = f^x, f^2 = f^y, f^3 = f^z$.

With all notation established, we can for example take the momentum equation,

$$\frac{\partial \rho \mathbf{v}}{\partial t} + \nabla \cdot (\rho \mathbf{v} \mathbf{v}) + \nabla p = \mathbf{f}, \tag{C.14}$$

and recast it more generally as (resisting the temptation to fold in time as a fourth coordinate, like we do in Chapter 8):

$$\frac{\partial \rho u^i}{\partial t} + \left(\rho u^j u^i\right)_{;j} + \left(p g^{ij}\right)_{;j} = f^i. \tag{C.15}$$

Writing out the covariant derivatives, Euler's equations in arbitrary coordinates therefore become:

$$\frac{\partial \rho}{\partial t} + \frac{\partial}{\partial x^i}\left(\rho u^i\right) + \Gamma^i_{ki}\rho u^k = 0, \tag{C.16}$$

$$\frac{\partial \rho u^i}{\partial t} + \frac{\partial}{\partial x^j}\left(\rho u^j u^i + pg^{ji}\right) + \Gamma^j_{mj}\left(\rho u^m u^i + pg^{mi}\right)$$
$$+ \Gamma^i_{mj}\left(\rho u^j u^m + pg^{jm}\right) = f^i, \tag{C.17}$$

$$\frac{\partial \mathscr{E}}{\partial t} + \frac{\partial}{\partial x^i}\left((\mathscr{E}+p)u^i\right) + \Gamma^i_{ki}(\mathscr{E}+p)u^k = u^i g_{ij} f^j. \tag{C.18}$$

The generalization to the Navier-Stokes equations is straightforward and requires the stress tensor in generalized coordinates. Using the same approach of replacing partial derivatives with covariant derivatives, we can define this as

$$\mathcal{T}^{ij} \equiv \eta \left(g^{jk}u^i_{;k} + g^{ik}u^j_{;k} - \frac{2}{3}g^{ij}u^k_{;k} \right) + \zeta g^{ij}u^k_{;k} - pg^{ij}, \tag{C.19}$$

where η the coefficient of shear viscosity and ζ the coefficient of bulk viscosity as defined in Chapter 9.

C.1   CYLINDRICAL COORDINATES

For cylindrical coordinates we take x^1 to be the projected distance from the z-axis in the xy-plane, $x^1 = \varsigma \equiv \sqrt{x^2 + y^2} = r\sin\theta$. For x^2 we take $x^2 = \phi$, the angle in the xy-plane starting at $\phi = 0$ along the positive x-axis. The third coordinates is just the z-coordinate from Cartesian coordinates, $x^3 = z$. The corresponding velocities are $u^1 = v^\varsigma$, $u^2 = \psi$, $u^3 = v^z$. The angular velocity ψ in the direction of increasing ϕ has cgs units of radians per second. According to basic mechanics, this velocity relates to linear velocity v^ϕ in the ϕ direction according to $v^\phi = \varsigma\psi$. The other velocities, v^ς and v^z still have cgs units of centimetres per second.

The relevant metric tensors for cylindrical coordinates can be computed to be

$$[g_{ab}] = \begin{bmatrix} 1 & 0 & 0 \\ 0 & \varsigma^2 & 0 \\ 0 & 0 & 1 \end{bmatrix}_{\varsigma,\phi,z}, \qquad [g^{ab}] = \begin{bmatrix} 1 & 0 & 0 \\ 0 & \varsigma^{-2} & 0 \\ 0 & 0 & 1 \end{bmatrix}_{\varsigma,\phi,z}. \tag{C.20}$$

With this in hand, the connection coefficients can be summarized as

$$\Gamma^l_{ki} = \varsigma \left(g^{l2}\delta^2_i\delta^1_k + g^{l2}\delta^1_i\delta^2_k - g^{l1}\delta^2_i\delta^2_k \right). \tag{C.21}$$

The only non-zero elements are therefore $\Gamma^1_{22} = -\varsigma$ and $\Gamma^2_{12} = \Gamma^2_{21} = \varsigma^{-1}$.

We can now construct Euler's equations in cylindrical coordinates:

$$\frac{\partial \rho}{\partial t} + \frac{\partial}{\partial \varsigma}(\rho v^\varsigma) + \frac{\partial}{\partial \phi}(\rho \psi) + \frac{\partial}{\partial z}(\rho v^z) = -\frac{\rho v^\varsigma}{\varsigma}, \tag{C.22}$$

$$\frac{\partial \rho v^\varsigma}{\partial t} + \frac{\partial}{\partial \varsigma}(\rho v^\varsigma v^\varsigma) + \frac{\partial}{\partial \phi}(\rho \psi v^\varsigma) + \frac{\partial}{\partial z}(\rho v^z v^\varsigma) + \frac{\partial p}{\partial \varsigma} = f^\varsigma$$
$$- \frac{\rho v^\varsigma v^\varsigma}{\varsigma} + \rho \varsigma \psi \psi, \tag{C.23}$$

$$\frac{\partial \rho \psi}{\partial t} + \frac{\partial}{\partial \varsigma}(\rho v^\varsigma \psi) + \frac{\partial}{\partial \phi}(\rho \psi \psi) + \frac{\partial}{\partial z}(\rho v^z \psi) + \frac{\partial}{\partial \phi}\left(\frac{p}{\varsigma^2}\right) = \frac{f^\phi}{\varsigma}$$
$$- \frac{3\rho v^\varsigma \psi}{\varsigma}, \tag{C.24}$$

$$\frac{\partial \rho v^z}{\partial t} + \frac{\partial}{\partial \varsigma}(\rho v^\varsigma v^z) + \frac{\partial}{\partial \phi}(\rho \psi v^z) + \frac{\partial}{\partial z}(\rho v^z v^z) + \frac{\partial p}{\partial z} = f^z - \frac{\rho v^\varsigma v^z}{\varsigma}, \tag{C.25}$$

$$\frac{\partial \mathscr{E}}{\partial t} + \frac{\partial}{\partial \varsigma}([\mathscr{E} + p]v^\varsigma) + \frac{\partial}{\partial \phi}([\mathscr{E} + p]\psi) + \frac{\partial}{\partial z}([\mathscr{E} + p]v^z) = \mathbf{v} \cdot \mathbf{f}$$
$$- \frac{[\mathscr{E} + p]v^\varsigma}{\varsigma}. \tag{C.26}$$

In these expressions, we have left in actual force density f^i (dyn / cm^3 in cgs units) rather than generalized force $\mathfrak{f}^i$. The dot product $\mathbf{v} \cdot \mathbf{f}$ is a scalar and therefore invariant under a change in coordinates, meaning that $v^i \delta_{ij} f^j = u^i g_{ij} \mathfrak{f}^j$ between Cartesian coordinates on the LHS and cylindrical coordinates on the RHS.

The formulation above is as close as possible to get to expressions with the same structure as their Cartesian counterparts, but with the presence of geometric source terms. However, the pressure gradient in Equation C.24 is also still different in structure from its Cartesian counterpart, which means that for example a numerical implementation of Euler's equations in cylindrical coordinates cannot completely be dealt with merely by adding source terms.

Note that the covariant form of the generalized velocity u_ϕ is given by $u_2 = g_{2i}u^i = \varsigma^2 \psi = \varsigma v^\phi$ and is therefore equal to the angular momentum density l^ϕ (i.e. $l^\phi \equiv u_\phi$). Replacing u^ϕ by $u_\phi \varsigma^{-2}$ in Equation C.24 (or using the metric tensor g_{ij} to lower the index of u^i in the generalized momentum conservation law before specializing to cylindrical coordinates) turns this expression into a conservation law for angular momentum density:

$$\frac{\partial \rho l^\phi}{\partial t} + \frac{1}{\varsigma}\frac{\partial}{\partial \varsigma}(\varsigma \rho v^\varsigma l^\phi) + \frac{1}{\varsigma}\frac{\partial}{\partial \phi}(\rho v^\phi l^\phi) + \frac{\partial}{\partial z}(\rho v^z l^\phi) + \frac{\partial p}{\partial \phi} = \varsigma f^\phi. \tag{C.27}$$

Unlike ρu^i, which only yields physically conserved quantities for $i = 1, 3$, the covariant ρu_i actually describes a conserved quantity for all i.

The equations in cylindrical coordinates are usually given in a form using v^i rather than u^i. These expressions can easily be obtained from the above by substitution:

$$\frac{\partial \rho}{\partial t} + \frac{1}{\varsigma}\frac{\partial}{\partial \varsigma}\left(\varsigma \rho v^{\varsigma}\right) + \frac{1}{\varsigma}\frac{\partial}{\partial \phi}\left(\rho v^{\phi}\right) + \frac{\partial}{\partial z}\left(\rho v^z\right) = 0, \tag{C.28}$$

$$\frac{\partial \rho v^{\varsigma}}{\partial t} + \frac{1}{\varsigma}\frac{\partial}{\partial \varsigma}\left(\varsigma \rho v^{\varsigma} v^{\varsigma}\right) + \frac{1}{\varsigma}\frac{\partial}{\partial \phi}\left(\rho v^{\phi} v^{\varsigma}\right) + \frac{\partial}{\partial z}\left(\rho v^z v^{\varsigma}\right) + \frac{\partial p}{\partial \varsigma} = f^{\varsigma}$$
$$+ \frac{\rho v^{\phi} v^{\phi}}{\varsigma}, \tag{C.29}$$

$$\frac{\partial \rho v^{\phi}}{\partial t} + \frac{1}{\varsigma}\frac{\partial}{\partial \varsigma}\left(\varsigma \rho v^{\varsigma} v^{\phi}\right) + \frac{1}{\varsigma}\frac{\partial}{\partial \phi}\left(\rho v^{\phi} v^{\phi}\right) + \frac{\partial}{\partial z}\left(\rho v^z v^{\phi}\right) + \frac{1}{\varsigma}\frac{\partial p}{\partial \phi} = f^{\phi}$$
$$- \frac{\rho v^{\phi} v^{\varsigma}}{\varsigma}, \tag{C.30}$$

$$\frac{\partial \rho v^z}{\partial t} + \frac{1}{\varsigma}\frac{\partial}{\partial \varsigma}\left(\varsigma \rho v^{\varsigma} v^z\right) + \frac{1}{\varsigma}\frac{\partial}{\partial \phi}\left(\rho v^{\phi} v^z\right) + \frac{\partial}{\partial z}\left(\rho v^z v^z\right) + \frac{\partial p}{\partial z} = f^z, \tag{C.31}$$

$$\frac{\partial \mathscr{E}}{\partial t} + \frac{1}{\varsigma}\frac{\partial}{\partial \varsigma}\left(\varsigma \left[\mathscr{E} + p\right] v^{\varsigma}\right) + \frac{1}{\varsigma}\frac{\partial}{\partial \phi}\left(\left[\mathscr{E} + p\right] v^{\phi}\right) + \frac{\partial}{\partial z}\left(\left[\mathscr{E} + p\right] v^z\right) = \mathbf{v}\cdot\mathbf{f}. \tag{C.32}$$

C.2 SPHERICAL COORDINATES

Spherical coordinates are defined as usual, with $x^1 = r$, $x^2 = \theta$ and $x^3 = \phi$. The generalized velocities become $u^1 = v^r$, $u^2 = \omega$ and $u^3 = \psi$, where the newly introduced ω is the angular velocity in the θ-direction, linked to linear velocity v^{θ} according to $v^{\theta} = r\omega$. The relevant metric tensors are given by

$$[g_{ij}] = \begin{bmatrix} 1 & 0 & 0 \\ 0 & r^2 & 0 \\ 0 & 0 & r^2\sin^2\theta \end{bmatrix}_{r,\theta,\phi}, \qquad [g^{ij}] = \begin{bmatrix} 1 & 0 & 0 \\ 0 & r^{-2} & 0 \\ 0 & 0 & \frac{1}{r^2\sin^2\theta} \end{bmatrix}_{r,\theta,\phi}. \tag{C.33}$$

Using these, we can work out that the only non-zero connection coefficients are

$$\begin{array}{lll} \Gamma^1_{22} = -r, & \Gamma^1_{33} = -r\sin^2\theta, & \Gamma^2_{33} = -r\sin\theta\cos\theta, \\ \Gamma^2_{12} = r^{-1}, & \Gamma^2_{21} = r^{-1}, & \Gamma^3_{13} = r^{-1}, \\ \Gamma^3_{31} = r^{-1}, & \Gamma^3_{23} = \frac{\cos\theta}{\sin\theta}, & \Gamma^3_{32} = \frac{\cos\theta}{\sin\theta}. \end{array} \tag{C.34}$$

Euler's equations in spherical coordinates follow as

$$\frac{\partial \rho}{\partial t} + \frac{\partial}{\partial r}(\rho v^r) + \frac{\partial}{\partial \theta}(\rho \omega) + \frac{\partial}{\partial \phi}(\rho \psi) = -\frac{2\rho v^r}{r} - \frac{\cos\theta}{\sin\theta}\rho\omega, \qquad (\text{C.35})$$

$$\frac{\partial \rho v^r}{\partial t} + \frac{\partial}{\partial r}(\rho v^r v^r) + \frac{\partial}{\partial \theta}(\rho \omega v^r) + \frac{\partial}{\partial \phi}(\rho \psi v^r) + \frac{\partial p}{\partial r}$$
$$= f^r - \frac{2\rho v^r v^r}{r} - \frac{\cos\theta}{\sin\theta}\rho\omega v^r + r\rho\omega\omega + r\rho\sin^2\theta\,\psi\psi, \qquad (\text{C.36})$$

$$\frac{\partial \rho\omega}{\partial t} + \frac{\partial}{\partial r}(\rho v^r \omega) + \frac{\partial}{\partial \theta}(\rho \omega\omega) + \frac{\partial}{\partial \phi}(\rho \psi\omega) + \frac{\partial}{\partial \theta}\left(\frac{p}{r^2}\right)$$
$$= \frac{f^\phi}{r} - \frac{4\rho v^r \omega}{r} - \frac{\cos\theta}{\sin\theta}\rho\omega\omega + \cos\theta\sin\theta\rho\psi\psi \qquad (\text{C.37})$$

$$\frac{\partial \rho\psi}{\partial t} + \frac{\partial}{\partial r}(\rho v^r \psi) + \frac{\partial}{\partial \theta}(\rho \omega\psi) + \frac{\partial}{\partial \phi}(\rho \psi\psi) + \frac{\partial}{\partial \phi}\left(\frac{p}{r^2\sin^2\theta}\right)$$
$$= \frac{f^\phi}{r\sin\theta} - \frac{4\rho v^r \psi}{r} - \frac{3\cos\theta}{\sin\theta}\rho\omega\psi, \qquad (\text{C.38})$$

$$\frac{\partial \mathscr{E}}{\partial t} + \frac{\partial}{\partial r}\left([\mathscr{E}+p]v^r\right) + \frac{\partial}{\partial \theta}\left([\mathscr{E}+p]\omega\right) + \frac{\partial}{\partial \phi}\left([\mathscr{E}+p]\psi\right)$$
$$= \mathbf{v}\cdot\mathbf{f} - \frac{2[\mathscr{E}+p]v^r}{r} - \frac{\cos\theta}{\sin\theta}[\mathscr{E}+p]\omega. \qquad (\text{C.39})$$

Again, we left in the force density f^i rather than the generalized f^i. As with cylindrical coordinates, the above will not be the most familiar way to express the conservation laws in spherical coordinates. Instead, we can rewrite these expressions in terms of actual linear velocities v^i:

$$\frac{\partial \rho}{\partial t} + \frac{1}{r^2}\frac{\partial}{\partial r}(r^2\rho v^r) + \frac{1}{r\sin\theta}\frac{\partial}{\partial \theta}\left(\sin\theta\rho v^\theta\right) + \frac{1}{r\sin\theta}\frac{\partial}{\partial \phi}(\rho v^\phi) = 0, \quad (\text{C.40})$$

$$\frac{\partial \rho v^r}{\partial t} + \frac{1}{r^2}\frac{\partial}{\partial r}(r^2\rho v^r v^r) + \frac{1}{r\sin\theta}\frac{\partial}{\partial \theta}\left(\sin\theta\rho v^\theta v^r\right) + \frac{1}{r\sin\theta}\frac{\partial}{\partial \phi}(\rho v^\phi v^r)$$
$$+ \frac{\partial p}{\partial r} = f^r + \frac{1}{r}\left(\rho v^\theta v^\theta\right) + \frac{1}{r}\left(\rho v^\phi v^\phi\right), \qquad (\text{C.41})$$

$$\frac{\partial \rho v^\theta}{\partial t} + \frac{1}{r^2}\left(r^2\rho v^r v^\theta\right) + \frac{1}{r\sin\theta}\frac{\partial}{\partial \theta}\left(\sin\theta\rho v^\theta v^\theta\right)$$
$$+ \frac{1}{r\sin\theta}\frac{\partial}{\partial \phi}\left(\rho v^\phi v^\theta\right) + \frac{\partial p}{\partial \theta} = f^\phi - \frac{\rho v^\theta v^r}{r} + \frac{\cos\theta}{r\sin\theta}\rho v^\phi v^\phi, \qquad (\text{C.42})$$

$$\frac{\partial \rho v^\phi}{\partial t} + \frac{1}{r^2}\frac{\partial}{\partial r}\left(r^2 \rho v^r v^\phi\right) + \frac{1}{r\sin\theta}\frac{\partial}{\partial \theta}\left(\sin\theta \rho v^\theta v^\phi\right) + \frac{1}{r\sin\theta}\frac{\partial}{\partial \phi}\left(\rho v^\phi v^\phi\right)$$

$$+ \frac{1}{r\sin\theta}\frac{\partial p}{\partial \phi} = f^\phi - \frac{\rho v^\phi v^r}{r} - \frac{\cos\theta}{r\sin\theta}\rho v^\phi v^\theta, \tag{C.43}$$

$$\frac{\partial \mathcal{E}}{\partial t} + \frac{1}{r^2}\left(r^2\left[\mathcal{E} + p\right]v^r\right) + \frac{1}{r\sin\theta}\frac{\partial}{\partial \theta}\left(\sin\theta\left[\mathcal{E} + p\right]v^\theta\right)$$

$$+ \frac{1}{r\sin\theta}\frac{\partial}{\partial \phi}\left(\left[\mathcal{E} + p\right]v^\phi\right) = \mathbf{v}\cdot\mathbf{f}. \tag{C.44}$$

D List of Symbols

a Acceleration.

c_p Specific heat capacity at constant pressure.

c_V Specific heat capacity at constant volume. Equivalent to $c_{\mathscr{V}}$ and c_ρ.

c Light speed.

c_s Sound speed.

$\mathscr{E}$ Total energy density (excluding rest mass term), $\mathscr{E} = e + \rho v^2/2$.

E Total energy.

E_g Gravitational energy.

E_{int} Internal energy.

e Internal energy density (excluding rest mass term).

e Internal energy density (including rest mass term), $e = e + \rho c^2$.

$\mathscr{F}$ Distribution function.

F Force.

f Force density.

f^i Components of generalized force density, $f^i = \rho \dot{u}^i$ for some acceleration $\dot{u}^i$ in generalized coordinates.

G Newtonian constant of gravitation.

g Gravitational acceleration.

g Specific Gibbs free energy, $g = \varepsilon = Ts + p\mathscr{V}$.

$\hbar$ Specific Helmholtz free energy, $\hbar = \varepsilon - Ts$.

H Scale height.

h Specific enthalpy (excluding rest mass term), $h = \varepsilon + p\mathscr{V}$.

I Intensity.

$\mathscr{K}$ Coefficient of thermal conductivity.

K Some constant of proportionality.

Kn Knudsen number.

k_B Boltzmann's constant.

k Wavenumber.

L Length scale.

L Luminosity.

$\mathscr{L}$ Angular momentum.

l^ϕ angular momentum density in the ϕ-direction.

$\mathscr{M}$ Mach number, $\mathscr{M} = |v|/c_s$.

M Mass (e.g. total stellar mass or total fluid element mass).

m Constituent particle mass *or* atomic mass unit.

$m(r)$ Cumulative mass up to radius r.

N Number of particles.

n Polytropic index, $\hat{\gamma} = 1 + 1/n$.

n Number density, $n = \rho/m$ or $n = \rho/(\mu m)$.

p Pressure.

DOI: 10.1201/9781003095088-D

q Specific heat.

q Heat flux vector.

R Radius (e.g. stellar radius or shock radius).

S^i $0i$-entry energy-momentum tensor, $S^i = \gamma^2 w \beta^i$.

s Specific entropy.

T Temperature.

T Time scale.

$T^{\alpha\beta}$ Energy-momentum tensor, $T^{\alpha\beta} = w \frac{V^\alpha}{c} \frac{V^\beta}{c} + p g^{\alpha\beta}$.

T^{ij} Stress tensor.

$\mathcal{T}^{ij}$ Stress tensor in generalized coordinates.

$\mathbf{U}$ Shock velocity.

$\mathbf{u}$ Total constituent particle velocity.

u^i Components of generalized velocity coordinate.

$\tilde{\mathbf{u}}$ Constituent particle peculiar velocity, $\tilde{\mathbf{u}} = \mathbf{u} - \mathbf{v}$.

$\mathscr{V}$ Specific volume, $\mathscr{V} = 1/\rho$.

V Volume.

V Velocity scale.

V^α relativistic Four-velocity.

$\mathbf{v}$ Bulk fluid flow velocity.

w Enthalpy density including rest mass, $w = \rho c^2 + e + p$.

w Specific work.

x^i Components of generalized coordinates.

β Flow speed in units of c.

Γ Lorentz factor of the shock.

$\hat{\gamma}$ Polytropic exponent, $p \propto \rho^{\hat{\gamma}}$.

$\hat{\gamma}_{ad}$ Adiabatic exponent, $p = (\hat{\gamma}_{ad} - 1) e$.

γ Fluid Lorentz factor.

ε Specific internal energy.

ζ Coefficient of bulk viscosity.

η Coefficient of shear viscosity.

λ Mean free path.

μ Mean molecular weight.

ρ Mass density in the comoving frame.

Σ vertically integrated density.

ς distance to z-axis (cylindrical coordinate).

σ Cross section.

τ Proper time.

τ 00-entry energy-momentum tensor minus rest mass, $\tau = \gamma^2 w - p - D c^2$.

Φ Velocity potential.

Ψ Gravitational potential.

ψ Angular velocity in the ϕ-direction.

ω Angular velocity in the θ-direction, angular frequency in general.

ω Vorticity.

E Abbreviated Answers to Selected Problems

This appendix contains abbreviated answers to a limited list of selected problems in the book. These worked solutions are selected because they introduce concepts not part of the main text or are of particular importance for understanding a chapter.

Answer to Problem 1.9

Start from

$$ds = \left(\frac{\partial s}{\partial p}\right)_T dp + \left(\frac{\partial s}{\partial T}\right)_p dT.$$

Differentiate with respect to p at constant s. In this case $ds = 0$, while dp/dp will always be 1 regardless of whether s is kept constant. Rearrange the outcome to arrive at the requested identity.

Answer to Problem 2.1

We start with the momentum equation:

$$\frac{\partial}{\partial t}(\rho \mathbf{v}) + \nabla \cdot (\rho \mathbf{v} \mathbf{v} + p \mathbb{1}) = \mathbf{f},$$

$$\frac{\partial \rho}{\partial t}\mathbf{v} + \rho \frac{\partial \mathbf{v}}{\partial t} + \nabla \cdot (\rho \mathbf{v} \mathbf{v} + p \mathbb{1}) = \mathbf{f}.$$

Applying the continuity equation and moving the pressure term to the RHS, we get

$$-\nabla \cdot (\rho \mathbf{v})\mathbf{v} + \rho \frac{\partial \mathbf{v}}{\partial t} + \nabla \cdot (\rho \mathbf{v} \mathbf{v}) = \mathbf{f} - \nabla p.$$

Dividing left and right by ρ, we get

$$\frac{\partial \mathbf{v}}{\partial t} - \frac{1}{\rho}\nabla \cdot (\rho \mathbf{v})\mathbf{v} + \frac{1}{\rho}\nabla \cdot (\rho \mathbf{v} \mathbf{v}) = \frac{\mathbf{f}}{\rho} - \frac{1}{\rho}\nabla p.$$

Since we are trying to derive the velocity equation

$$\frac{\partial \mathbf{v}}{\partial t} + \mathbf{v} \cdot \nabla \mathbf{v} = \frac{\mathbf{f}}{\rho} - \frac{1}{\rho}\nabla p,$$

it remains to be shown that

$$-\nabla \cdot (\rho \mathbf{v})\mathbf{v} + \nabla \cdot (\rho \mathbf{v} \mathbf{v}) = \rho \mathbf{v} \cdot \nabla \mathbf{v}.$$

DOI: 10.1201/9781003095088-E

Let us use index notation for clarity:

$$[-\nabla \cdot (\rho \mathbf{v})\mathbf{v} + \nabla \cdot (\rho \mathbf{vv})]^i = -\frac{\partial}{\partial x^j}\left(\rho v^j\right)v^i + \frac{\partial}{\partial x^j}\left(\rho v^j v^i\right),$$

$$= -\frac{\partial}{\partial x^j}\left(\rho v^j\right)v^i + \frac{\partial}{\partial x^j}\left(\rho v^j\right)v^i + \rho v^j \frac{\partial v^i}{\partial x^j},$$

$$= \rho \mathbf{v} \cdot \nabla v^i.$$

Answer to Problem 2.2

We are first asked to derive the equation for kinetic energy density. Our starting point is the velocity equation multiplied by $\mathbf{v}$:

$$\mathbf{v} \cdot \frac{\partial \mathbf{v}}{\partial t} + \mathbf{v} \cdot (\mathbf{v} \cdot \nabla \mathbf{v}) = -\frac{\mathbf{v}}{\rho} \cdot \nabla p + \mathbf{v} \cdot \frac{\mathbf{f}}{\rho}.$$

If we multiply this by ρ and shift the pressure term to the LHS, we get

$$\rho \mathbf{v} \cdot \frac{\partial \mathbf{v}}{\partial t} + \rho \mathbf{v} \cdot (\mathbf{v} \cdot \nabla \mathbf{v}) + \mathbf{v} \cdot \nabla p = \mathbf{v} \cdot \mathbf{f}.$$

The pressure term can also be expressed as $\mathbf{v} \cdot \nabla \cdot (p\mathbb{1})$. The partial velocity derivative can be treated as follows:

$$v_i \frac{\partial v^i}{\partial t} = \sum_i v^i \frac{\partial v^i}{\partial t} = \sum_i \frac{1}{2}\frac{\partial}{\partial t}(v^i v^i) = \frac{1}{2}\frac{\partial}{\partial t}(v^i v_i) = \frac{1}{2}\frac{\partial v^2}{\partial t}.$$

At this point we have the intermediate result

$$\rho \frac{\partial}{\partial t}\left(\frac{1}{2}v^2\right) + \rho \mathbf{v} \cdot (\mathbf{v} \cdot \nabla \mathbf{v}) + \mathbf{v} \cdot \nabla \cdot (p\mathbb{1}) = \mathbf{v} \cdot \mathbf{f},$$

which means that we still need to check whether

$$\rho \frac{\partial}{\partial t}\left(\frac{1}{2}v^2\right) + \rho v^i\left(v^j \frac{\partial}{\partial x^j}v_i\right) \overset{?}{=} \frac{\partial}{\partial t}\left(\frac{1}{2}\rho v^2\right) + \frac{\partial}{\partial x^j}\left(\frac{1}{2}\rho v^j v^2\right),$$

having switched to index notation for convenience. We can treat the second term on the LHS in the same manner as we treated the partial time derivative above:

$$v^i \frac{\partial}{\partial x^j}v_i = \frac{1}{2}\frac{\partial}{\partial x^j}(v^i v_i) = \frac{\partial}{\partial x^j}(\frac{1}{2}v^2),$$

and substitute the result, so it now remains to show

$$\rho \frac{\partial}{\partial t}\left(\frac{1}{2}v^2\right) + \rho v^j \frac{\partial}{\partial x^j}\left(\frac{1}{2}v^2\right) \overset{?}{=} \frac{\partial}{\partial t}\left(\frac{1}{2}\rho v^2\right) + \frac{\partial}{\partial x^j}\left(\frac{1}{2}\rho v^j v^2\right).$$

Both terms on the RHS can be expanded using the product rule. This yields the LHS above, plus $\frac{1}{2}v^2$ times the LHS of the continuity equation. Because the RHS of the continuity equation is zero, this means we have completed our derivation of the kinetic energy density equation. Adding to this the internal energy density equation to arrive at the full energy conservation law is straightforward.

Answer to Problem 2.4

Some intermediate steps (using the differential for specific enthalpy $dh = T ds + \rho^{-1} dp$ near the end):

$$\frac{\partial}{\partial t}(\rho h - p) + \nabla \cdot (\rho h \mathbf{v}) - \mathbf{v} \cdot \nabla p \;=\; \mathscr{K} \nabla^2 T,$$

$$-\left(\frac{\partial p}{\partial t} + \mathbf{v} \cdot \nabla p\right) + \frac{\partial}{\partial t}(\rho h) + \nabla \cdot (\rho h \mathbf{v}) \;=\; \mathscr{K} \nabla^2 T,$$

$$-\frac{dp}{dt} + \frac{\partial}{\partial t}(\rho h) + \mathbf{v} \cdot \nabla (\rho h) + \rho h \nabla \cdot \mathbf{v} \;=\; \mathscr{K} \nabla^2 T,$$

$$-\frac{dp}{dt} + \rho \frac{dh}{dt} + h \frac{d\rho}{dt} + h \rho \nabla \cdot \mathbf{v} \;=\; \mathscr{K} \nabla^2 T,$$

$$-\frac{dp}{dt} + \rho T \frac{ds}{dt} + \frac{dp}{dt} + \left(\frac{d\rho}{dt} + \rho \nabla \cdot \mathbf{v}\right) \;=\; \mathscr{K} \nabla^2 T.$$

At this point, the two pressure derivatives cancel and the term in parentheses is equal to zero according to the continuity equation. What remains is the requested heat equation. This can be rewritten using $C_{\mathscr{V}} = T (\partial s / \partial T)_V$ to provide the heat equation in terms of a temperature time-derivative under the condition that $\mathscr{V}$ is not changing.

Answer to Problem 2.5

The energy equation results from integrating over the individual particle velocities when computing their average kinetic energy. Our starting point Equation 2.37 without heating term,

$$\frac{\partial \mathscr{F}}{\partial t} + u^i \frac{\partial \mathscr{F}}{\partial x^i} + \frac{F^i}{m} \frac{\partial \mathscr{F}}{\partial u^i} = 0,$$

can therefore be written as

$$\int \frac{1}{2} m u^2 \frac{\partial \mathscr{F}}{\partial t} d\mathbf{u} + \int \frac{1}{2} m u^2 u^i \frac{\partial \mathscr{F}}{\partial x^i} d\mathbf{u} + \int \frac{1}{2} m u^2 \frac{F^i}{m} \frac{\partial \mathscr{F}}{\partial u^i} d\mathbf{u} = 0.$$

From this expression we are supposed to derive Equation 2.9:

$$\frac{\partial \mathscr{E}}{\partial t} + \nabla \cdot ([\mathscr{E} + p] \mathbf{v}) - \mathbf{v} \cdot \mathbf{f} = 0.$$

We treat the three LHS terms separately in order of appearance. The partial time derivative can be taken outside of the integral completely, since $\mathbf{u}$ and t are independent variables by definition of this partial derivative. Expanding the velocity in terms of bulk and peculiar velocities (and taking some terms out of the integral that are not a function of $\mathbf{u}$) yields

$$\frac{\partial}{\partial t} \int \frac{1}{2} m \tilde{u}^2 \mathscr{F} d\mathbf{u} + \frac{\partial}{\partial t} m v_i \int \tilde{u}^i \mathscr{F} d\mathbf{u} + \frac{\partial}{\partial t} \frac{1}{2} v^2 \int m \mathscr{F} d\mathbf{u},$$

which evaluates to

$$\frac{\partial}{\partial t} \left(\rho \varepsilon + 0 + \frac{1}{2} \rho v^2\right) = \frac{\partial \mathscr{E}}{\partial t}.$$

Here we identified the internal energy density as the averaged peculiar velocity contribution to the individual kinetic energies. The middle term is zero by definition of the integral over peculiar velocities, and the third term is integral is equal by definition to density.

Expanding the partial spatial derivative in a similar manner, yields

$$\frac{\partial}{\partial x^i} \int \frac{1}{2} m \left(\bar{u}^2 \bar{u}^i + \bar{u}^2 v^i + v^2 \bar{u}^i + v^2 v^i + 2 v^j \bar{u}_j \bar{u}^i + 2 v^j \bar{u}_j v^i \right) \mathscr{F} d\mathbf{u},$$

$$\frac{\partial}{\partial x^i} \left(0 + v^i \rho \varepsilon + 0 + \frac{v^2 \rho v^i}{2} + v_j p \delta^{ij} + 0 \right),$$

$$\frac{\partial}{\partial x^i} \left(v^i \rho \varepsilon + \frac{\rho v^2}{2} v^i + p v^i \right).$$

Note again that bulk velocity terms can be taken out of the integral. We have also identified the microphysical definition of pressure.

The force term can be computed using partial integration:

$$\frac{F^i}{m} \int \frac{m u^2}{2} \frac{\partial \mathscr{F}}{\partial u^i} d\mathbf{u} = -\frac{F^i}{m} \int m u_i \mathscr{F} d\mathbf{u} = -\frac{F^i}{m} \rho v_i = -f^i v_i.$$

In the last step, we switched from force $\mathbf{F} = m\mathbf{a}$ per particle of mass m, to a force density $\mathbf{f} = \rho\mathbf{a}$. For those not yet too used to the index notation approach, the partial integration step goes as follows:

$$\frac{F^i}{m} \int \frac{m u^2}{2} \frac{\partial \mathscr{F}}{\partial u^i} d\mathbf{u} = -\frac{F^i}{m} \int \frac{\partial}{\partial u^i} \left(\frac{m u^j u_j}{2} \right) \mathscr{F} d\mathbf{u} + 0,$$

where the zero term has its zero value because it represents the population values at the boundaries of velocity-space, where the population density is presumably zero (i.e. we assume only finite velocities are realized). The above RHS term can be expanded according to

$$-\frac{F^x}{m} \frac{m}{2} \int \frac{\partial}{\partial u^x} \left(u^x u^x + u^y u^y + u^z u^z \right) \mathscr{F} d\mathbf{u} +$$

$$-\frac{F^y}{m} \frac{m}{2} \int \frac{\partial}{\partial u^y} \left(u^x u^x + u^y u^y + u^z u^z \right) \mathscr{F} d\mathbf{u} +$$

$$-\frac{F^z}{m} \frac{m}{2} \int \frac{\partial}{\partial u^z} \left(u^x u^x + u^y u^y + u^z u^z \right) \mathscr{F} d\mathbf{u} =$$

$$-\frac{F^x}{m} m \int u^x \mathscr{F} d\mathbf{u} - \frac{F^y}{m} m \int u^y \mathscr{F} d\mathbf{u} - \frac{F^z}{m} m \int u^z \mathscr{F} d\mathbf{u}.$$

Answer to Problem 2.6

Some intermediate steps:

$$\frac{1}{2}m\int \tilde{u}^2 n\left(\frac{m}{2\pi k_B T}\right)^{\frac{3}{2}}\exp\left[-\frac{m(\mathbf{u}-\mathbf{v})^2}{2k_B T}\right]d\mathbf{u} =$$

$$\frac{1}{2}mn\left(\frac{m}{2\pi k_B T}\right)^{\frac{3}{2}}\int_{u_x,u_y,u_z=-\infty}^{\infty}\tilde{u}^2\exp\left[-\frac{m\tilde{u}^2}{2k_B T}\right]d\left(\tilde{u}_x+v_x\right)d\left(\tilde{u}_y+v_y\right)d\left(\tilde{u}_z+v_z\right) =$$

$$\frac{1}{2}mn\left(\frac{m}{2\pi k_B T}\right)^{\frac{3}{2}}\int_{\tilde{u}=0}^{\infty}\tilde{u}^4 4\pi\exp\left[-\frac{m\tilde{u}^2}{2k_B T}\right]d\tilde{u} =$$

$$\frac{1}{2}mn\pi^{-\frac{3}{2}}(4\pi)\left(\frac{2k_B T}{m}\right)\int_0^{\infty}x^4\exp\left[-x^2\right]dx.$$

The dimensionless integral equals $\frac{3}{8}\sqrt{\pi}$, so this altogether indeed evaluates to $\frac{3}{2}nk_B T$. Note that in the above the change in integration variables $u_x \to \tilde{u}_x$ etc. had no effect since the integration domain is $[-\infty,\infty]$ in both cases.

Answer to Problem 4.3

i) $$\frac{dp}{dr}=-\frac{MG\rho}{r^2}\Rightarrow\frac{d\rho}{\rho}=-\frac{MG}{A}\frac{dr}{r^2}\Rightarrow\frac{\rho}{\rho_0}=\exp\left(\frac{MG}{Ar}-\frac{MG}{Ar_0}\right)$$

ii) This can be shown by considering small distances $\delta r \equiv r-r_0$, such that $\delta r/r_0 \ll 1$. We have

$$\frac{\rho}{\rho_0}=\exp\left[\frac{MG}{A(\delta r-r_0)}-\frac{MG}{Ar_0}\right]\approx\exp\left[\frac{1}{\rho_0}\frac{MG}{A}\left(1-\frac{\delta r}{r_0}\right)-\frac{MG}{Ar_0}\right],$$

leading to Equation 4.5 once the cancelling terms are taken out.

Answer to Problem 4.4

i) $$\frac{dp}{dz}=-\rho g\Rightarrow\rho^{\hat{\gamma}-2}d\rho=-\frac{g}{A\hat{\gamma}}dz\Rightarrow\rho^{\hat{\gamma}-1}=\rho_0^{\hat{\gamma}-1}-\frac{g(\hat{\gamma}-1)}{\hat{\gamma}A}z,$$

which sets $\rho=\rho_0$ at $z_0=0$. Normalizing the expression yields the requested answer.

ii) Start from the total column depth $\bar{m}=\int\rho/\rho_0 dz$ for both (normalized) atmospheric profiles:

$$\bar{m} = \int_0^{\infty}\exp\left(-z/H\right)dz=H,$$

$$\bar{m} = \int_0^{z_{max}}\left(1-\frac{z}{z_{max}}\right)^{\frac{1}{\hat{\gamma}-1}}dz=z_{max}\frac{\hat{\gamma}-1}{\hat{\gamma}}.$$

Equating the two yields the requested answer.

iii) Figure E.1 shows the requested comparison on a logarithmic plot.

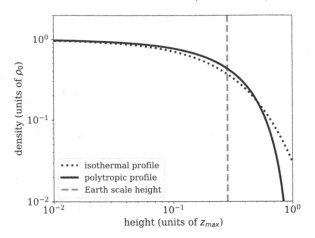

Figure E.1 A comparison between isothermal and polytropic plane-parallel atmosphere profiles.

Answer to Problem 5.2

To derive the wave equation for Φ, start from the continuity equation

$$\frac{\partial \rho_1}{\partial t} + \rho_0 \nabla^2 \Phi = 0 \Rightarrow c_s^2 \frac{\partial \rho_1}{\partial t} = -\rho_0 c_s^2 \nabla^2 \Phi,$$

which we can file away now for further use later on. Taking the partial time derivative of the linearized velocity perturbation equation yields

$$\nabla \left(\rho_0 \frac{\partial^2 \Phi}{\partial t^2} + c_s^2 \frac{\partial \rho_1}{\partial t} \right) = 0.$$

Combining the two results therefore gives us

$$\nabla \left(\rho_0 \frac{\partial^2 \Phi}{\partial t^2} - \rho_0 c_s^2 \nabla^2 \Phi \right) = 0,$$

which has to hold throughout the fluid including far away from the perturbation. Integrating therefore yields constant of integration equal to zero, and we have our wave equation for Φ as a result (after dividing by ρ_0).

The scalar potential implies that the vorticity of the velocity is equal to zero, i.e. $\nabla \times \nabla \Phi = 0 \Rightarrow \nabla \times \mathbf{v}' = 0$, as for any cross product between two vectors pointing in the same direction. This also follows from

$$\nabla \times \frac{\partial \mathbf{v}'}{\partial t} + \frac{1}{\rho_0} \nabla \times \nabla p' = 0 \Rightarrow \frac{\partial}{\partial t} \nabla \times \mathbf{v}' = 0.$$

Since the vorticity was zero to begin with (fluid at rest), it will remain zero, and the use of a scalar potential is appropriate.

Answer to Problem 5.3

Starting from the conservation law for energy density, the energy flux associated with the perturbation is given by

$$h\rho\mathbf{v}_1 = h_0\rho\mathbf{v}_1 + h_1\rho\mathbf{v}_1 \approx h_0\rho\mathbf{v}_1 + h_1\rho_0\mathbf{v}_1.$$

The thermodynamic identity for the specific enthalpy is given by $dh = Tds + \rho^{-1}dp$, from which it follows that we can take

$$h_1 = \left(\frac{\partial h}{\partial p}\right)_s p_1 = p_1/\rho_0.$$

We now have a flux

$$h\rho\mathbf{v}_1 = h_0\rho\mathbf{v}_1 + p_1\mathbf{v}_1.$$

The first term on the RHS, however, will be zero when integrated over a surface covering the perturbation in the same way that the volume-integrated density perturbation ends up zero, as it describes a mass flux. The remaining term can be rewritten using $p_1 = \rho_0 c_s v_1$ (Equation 5.17), which means that

$$p_1\mathbf{v}_1 = \rho_0 c_s v_1^2 \hat{\mathbf{n}} = \mathscr{E}_1 c_s \hat{\mathbf{n}}.$$

Answer to Problem 5.4

i) At some point, you will need to use an identity of the type discussed in exercise 1.9:

$$ds = \left(\frac{\partial s}{\partial \rho}\right)_{p,0} d\rho + \left(\frac{\partial s}{\partial p}\right)_{p,0} dp,$$

$$ds = \left(\frac{\partial s}{\partial \rho}\right)_{p,0}\left(d\rho + \left(\frac{\partial s}{\partial p}\right)_{p,0}\left(\frac{\partial \rho}{\partial s}\right)_{p,0} dp\right),$$

$$ds = \left(\frac{\partial s}{\partial \rho}\right)_{p,0}\left(d\rho - \left(\frac{\partial \rho}{\partial p}\right)_{s,0} dp\right).$$

Substituting the sound speed in the final expression yields the requested result.

ii) The answer is obtained by both dropping all non-linear terms in the perturbation and realizing that $\nabla s_0 \approx \nabla^2 T_0 \approx 0$ (the latter is a restatement of the assumption from the beginning of the chapter that the fluid at rest varies over scales far larger than the perturbation that is introduced).

iii) The expressions evaluated for the rest state are not affected by the partial derivative operators, so we have

$$\rho_0 T_0 \left(\frac{\partial s}{\partial \rho}\right)_{p,0} \frac{\partial}{\partial t}\left(\rho_1 - \frac{p_1}{c_s^2}\right) = \mathscr{K}\left(\frac{\partial T}{\partial \rho}\right)_{p,0} \nabla^2\left(\rho_1 - \frac{p_1}{c_T^2}\right).$$

Applying the identity

$$\left(\frac{\partial s}{\partial \rho}\right)_{p,0} = \left(\frac{\partial s}{\partial T}\right)_{p,0}\left(\frac{\partial T}{\partial \rho}\right)_{p,0},$$

allows us to cancel the derivatives of temperature and to introduce the heat capacity at constant pressure on the LHS.

iv) Take two times the partial derivative with respect to time on both sides of the result from (iii). Then eliminate $\partial^2 \rho_1/\partial t^2$ in favour of $\nabla^2 p_1$.

v) This follows directly from applying the requested derivatives, leading to $\frac{\partial}{\partial t}\frac{\partial^2}{\partial x^2}p = ik^2\omega p$, and so forth.

vi) For an adiabatic sound wave we have that $k^2 \approx \omega^2/c_s^2$ so the LHS of the expression from (v) reduces to zero. The magnitude of the RHS should therefore also be close to zero, which implies that $\omega \ll \omega_{TC}$.

vii) For an isothermal sound wave we have instead that $k^2 \approx \omega^2/c_T^2$, so the RHS needs to have a very large magnitude instead to keep up. This implies $\omega \gg \omega_{TC}$

viii)
$$\nu_{TC} = \frac{\rho c_p c_s^2}{2\pi\mathcal{K}} = \frac{c_p \hat\gamma p}{2\pi\mathcal{K}} = \frac{\hat\gamma^2}{\hat\gamma - 1}\frac{k_B}{\mu m}\frac{p}{2\pi\mathcal{K}} \approx 10^9 \text{ Hz.}$$

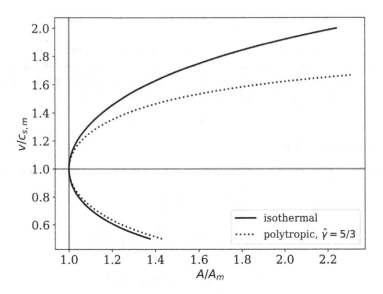

Figure E.2 Velocity of an isothermal and polytropic De Laval nozzle versus cross-section.

Answer to Problem 6.2

i) Some intermediate steps starting from the pre-cursor to the Bernoulli equation:

$$d\left(\frac{1}{2}v^2\right) + \rho^{-1}dp(\rho) = 0,$$

$$d\left(\frac{1}{2}v^2\right) + \frac{dp}{d\rho}\frac{d\rho}{\rho} = 0,$$

$$\frac{1}{2}v^2 - \frac{1}{2}c^2 + c_s^2\ln\frac{\rho}{\rho_m} = 0,$$

$$\frac{1}{2}v^2 - \frac{1}{2}c^2 + c_s^2\ln\frac{\dot M}{vA\rho_m} = 0,$$

$$\frac{1}{2}v^2 - \frac{1}{2}c^2 + c_s^2\ln\frac{c_s A_m\rho_m}{vA\rho_m} = 0.$$

Here, we have used the definition of the sound speed $c_s^2 = dp/d\rho$ and substituted $\dot{M} = \rho v A$. Subscript m refers to the narrowest point, where $v = c_s$ and used that as a boundary condition when integrating the differential expression.

ii) The most efficient approach is to solve analytically for A/A_m rather than try to find a (semi-)analytical solution for $\mathcal{M}$:

$$\frac{A}{A_m} = \mathcal{M}^{-1} \exp\left(\frac{\mathcal{M}^2 - 1}{2}\right).$$

A plot of $\mathcal{M}$ versus A/A_m is shown in Figure E.2.

iii) Thanks to $\dot{M} = \rho v A$ and the assumption of a polytropic gas, the sound speed can be rewritten in the forms

$$c_s^2 = \hat{\gamma} K \rho^{\hat{\gamma}-1} = \hat{\gamma} K \dot{M}^{\hat{\gamma}-1} v^{1-\hat{\gamma}} A^{1-\hat{\gamma}}.$$

This comes in useful in deriving the requested result. Some intermediate steps:

$$d\left(\frac{1}{2}v^2\right) + \rho^{-1} dp(\rho) = 0,$$

$$d\left(\frac{1}{2}v^2\right) + K\hat{\gamma}\rho^{\hat{\gamma}-2} d\rho = 0,$$

$$\frac{1}{2}v^2 - \frac{1}{2}c_{s,m}^2 + \frac{K\hat{\gamma}}{\hat{\gamma}-1}\rho^{\hat{\gamma}-1} - \frac{K\hat{\gamma}}{\hat{\gamma}-1}\rho_m^{\hat{\gamma}-1} = 0,$$

$$\frac{1}{2}v^2 - \frac{1}{2}c_{s,m}^2 + \frac{c_s^2}{\hat{\gamma}-1} - \frac{c_{s,m}^2}{\hat{\gamma}-1} = 0,$$

$$\frac{1}{2}\left(\frac{v}{c_{s,m}}\right)^2 - \frac{1}{2} + \frac{1}{\hat{\gamma}-1}\left(\frac{v}{c_{s,m}}\right)^{1-\hat{\gamma}}\left(\frac{A}{A_m}\right)^{1-\hat{\gamma}} - \frac{1}{\hat{\gamma}-1} = 0.$$

iv) Again, the best approach is to isolate the cross-section. In this case, it is straightforward to derive that

$$\frac{A}{A_m} = \left[\frac{\hat{\gamma}+1}{2}\left(\frac{v}{c_{s,m}}\right)^{\hat{\gamma}-1} - \frac{\hat{\gamma}-1}{2}\left(\frac{v}{c_{s,m}}\right)^{\hat{\gamma}+1}\right]^{\frac{1}{1-\hat{\gamma}}}.$$

The resulting profile is included in Figure E.2.

Answer to Problem 7.1

For the velocity squared, directly combine Equations 7.12, 7.13:

$$\frac{1}{2}(v_{1,s})^2 - \frac{1}{2}(v_{2,s})^2 = \frac{1}{2}(p_2 - p_1)\frac{\mathcal{V}_2^2 - \mathcal{V}_2^2}{\mathcal{V}_1 - \mathcal{V}_2}$$

$$\Rightarrow \quad \frac{1}{2}(v_{1,s})^2 - \frac{1}{2}(v_{2,s})^2 = \frac{1}{2}(p_2 - p_1)\frac{(\mathcal{V}_1 - \mathcal{V}_2)(\mathcal{V}_1 + \mathcal{V}_2)}{\mathcal{V}_1 - \mathcal{V}_2},$$

which is the desired outcome once cancelling terms are accounted for. The velocity equations can be derived completely analogously. The equation for specific internal energies is more involved. We can start by rewriting the jump-condition for total energy, and make use of (in this order), the mass conservation jump condition and the velocity-squared jump condition:

$$(\mathscr{E}_2 + p_2)\, v_{2,S}^x - (\mathscr{E}_1 + p_1)\, v_{1,S}^x = 0$$

$$\Rightarrow \quad (\mathscr{E}_2 + p_2)\, \rho_2^{-1} (\rho_2 v_{2,S}^x) - (\mathscr{E}_1 + p_1)\, \rho_1^{-1} (\rho_1 v_{1,S}^x) = 0$$

$$\Rightarrow \quad \left(\varepsilon_2 + \frac{1}{2} \left(v_{2,S}^x\right)^2 + p_2 \rho_2^{-1}\right) - \left(\varepsilon_1 + \frac{1}{2} \left(v_{1,S}^x\right)^2 + p_1 \rho_1^{-1}\right) = 0$$

$$\Rightarrow \quad \left(\varepsilon_2 + p_2 \rho_2^{-1}\right) - \left(\varepsilon_1 + p_1 \rho_1^{-1}\right) = \frac{1}{2} (p_2 - p_1)(\mathscr{V}_1 + \mathscr{V}_2)$$

$$\Rightarrow \quad \varepsilon_2 + p_2 \mathscr{V}_2 - \varepsilon_1 - p_1 \mathscr{V}_1 = \frac{1}{2} p_2 \mathscr{V}_1 + \frac{1}{2} p_2 \mathscr{V}_2 - \frac{1}{2} p_1 \mathscr{V}_1 - \frac{1}{2} p_1 \mathscr{V}_2$$

$$\Rightarrow \quad \varepsilon_2 - \varepsilon_1 = \frac{1}{2} (p_2 + p_1)(\mathscr{V}_1 - \mathscr{V}_2).$$

Answer to Problem 7.2

Some intermediate steps in deriving Equation 7.18:

$$\frac{p_2 \mathscr{V}_2}{\hat{\gamma} - 1} - \frac{p_1 \mathscr{V}_1}{\hat{\gamma} - 1} = \frac{1}{2} (p_2 + p_1)(\mathscr{V}_1 - \mathscr{V}_2)$$

$$\Rightarrow \quad p_2 \frac{\mathscr{V}_2}{\hat{\gamma} - 1} - \frac{1}{2} p_2 (\mathscr{V}_1 - \mathscr{V}_2) = p_1 \frac{\mathscr{V}_1}{\hat{\gamma} - 1} + \frac{1}{2} p_1 (\mathscr{V}_1 - \mathscr{V}_2)$$

$$\Rightarrow \quad p_2 \frac{2\mathscr{V}_2 - (\hat{\gamma} - 1)\mathscr{V}_1 + (\hat{\gamma} - 1)\mathscr{V}_2}{2(\hat{\gamma} - 1)} = p_1 \frac{2\mathscr{V}_1 + (\hat{\gamma} - 1)\mathscr{V}_1 - (\hat{\gamma} - 1)\mathscr{V}_2}{2(\hat{\gamma} - 1)}.$$

In order to derive Equation 7.19, we could take the second line above and proceed to instead collect $\mathscr{V}_1$ terms on the LHS, $\mathscr{V}_2$ terms on the RHS. The remaining equal signs follow from the definition of $\mathscr{V}$ and application of the mass conservation jump-condition.

Answer to Problem 7.3

i) Multiplying our starting equation by ρ, and expanding the comoving derivative in terms of partial derivatives, we obtain:

$$\rho \frac{d(p/\rho)}{dt} = 0 \Rightarrow \rho \frac{\partial (p/\rho)}{\partial t} + \rho \mathbf{v} \cdot \nabla (p/\rho) = 0.$$

The continuity equation, multiplied by (p/ρ), reads

$$(p/\rho) \frac{\partial \rho}{\partial t} + (p/\rho)\nabla \cdot (\rho \mathbf{v}) = 0.$$

Summing these two equations yields the requested result.

ii) The jump condition follows completely analogously to the other derivations of jump conditions and is therefore

$$p_1 v_{1,S}^x = p_2 v_{2,S}^x.$$

iii) Using mass conservation, the jump condition can be rewritten in the form $p_1/\rho_1 = p_2/\rho_2$. Since according to the ideal gas law $(p/\rho) \propto T$, this implies $T_2 = T_1$.

iv) Since density is linearly proportional to density in the case of an isothermal fluid, the density jump can become arbitrarily high. This stands in contrast to strong adiabatic shock waves.

v) Since the unshocked medium is at rest, we have $v_{1,S}^2 = v_S^2$. We can derive

$$U^2 = \rho_1^{-2}\frac{p_2 - p_1}{\rho_1^{-1} - \rho_2^{-1}} = \frac{p_1}{\rho_1^2}\frac{(\rho_2/\rho_1) - 1}{\rho_1^{-1} - \rho_2^{-1}} = \frac{p_1}{\rho_1^2}\frac{(\rho_2/\rho_1) - 1}{(\rho_2/\rho_1) - 1}p_2 \Rightarrow p_2 = (U/c_{s,1})^2\rho_1,$$

where the last expression includes the sound speed of the unshocked medium, given by $c_{s,1}^2 = p_1/\rho_1$. We can use the ideal gas law to derive the pressure in the unshocked medium using $p_1 = n_1 k_B T_1 = 1.38 \times 10^{-9}$ Ba, and therefore the unshocked medium sound speed ($c_{s,1} = 9.08 \times 10^5$ cm s^{-1}). We find that $\rho_2 = 1.21 \times 10^5 m$ cm^{-3}. Using our original jump condition, which implies that $p_2/p_1 = \rho_2/\rho_1 = 121$, we also find $p_2 = 1.67 \times 10^{-7}$ Ba. In the frame of the shock $|v_{2,S}| = |v_{1,S}| \times \rho_1/\rho_2 = 8.24 \times 10^4$ cm s^{-1}. In the frame of the shock, this fluid is moving *away* from the shock. The corresponding lab frame velocity is therefore $v_2 = U - |v_{2,S}| \approx U = 10^7$ cm s^{-1}. The shocked fluid will move at a velocity very close to that of the shock.

Answer to Problem 7.4

If fluid 1 is at rest, we immediately have $v_{1,S} = -U$. We also have the identity $v_{2,S} = -Uv_{2,S}/v_{1,S}$. We can apply this to the shock jump condition for velocity-squared:

$$v_{1,S}^2 - v_{2,S}^2 = (p_2 - p_1)(\mathcal{V}_1 + \mathcal{V}_2)$$

$$\Rightarrow \quad U^2 - U^2\left(\frac{v_{2,S}}{v_{1,S}}\right)^2 = (p_2 - p_1)(\mathcal{V}_1 + \mathcal{V}_2),$$

which evaluates to the requested Equation 7.27.

To address the point about sound speeds in the question, we proceed as follows. Compute p_1 using $p_1 = \rho_1 k_B T/m_p$ and p_2 from the provided ratio. Use Equation 7.19 to compute $\mathcal{V}_2$, using $\mathcal{V}_1 \equiv 1/\rho_1$. This translates to $\rho_2 = 3.86\rho_1$ in the strong shock case. The ratio between velocities $v_{1,S}$ and $v_{2,S}$ follows from the ratio between densities. This ratio can be used directly in the equation provided in the exercise, along with the results for pressure and density, in order to compute the shock velocity U. The sound speeds follow from $c_S^2 = \hat{\gamma}p/\rho$, using $\hat{\gamma} = 5/3$. For the upstream fluid velocity, we have $v_{1,S} = -U$, given that the medium is at rest. The downstream fluid velocity then follows from the velocity ratio. Note that $v_{2,S}$ and $v_{1,S}$ will need to have the same sign if they are to obey the shock jump condition that equates them to a density ration (which is positive by definition).

For the strong compression shock, it will follow that

$$c_{s,2} + v_2 = 4.3 \times 10^5 \text{ cm s}^{-1} > U = 3.2 \times 10^5 \text{ cm s}^{-1},$$

$$U = 3.2 \times 10^5 \text{ cm s}^{-1} > c_{s,1} = 3.7 \times 10^4 \text{ cm s}^{-1}.$$

For the 'rarefaction shock', it will follow that

$$c_{s,2} + v_2 = -4.1 \times 10^4 \text{ cm s}^{-1} < U = 1.7 \times 10^4 \text{ cm s}^{-1}$$

$$U = 1.7 \times 10^4 \text{ cm s}^{-1} > c_{s,1} = 3.7 \times 10^4 \text{ cm s}^{-1}.$$

For the compression shock, these numbers are fine. The upstream sound speed is smaller than the shock velocity, so the upstream medium has no means of receiving advance warning of the shock (a key property of shocks). At the same time, the downstream medium can communicate with the shock, and in particular, deliver the energy to sustain the shock.

For the rarefaction shock, the numbers are the wrong way round (at least when attempting to let the shock run *into* medium 1, coming from medium 2, note that U is determined from U^2 so it becomes possible to flip the whole situation by picking the negative solution for U). The upstream medium is in contact already with the shock before it arrives, with the perturbation in the upstream medium outrunning the shock itself. As a result, the shock will diffuse, as the upstream medium gets disturbed more, and more energy gets carried away from the shock front. At the same time, the shock will outpace causal contact with the downstream medium and cannot be sustained.

Answer to Problem 8.5

For the non-relativistic limit, we will need approximations for γ^2 and γ:

$$\gamma^2 = \frac{1}{1 - v^2/c^2} = 1 + \frac{v^2}{c^2} + O\left(\frac{v^4}{c^4}\right) \approx 1 + \frac{v^2}{c^2}, \qquad \gamma = \frac{1}{\sqrt{1 - v^2/c^2}} \approx 1 + \frac{1}{2}\frac{v^2}{c^2}.$$

The other thing to keep in mind is that we will want to use a density measure in the lab frame, i.e. $D \equiv \gamma\rho$, in order to later subtract away the continuity equation. This does not apply to p or $e = \rho\varepsilon$, and you can check from the microphysics chapter that they were set up in terms of *peculiar* velocities of constituent particles, i.e., velocities in the frame comoving with the fluid parcel). Let us first take the limits for the components T^{00}, $T^{0i} = T^{i0}$ and $T^{ij} = T^{ji}$ separately.

$$T^{00} = \gamma^2 \rho c^2 + \gamma^2 \rho\varepsilon + \gamma^2 p - p,$$
$$= \gamma D c^2 + \gamma^2 \rho\varepsilon + \gamma^2 p - p,$$
$$\approx D c^2 + \frac{1}{2} D v^2 + \rho\varepsilon + \rho\varepsilon\frac{v^2}{c^2} + p + p\frac{v^2}{c^2} - p.$$

We can discard a lot of terms here. The pressure terms cancel even at lowest order, while some terms are extremely small (v^2/c^2), leaving

$$T^{00} \approx D c^2 + \mathscr{E},$$

which can be recognized as the energy density (plus a rest mass term). In a similar manner, we can show

$$T^{0i} \approx \left(D c^2 + \frac{1}{2} D v^2 + \rho\varepsilon + p\right)\frac{v^i}{c}.$$

It then follows that

$$\frac{\partial}{\partial X^\alpha} T^{\alpha 0} = \frac{\partial}{\partial ct} T^{00} + \frac{\partial}{\partial x^i} T^{i0},$$

is identical to conservation of energy plus conservation of mass (the latter multiplied by c^2). Conservation of momentum follows from considering

$$T^{ij} = \left(\rho c^2 + \rho \varepsilon + p\right) \gamma^2 \frac{v^i v^j}{c^2} + p\delta^{ij},$$

$$= \left(\gamma D c^2 + \gamma^2 \rho \varepsilon + \gamma^2 p\right) \frac{v^i v^j}{c^2} + p\delta^{ij},$$

$$\approx \left(Dc^2 + \frac{1}{2}Dv^2 + \rho\varepsilon + \rho\varepsilon\frac{v^2}{c^2} + p + p\frac{v^2}{c^2}\right) \frac{v^i v^j}{c^2} + p\delta^{ij},$$

$$\approx D v^i v^j + p\delta^{ij}.$$

As a result (we can now drop also almost everything in T^{0i}),

$$\frac{\partial}{\partial X^\alpha} T^{\alpha i} = \frac{\partial}{\partial ct} \left(Dv^i c\right) + \frac{\partial}{\partial x^i} \left(Dv^i v^j + p\delta^{ij}\right).$$

Answer to Problem 8.6

The jump condition for internal energy can take the forms $e_2 - e_1 = w_2 \gamma_{2,s}^2 (1 - \beta_{2,s}/\beta_{1,s})$ and $e_1 - e_2 = w_1 \gamma_{1,s}^2 (1 - \beta_{1,s}/\beta_{2,s})$. Using this when working backwards, we get

$$\frac{e_1 + p_2}{e_2 + p_1} = \frac{e_2 - w_2 \gamma_{2,s}^2 (1 - \beta_{2,s}/\beta_{1,s}) + p_2}{e_1 - w_1 \gamma_{1,s}^2 (1 - \beta_{1,s}/\beta_{2,s}) + p_1},$$

$$\frac{e_1 + p_2}{e_2 + p_1} = \frac{-\left(\gamma_{2,s}^2 - 1\right) w_2 + w_2 \gamma_{2,s}^2 \beta_{2,s}/\beta_{1,s}}{-\left(\gamma_{1,s}^2 - 1\right) w_1 + w_1 \gamma_{1,s}^2 \beta_{1,s}/\beta_{2,s}},$$

$$\frac{e_1 + p_2}{e_2 + p_1} = \frac{-w_2 \gamma_{2,s}^2 \beta_{2,s}^2 + w_2 \gamma_{2,s}^2 \beta_{2,s}/\beta_{1,s}}{-w_1 \gamma_{1,s}^2 \beta_{1,s}^2 + w_1 \gamma_{1,s}^2 \beta_{1,s}/\beta_{2,s}},$$

$$\frac{e_1 + p_2}{e_2 + p_1} = \frac{w_2 \gamma_{2,s}^2}{w_1 \gamma_{1,s}^2} \frac{\beta_{2,s}/\beta_{1,s} - \beta_{2,s}^2}{\beta_{1,s}/\beta_{2,s} - \beta_{1,s}^2},$$

$$\frac{e_1 + p_2}{e_2 + p_1} = \frac{\beta_{1,s}}{\beta_{2,s}} \frac{\beta_{2,s}}{\beta_{1,s}} \frac{1/\beta_{1,s} - \beta_{2,s}}{1/\beta_{2,s} - \beta_{1,s}} = \frac{\beta_{2,s}}{\beta_{1,s}}.$$

Answer to Problem 8.7

i) Our suggested starting point can quickly be rewritten as

$$\frac{p_2}{p_1} \left(\rho_1 c^2 + e_1 + p_2\right) = \left(\rho_2 c^2 + \frac{\hat{\gamma}_{ad,2}}{\hat{\gamma}_{ad,2} - 1} p_2\right) \gamma_{2,1}.$$

If we use the strong shock assumption $e_1 \ll p_2$ and collect terms of p_2 on the RHS, we obtain

$$\frac{\rho_2}{\rho_1} p_2 = (\gamma_{2,1} - 1) \rho_2 c^2 + \frac{\hat{\gamma}_{ad,2}}{\hat{\gamma}_{ad,2} - 1} p_2 \gamma_{2,1}.$$

Dividing left and right by p_2 yields Equation 8.88. A further application of Equation 8.84 yields the requested Equation 8.89.

ii) To make this worked solution easier on the eye, we will temporarily use the symbols $y \equiv \gamma_{2,1}$ and $x \equiv \hat{\gamma}_{ad,2}$. We first work backwards a bit from Equation 8.90, as suggested:

$$\Gamma^2 = \frac{(y+1)(xy-x+1)^2}{(2x-x^2)(y-1)+2},$$

$$= \frac{(y+1)\left[y^2x^2 + y(-2x^2+2x) + x^2 - 2x + 1\right]}{y(-x^2+2x) + x^2 - 2x + 2},$$

$$= \frac{y^3x^2 + y^2(-x^2+2x) + y(-x^2+1) + x^2 - 2x + 1}{y(-x^2+2x) + x^2 - 2x + 2}.$$

Now, we move forwards from Equation 8.76, to see if we can arrive at the expression above. We first specialize to the case of a cold external medium:

$$\Gamma^2 = \frac{(\rho_2 c^2 + e_2 - \rho_1 c^2)(\rho_1 c^2 + p_2)}{(\rho_2 c^2 + e_2 - \rho_1 c^2)(\rho_1 c^2 + p_2) - p_2(\rho_2 c^2 + e_2)}.$$

The rest is a matter of applying Equation 8.84 and Equation 8.89, applicable to strong shocks running into a cold environment. Some steps to get you started:

$$\Gamma^2 = \frac{\left[1 + y - 1 - \frac{x-1}{yx+1}\right]\left[\frac{x-1}{yx+1} + (x-1)(y-1)\right]}{\left[1 + y - 1 - \frac{x-1}{yx+1}\right]\left[\frac{x-1}{yx+1} + (x-1)(y-1)\right] - (y-1)(x-1)(1+y-1)},$$

$$= \frac{\left[y - \frac{x-1}{yx+1}\right]\left[\frac{1}{yx+1} + (y-1)\right]}{\left[y - \frac{x-1}{yx+1}\right]\left[\frac{1}{yx+1} + (y-1)\right] - (y-1)y},$$

$$= \frac{(y^2x + y - x + 1)(1 + y^2x + y - yx - 1)}{(y^2x + y - x - 1)(1 + y^2x + y - yx - 1) - (yx+1)^2(y-1)y}.$$

Expanding the products yields the midway solution we found above by working backwards.

Answer to Problem 8.10

i) We start from

$$\frac{d\chi}{dt} = \left(\frac{\partial \chi}{\partial t}\right) + c\beta_m \frac{\partial \chi}{\partial r} = 0.$$

Plugging in the expressions for the partial derivatives in terms of Γ^2 and χ, we obtain

$$4\left(2\Gamma^2 - \chi\right) + 1 - \beta_m\left(1 + 8\Gamma^2\right) = 0, \tag{E.1}$$

which can be further manipulated (dropping terms that are too small) into

$$\beta_m = \left(1 - 4\chi + 8\Gamma^2\right)\frac{1}{8\Gamma^2}\left(1 - \frac{1}{8\Gamma^2}\right),$$

$$\beta_m = \frac{8\Gamma^2 - 4\chi}{8\Gamma^2},$$

which reduces further to the requested result.

ii) The derivation is analogous to the non-relativistic case, which means that we are equating energy flux to additional energy captured in the increasing sphere at fixed χ,

$$4\pi r^2 \beta_m dt \left(T^{00} - \rho c^2\right) = 4\pi r^2 dt \left(w\gamma^2\beta\right).$$

In the relativistically hot limit, rest mass energy can be neglected and the above can be rewritten as

$$\left(1 - \frac{\chi}{2\Gamma^2}\right)\left(\gamma^2 e + \gamma^2 p - p\right) = (e+p)\left(\gamma^2 - \frac{1}{2}\right),$$

where we also approximated $\beta \approx 1 - \gamma^{-2}/2$. Plugging in the ultra-relativistic EOS $p = e/3$ gets us eventually to

$$\left(1 - \frac{\chi}{2\Gamma^2}\right)\left(4\gamma^2 - 1\right) = 4\gamma^2 - 2,$$

$$4\gamma^2 - 1 - \frac{\chi\gamma^2 2}{\Gamma^2} + \frac{\chi}{2\Gamma^2} = 4\gamma^2 - 2.$$

The first terms on either side cancel, and the last LHS term can be neglected. The result is $\gamma^2 = \Gamma^2\chi^{-1}/2$, which contains the requested expression $g = \chi^{-1}$.

Answer to Problem 9.1

(i) The bulk viscosity terms drop out of σ, leaving

$$\sigma^{ij} = \eta\left(\frac{\partial v^i}{\partial x_j} + \frac{\partial v^j}{\partial x_i} - \frac{2}{3}\frac{\partial v^k}{\partial x^k}\delta^{ij}\right) \Rightarrow \frac{\partial}{\partial x^i}\sigma^{ij} = \eta\left(\frac{\partial^2 v^i}{\partial x^i \partial x_j} + \frac{\partial^2 v^j}{\partial x^i \partial x_i} - \frac{2}{3}\frac{\partial^2 v^k}{\partial x_j \partial x^k}\right),$$

after applying the δ-function in the rightmost term. The order of partial differentiation operators is interchangeable, hence

$$\nabla \cdot \sigma = \eta\left(\nabla(\nabla \cdot \mathbf{v}) + \nabla^2\mathbf{v} - \frac{2}{3}\nabla(\nabla \cdot \mathbf{v})\right) = \eta\left(\frac{1}{3}\nabla(\nabla \cdot \mathbf{v}) + \nabla^2\mathbf{v}\right).$$

(ii) We have

$$\nabla \times \left(\frac{1}{\rho}\nabla p\right) = \left(\nabla\rho^{-1}\right) \times \nabla p + \rho^{-1}(\nabla \times \nabla)p = \left(\nabla\rho^{-1}\right) \times \nabla p + 0.$$

For a barotropic gas we further have $\nabla\rho^{-1} = (d\rho^{-1}/dp)\nabla p$, so the remaining RHS term also evaluates to zero.

(iii) Starting from the full equation provided under question (ii), a few terms can be discarded immediately. This includes the pressure term (as shown above) and all terms containing $\nabla \times \nabla$. This leaves us with

$$\frac{\partial \omega}{\partial t} + \nabla \times (\mathbf{v} \cdot \nabla \mathbf{v}) = \nabla \times \left(\frac{\eta}{\rho} \nabla^2 \mathbf{v} \right).$$

We can plug in the provided vector identity, and obtain

$$\frac{\partial \omega}{\partial t} + \nabla \times \left(\frac{1}{2} \nabla v^2 - \mathbf{v} \times (\nabla \times \mathbf{v}) \right) = \nabla \times \left(\frac{\eta}{\rho} \nabla^2 \mathbf{v} \right).$$

Again we can discard the ∇ cross product, which leads to

$$\frac{\partial \omega}{\partial t} - \nabla \times (\mathbf{v} \times \omega) = \nabla \times \left(\frac{\eta}{\rho} \nabla^2 \mathbf{v} \right).$$

This can then be recast in the requested format by applying the product rule to the RHS.

(iv) The difference between viscous and inviscid fluids is that for the latter Kelvin's theorem holds and vorticity is preserved along flow lines. In the viscous fluid case, the equation shows additional terms that can lead to the decay *and* emergence of eddies. The physical reason for this is that the viscous shear forces do not act perpendicular to the fluid parcel surfaces and can therefore spin parcels up or down.

Answer to Problem 9.2

$$\frac{\partial E_{kin}}{\partial t} = \ldots \int_V v_i \frac{\partial}{\partial x_j} \left(\zeta \delta^{ji} \frac{\partial v^k}{\partial x^k} \right) dV,$$

$$= \ldots \int_V v_i \frac{\partial}{\partial x_i} \left(\zeta \frac{\partial v^k}{\partial x^k} \right) dV,$$

$$= \ldots \int_V \frac{\partial}{\partial x_i} \left(v_i \zeta \frac{\partial v^k}{\partial x^k} \right) dV - \int_V \zeta \frac{\partial v^i}{\partial x^i} \frac{\partial v^k}{\partial x^k},$$

$$= \ldots \oint_S \zeta (\nabla \cdot \mathbf{v}) \mathbf{v} \cdot d\mathbf{S} - \int_V \zeta (\nabla \cdot \mathbf{v})^2 dV$$

Here the resulting left term on the RHS represents a flux through a boundary. This will only move things around within the fluid and cause no net gain or dissipation of kinetic energy. If $\zeta > 0$, the right term *must* altogether be negative, and dissipation is inevitable no matter the velocity profile of the fluid.

Answer to Problem 10.4

i) If wavenumber $k \propto L^{-1}$, dimensional analysis dictates that $E_k \propto \varepsilon_{rate} T L \propto \varepsilon_{rate} L^2 / V$. Since the energy transfer rate is the same across scales, we have

$$\frac{E_k' V'}{(L')^2} = \frac{E_k V}{L^2} \Rightarrow E_k' V \left(\frac{L'}{L} \right)^{\frac{1}{3}} = E_k V \left(\frac{L'}{L} \right)^2 \rightarrow E_k' = E_k \left(\frac{L'}{L} \right)^{\frac{5}{3}} = E_k \left(\frac{k'}{k} \right)^{-\frac{5}{3}}.$$

ii) This time we have $E_{kkk} \propto \varepsilon_{rate} T L^3 \propto \varepsilon_{rate} L^4 / V$, so

$$\frac{E'_{kkk} V'}{(L')^4} = \frac{E_{kkk} V}{L^4} \Rightarrow E'_{kkk} = E_{kkk} \left(\frac{L'}{L}\right)^{\frac{11}{3}} = E_{kkk} \left(\frac{k'}{k}\right)^{-\frac{11}{3}}$$

iii) This follows from $\Delta n \propto V$, according to $\left\langle (\Delta n)^2 \right\rangle \propto P_{3N} L^{-3} \propto V^2$. We get

$$\frac{P'_{3N}}{P_{3N}} = \left(\frac{V'}{V}\right)^2 \left(\frac{L'}{L}\right)^3 = \left(\frac{L'}{L}\right)^{\frac{11}{3}} = \left(\frac{k'}{k}\right)^{-\frac{11}{3}}.$$

Answer to Problem 11.1

Getting rid of pressure first, yields

$$v \frac{dv}{dr} + \rho^{-1} c_s^2 \frac{d\rho}{dr} = -\frac{GM}{r^2}.$$

As stated in the exercise, we need to apply the continuity equation for steady flow here. This equation tells us:

$$\frac{1}{r^2} \frac{d}{dr} \left(r^2 \rho v\right) = 0 \Rightarrow v \frac{d\rho}{dr} + \frac{2\rho v}{r} + \rho \frac{dv}{dr} = 0 \Rightarrow c_s^2 \rho^{-1} \frac{d\rho}{dr} = -\frac{2 c_s^2}{r} - \frac{c_s^2}{v} \frac{dv}{dr}.$$

Plugging this in and using $v dv/dr = (1/2) dv^2/dr$ completes the derivation.

Answer to Problem 11.2

We have three terms to eliminate. First we get rid of r_s using its definition, leading to

$$\dot{M} = 4\pi \frac{G^2 M^2}{4 c_s^4 (r_s)} \rho(r_s) c_s(r_s) = \pi \frac{G^2 M^2}{c_s^3 (r_s)} \rho(r_s).$$

The we get rid of $\rho(r_s)$ by using the polytropic gas law, and obtain:

$$\dot{M} = \pi \frac{G^2 M^2}{c_s^3 (r_s)} \rho(\infty) \left(\frac{c_s(r_s)}{c_s(\infty)}\right)^{\frac{2}{\hat{\gamma}-1}}$$

The requested answer can be obtained by plugging in

$$c_s(r_s) = c_s(\infty) \left(\frac{2}{5 - 3\hat{\gamma}}\right)^{1/2}.$$

References

1. R.D. Blandford and C.F. McKee. Fluid dynamics of relativistic blast waves. *The Physics of Fluids*, 19(8):1130–1138, 1976.

2. S.J. Blundell and K. Blundell. *Concepts in Thermal Physics (Second Edition)*. Oxford University Press, 2010.

3. P. Bodenheimer, G.P. Laughlin, M. Rozyczka, and H.W. Yorke. *Numerical Methods in Astrophysics: An Introduction*. Taylor & Francis, 2007.

4. D. Chandler. *Introduction to Modern Statistical Mechanics*. Oxford University Press, 1987.

5. S. Chandrasekhar. *An Introduction to the Study of Stellar Structure*. Dover Publications, 1958.

6. S. Chandrasekhar. *Hydrodynamic and Hydromagnetic Stability*. Dover Publications, 1981.

7. A. Chepurnov and A. Lazarian. Extending the big power law in the sky with turbulence spectra from Wisconsin Hα mapper data. *Astrophysical Journal*, 710(1):853–858, February 2010.

8. T.J. Chung. *Computational Fluid Dynamics*. Cambridge University Press, 2002.

9. C.J. Clarke and R.F. Carswell. *Astrophysical Fluid Dynamics*. Cambridge University Press, 2007.

10. B.T. Draine. *Physics of the Interstellar and Intergalactic Medium*. Princeton University Press, 2011.

11. E. Fermi. On the origin of the cosmic radiation. *Physical Review*, 75:1169–1174, 1949.

12. G.B. Field. Thermal Instability. *Astrophysical Journal*, 142:531, August 1965.

13. J. Foster and J.D. Nightingale. *A Short Course in General Relativity (Second Edition)*. Springer, 1995.

14. J. Frank, A. King, and D. Raine. *Accretion Power in Astrophysics (Third Edition)*. Cambridge University Press, 2002.

15. B. Fryxell, K. Olson, P. Ricker, F.X. Timmes, M. Zingale, D.Q. Lamb, P. MacNeice, R. Rosner, J. W. Truran, and H. Tufo. FLASH: An adaptive mesh hydrodynamics code for modeling astrophysical thermonuclear flashes. *Astrophysical Journal Supplement Series*, 131(1):273–334, November 2000.

16. D.J. Griffiths. *Introduction to Electrodynamics (Third Edition)*. Prentice Hall, 1999.

17. M. Guidry. *Modern General Relativity: Black Holes, Gravitational Waves and Cosmology*. Cambridge University Press, 2019.

18. T.E. Holzer and W.I. Axford. The theory of stellar winds and related flows. *Annual Review of Astronomy and Astrophysics*, 8:31, January 1970.

19. F. Jüttner. Das maxwellsche gesetz der geschwindigkeitsverteilung in der relativtheorie. *Annalen der Physik*, 34:856, 1911.

20. R. Kippenhahn, A. Weigert, and A. Weiss. *Stellar Structure and Evolution (Second Edition)*. Springer, 2012.

21. C. Kittel and H. Kroemer. *Thermal Physics (Second Edition)*. Freeman and co., 1980.

22. E. Kreyszig. *Advanced Engineering Mathematics (Tenth Edition)*. John Wiley & Sons, Inc., 2011.

23. L.D. Landau and E.M. Lifshitz. *Fluid Mechanics (Second Edition)*. Pergamon Press, 1987.

24. K.H. Lee and L.C. Lee. Interstellar turbulence spectrum from in situ observations of Voyager 1. *Nature Astronomy*, 3:154–159, January 2019.

25. R.J. Leveque. *Finite-Volume Methods for Hyperbolic Problems*. Cambridge University Press, 2004.

26. M.J. Lighthill and F.B. Whitham. On Kinematic Waves. II. A Theory of Traffic Flow on Long Crowded Roads. *Proc. of the Royal Society of London. Series A*, pages 317–345, 1956.

27. M. Longair. *High Energy Astrophysics*. Cambridge University Press, 2011.

28. R. Margutti et al. The signature of the central engine in the weakest relativistic explosions: GRB 100316D. *Astrophysical Journal*, 778(18), 2103.

29. J.M. Marti and E. Muller. Analytical solution of the Riemann problem in relativistic hydrodynamics. *Journal of Fluid Mechanics*, 258:317–333, January 1994.

30. F. Melia. *High-Energy Astrophysics*. Princeton University Press, 2009.

31. B.D. Metzger. Kilonovae. *Living Reviews in Relativity*, 23(1):1, December 2019.

32. A. Mignone. High-order conservative reconstruction schemes for finite volume methods in cylindrical and spherical coordinates. *Journal of Computational Physics*, 270:784–814, August 2014.

33. A. Mignone and G. Bodo. An HLLC Riemann solver for relativistic flows - I. Hydrodynamics. *Monthly Notices of the Royal Astronomical Society*, 364(1):126–136, November 2005.

34. A. Mignone, G. Bodo, S. Massaglia, T. Matsakos, O. Tesileanu, C. Zanni, and A. Ferrari. PLUTO: A numerical code for computational astrophysics. *Astrophysical Journal Supplement Series*, 170(1):228–242, May 2007.

35. A. Mignone, T. Plewa, and G. Bodo. The piecewise parabolic method for multidimensional relativistic fluid dynamics. *Astrophysical Journal Supplement Series*, 160(1):199–219, September 2005.

36. D. Mihalas and B. Weibel-Mihalas. *Foundations of Radiation Hydrodynamics*. Dover Publications, 1999.

37. C.E. Mungan. The Bernoulli equation in a moving reference frame. *European Journal of Physics*, pages 517–520, 2011.

38. L. Nava, L. Sironi, G. Ghisellini, A. Celotti, and G. Ghirlanda. Afterglow emission in gamma-ray bursts - I. Pair-enriched ambient medium and radiative blast waves. *Monthly Notices of the Royal Astronomical Society*, 433(3):2107–2121, August 2013.

39. E.N. Parker. The Hydrodynamic Theory of Solar Corpuscular Radiation and Stellar Winds. *Astrophysical Journal*, 132:821, November 1960.

40. A.D. Pierce. *Acoustics: An Introduction to Its Physical Principles and Applications*. Acoustical Society of America, 1991.

41. A. Pillepich et al. Simulating galaxy formation with the IllustrisTNG model. *Monthly Notices of the Royal Astronomical Society*, 473(3):4077–4106, January 2018.

42. W.H. Press, S.A. Teukolsky, W.T. Vetterinling, and B.P. Flannery. *Numerical Recipes: the Art of Scientific Computing*. Cambridge University Press, 2007.

43. O. Regev, O.M. Umurhan, and P.A. Yecko. *Modern Fluid Dynamics for Physics and Astrophysics*. Springer, 2016.

44. P.I. Richards. Shock waves on the highway. *Operations Research*, 4(1):42–51, 1956.

45. S. Rosswog. Astrophysical smooth particle hydrodynamics. *New Astronomy Reviews*, 53(4-6):78–104, April 2009.

46. L.I. Sedov. *Similarity and Dimensional Methods in Mechanics*. Academic Press, 1959.

47. S. Shore. *Astrophysical Hydrodynamics*. Wiley-VCH, 2007.

48. F.H. Shu. *The Physics of Astrophysics Volume II: Gas Dynamics*. University Science Books, 1992.

49. G.A. Sod. A survey of several finite difference methods for systems of nonlinear hyperbolic conservation laws. *Journal of Computational Physics*, 27:1–31, 1978.

50. V. Springel. E pur si muove: Galilean-invariant cosmological hydrodynamical simulations on a moving mesh. *Monthly Notices of the Royal Astronomical Society*, 401(2):791–851, January 2010.

51. V. Springel, et al. Simulations of the formation, evolution and clustering of galaxies and quasars. *Nature*, 435(7042):629–636, June 2005.

52. J.L. Synge. *The Relativistic Gas*. North-Holland Publishing Co, 1957.

53. E.F. Toro. *Riemann Solvers and Numerical Methods for Fluid Dynamics (Third Edition)*. Springer, 2008.

54. H.J. Van Eerten. Gamma-ray burst afterglow theory. In *eConf Proceedings C1304143, 7th Huntsville Gamma-Ray Burst Symposium, GRB 2013*, page paper 24, 2013.

55. H.J. Van Eerten. Simulation and physical model based gamma-ray burst afterglow analysis. *JHEAp*, (7):22, 2015.

56. M.K. Vyas. Effect of fluid composition on a jet breaking out of a Cocoon in gamma-ray bursts: A relativistic de Laval Nozzle treatment. *Universe*, (8):294, 2022.

57. Ya.B. Zel'dovich and Yu.P. Raizer. *Physics of Shock Waves and High-Temperature Hydrodynamic Phenomena*. Dover Publications, 2002.

Index

Printed in the United States
by Baker & Taylor Publisher Services